BIOACTIVE FOOD PEPTIDES IN HEALTH AND DISEASE

Edited by **Blanca Hernández-Ledesma** and **Chia-Chien Hsieh**

Bioactive Food Peptides in Health and Disease
http://dx.doi.org/10.5772/3318
Edited by Blanca Hernández-Ledesma and Chia-Chien Hsieh

Contributors

Blanca Hernández-Ledesma, Ben O. de Lumen, Chia-Chien Hsieh, Alfonso Clemente, Maria del Carmen Marín-Manzano, Maria del Carmen Arques, Claire Domoney, Roseanne Norris, Richard J. FitzGerald, Tânia G. Tavares, F. Xavier Malcata, Adham M. Abdou, Mujo Kim, Kenji Sato, Anne Pihlanto, Sari Mäkinen, Maira R. Segura-Campos, Luis A. Chel-Guerrero, David A. Betancur-Ancona, Kuo-Chiang Hsu, Yu-Shan Tung, Shih-Li Huang, Chia-Ling Jao, Sameh Sarray, Jose Luis, Mohamed El Ayeb, Naziha Marrakchi, Claudia Muro, Francisco Riera, Ayoa Fernández

Published by InTech
Janeza Trdine 9, 51000 Rijeka, Croatia

Publishing Process Manager Petra Nenadic
Technical Editor InTech DTP team
Cover InTech Design team

Additional hard copies can be obtained from orders@intechopen.com

Bioactive Food Peptides in Health and Disease,
Edited by Blanca Hernández-Ledesma and Chia-Chien Hsieh
p. cm.
ISBN 978-953-51-0964-8

For EU product safety concerns:
IN TECH d.o.o., Prolaz Marije Krucifikse Kozulić 3, 51000 Rijeka, Croatia,
info@intechopen.com or visit our website at intechopen.com.

Meet the editors

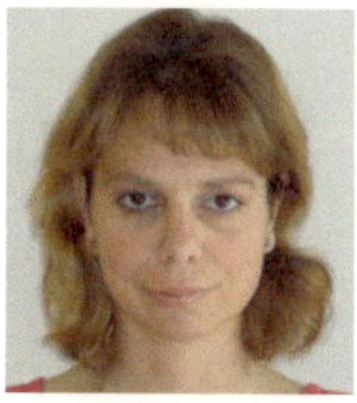

Dr. Blanca Hernandez-Ledesma is currently working as a Postdoctoral researcher under the Ramón & Cajal Programme at the Institute of Food Science Research (CSIC-UAM) in Spain. She was working (1998 to 2002) at the Institute of Industrial Fermentations (CSIC), earning her Ph.D. in Pharmacy by the Complutense University of Madrid, Spain in 2002. She continue working as postdoctoral fellow at the CSIC until 2007 when she moved to USA for a postdoctoral work at the University of California, Berkeley. In 2010, she went back to Spain to work as postdoctoral researcher under different contracts. Her work has been focused on the biological activities of peptides derived from food sources, evaluating their mechanisms of actions, bioavailability and technological aspects by in vitro experiments, cell culture and animal models.

Dr. Chia-Chien Hsieh is currently an Assistant Professor of Nutritional Sciences at the Department of Human Development and Family Studies, National Taiwan Normal University in Taiwan. She is an expert in the field of Nutrition Biochemistry with emphasis in Cancer Biology and Immunology. She received her Ph.D. degree from National Taiwan University in 2005. Her thesis focused on the effect of dietary factors in autoimmunity. Then, she worked as a postdoctoral fellow in the National Health Research Institute in Taiwan for two years. In 2007, she joined the Prof. de Lumen´s laboratory at the University of California Berkeley to work on the chemopreventive properties of lunasin peptide. She investigated its molecular mechanisms and conducted lunasin's efficacy studies in animals. She also studied cyclooxygenase-2 signaling in hepatocarcinoma at the University of Pittsburgh in 2009. Her works provide new knowledge on bioactive food components in diseases prevention and therapy.

Contents

Preface

In recent years, nutrition and food sciences have been focused on biologically active peptides present in the sequences of food proteins. Such peptides are inactive within the sequence of the precursor proteins and can be released by enzymatic proteolysis during gastrointestinal digestion or during food processing. Liberated peptides may act as regulatory compounds with hormone-like activity in the human body. To date, food-derived peptides with antihypertensive, antimicrobial, antioxidant, anti-carcinogenic, anti-inflammatory, opioid agonist or antagonist, antiviral, among others, have been characterized. These peptides possess properties that help to prevent and/or treat different disorders, maintaining the well-health status of humans.

Bioactive Food Peptides in Health and Disease provides a general overview of food-derived peptides for the promotion of human health and the prevention/management of chronic diseases. The book provides updated and interesting information on bioactive peptides obtained from both animal and plant food sources, emphasizing on important aspects related to their bioactivity, mechanism of action, and bioavailability. Also, the chapters describe the impact of bioactive peptides on the physiological absorption, regulation and disease prevention. The book also covers the recent technological advances for the production of food peptides.

The editors want to thank the authors for their important contribution to the success of this book. They are eminent researchers all over the world that have accepted to share and turn their ideas and work into a book that we hope to be an essential resource and reference for nutritional and food scientists, biochemists, industry producers and consumers.

Dr. Blanca Hernández-Ledesma
Institute of Food Science Research (CIAL, CSIC-UAM), Madrid, Spain

Dr. Chia-Chien Hsieh
Department of Nutritional Science and Toxicology.
University of California, Berkeley, California, USA

Section 1

Health Benefits of Food Peptides

Chapter 1

1997-2012: Fifteen Years of Research on Peptide Lunasin

Blanca Hernández-Ledesma, Ben O. de Lumen and Chia-Chien Hsieh

Additional information is available at the end of the chapter

http://dx.doi.org/10.5772/52368

1. Introduction

1.1. Chemopreventive role of food peptides

Cancer is a major killer in today's world accounting for around 13% of all deaths according to the World Health Organisation. It has been estimated that by 2020, approximately 17 million new cancer cases will be diagnosed, and 10 million cancer patients will die [1]. Epidemiological evidence has demonstrated that as many as 35% of these cancer cases may be related to dietary factors, and thus modifications of nutritional and lifestyle habits can prevent this disease [2]. Cell experiments, animal models and human trials have revealed that a large number of natural compounds present in the diet could lower cancer risk and even, sensitize tumor cells against anti-cancer therapies [3]. Therefore, knowledge on the effect of diet components on health will bring new opportunities for chemoprevention through intense alterations in diet regimens.

In the last few years, food proteins and peptides have become one group of nutraceuticals with demonstrated effects preventing the different stages of cancer, including initiation, promotion, and progression [4]. Certain advantages over alternative chemotherapy molecules, such as their high affinity, strong specificity for targets, low toxicity and good penetration of tissues, have made food proteins and peptides a new and promising anticancer strategy [5].

Protease inhibitors are found in plant tissues, particularly from legumes. One of the most extensively studied inhibitors in the field of carcinogenesis is the soybean derived Bowman-Birk protease inhibitor (BBI). It is a 71-amino acids polypeptides which chemopreventive properties have been demonstrated in both *in vitro* and *in vivo* systems [6]. As a result of this evidence, BBI acquired the status of "investigational new drug" from the Food and Drug Administration in 1992, and since then, large-scale human trials are being carried out to evaluate its use as an anticarcinogenic agent in the form of BBI concentrate (BBIC) [7-9]. These studies have shown that BBIC is well-tolerated by the patients and led to promising results for prostate and oral carcinomas.

Milk contains a number of proteins and peptides exhibiting chemopreventive properties. As an example, lactoferrin is a well-known whey protein for its inhibitory action on cancer cells proliferation, as well as for its antimicrobial, anti-inflammatory and antioxidant abilities [10]. The protective effects of orally administered lactoferrin against chemically induced carcinogenesis, tumor growth, and/or metastasis have been demonstrated in an increasing number of animal model studies, thereby suggesting its great potential therapeutic use in cancer disease prevention and/or treatment. Lactoferricin is a cationic peptide produced by acid-pepsin hydrolysis of lactoferrin. Similarly to its source protein, lactoferricin has been demonstrated, by cell culture and animal models, to exert anticarcinogenic properties against different types of cancer, such as leukemia, colon, breast, and lung cancer, among others [11]. This peptide acts through cell proliferation inhibition, apoptosis induction, angiogenesis suppression, and modulation of protein expression involved in different carcinogenesis pathways.

Recent studies have identified and characterized, peptides derived from animal and vegetal sources as promising chemopreventive agents [12-14]. One of these peptides, called lunasin, was identified in soybean and other plants and legumes. Studies performed in the last five years have revealed lunasin's properties in both cell culture and animal models, making it a potential strategy for cancer prevention and/or therapy. The purpose of this chapter is to summarize the evidence reported since lunasin's discovery in 1997 on its possible benefits as a chemopreventive agent as well as its demonstrated mechanisms of action.

2. Lunasin: Discovery and beyond

Lunasin has been described as a 43-amino acid peptide encoded within the soybean 2S albumin. Its sequence is SKWQHQQDSCRKQLQGVNLTPCEKHIMEKIQGRGDDDDD DDDD, containing 9 Asp residues, and an Arg-Gly-Asp cell adhesion motif [15]. Lunasin was first identified in the soybean seed, with variable concentrations ranged from 0.5 to 8.1 mg lunasin/g seed [16,17]. This variation has been found to mainly depend on the soybean genotype, suggesting the possibility of selecting and breeding varieties of soybean with higher lunasin contents [16]. The stages of seed development have also been found to affect lunasin's concentration, and a notable increase occurs during seed maturation [18]. However, sprouting leads to a continuing decrease of lunasin with soaking time. Recent studies have revealed the influence on lunasin content of environmental factors, such as temperature, soil moisture and germination time, as well as of processing conditions [19-21].

Presence of lunasin has been demonstrated in commercial and pilot plant produced soybean products, including soy milk, infant formula, high protein soy shake, tofu, bean curd, soybean cake, tempeh, and su-jae (Table 1) [22,23]. Results from these studies establish the influence on lunasin concentration in the food products of different parameters, such as the soy genotype, the environmental factors, the manufacturing process and the storage conditions. Thus, these parameters might be used to control the content of this bioactive peptide.

Type of sample	Composition-main ingredients	Lunasin (mg/100 g product)	Reference
Regular soymilk	Soybeans	15.7 ± 1.3	[22]
	Soybeans	12.3 ± 0.8	[22]
	Soybeans	11.8 ± 1.3	[22]
	Whole soybeans	9.3 ± 0.3	[23]
	Soybeans, soybean oil, soy lecithin	9.2 ± 1.7	[23]
	Soy flour, Stevia sweetened	7.0 ± 0.1	[23]
	Soy flour, Stevia sweetened	6.3 ± 0.2	[23]
	Whole soybeans, filtered water	7.9 ± 0.0	[23]
	Whole soybeans, filtered water	6.3 ± 0.1	[23]
	Whole soybeans, filtered water	6.1 ± 0.1	[23]
	Whole soybeans	6.0 ± 0.6	[23]
	Whole soybeans	2.2 ± 0.1	[23]
	Soybeans, soy lecithin	5.2 ± 0.7	[23]
	Soybeans, calcium fortified	1.8 ± 0.3	[23]
	Aqueous extract of soybeans	2.3 ± 0.4	[23]
Organic soymilk	Organic soybeans, malted wheat and barley extract	18.9 ± 2.6	[22]
	Organic soybeans, malted wheat and barley extract	14.2 ± 1.1	[22]
	Organic soybeans	13.8 ± 2.6	[22]
	Organic soybeans	14.4 ± 2.4	[22]
	Organic soybeans	14.7 ± 0.8	[22]
	Organic soybeans, rice syrup	13.7 ± 0.9	[22]
	Organic soybeans, soy protein isolate	13.9 ± 1.0	[22]
	Organic soybeans, malt syrup	18.3 ± 2.4	[22]
	Organic soybeans, barley extract	10.7 ± 0.8	[22]
	Whole organic soybeans, isoflavones	9.1 ± 0.1	[23]
	Organic soybeans	8.8 ± 0.0	[23]
	Organic soybeans, calcium fortified	8.2 ± 0.0	[23]
	Whole organic soybeans, calcium enriched	5.6 ± 0.7	[23]
	Whole organic soybeans, calcium enriched	5.4 ± 0.6	[23]
	Organic soybeans	3.8 ± 0.6	[23]
	Organic soybeans	2.9 ± 0.8	[23]
Soy formula	Corn syrup, soy protein isolate	4.1 ± 0.4	[22]
	Corn syrup, soy protein	2.8 ± 0.2	[22]
	Rice syrup, soy protein concentrate	1.5 ± 0.1	[22]
	Soy protein isolate, soy oil, iron fortified	8.9 ± 0.4	[23]

Type of sample	Composition-main ingredients	Lunasin (mg/100 g product)	Reference
	Soy protein isolate, soy oil, iron fortified	8.1 ± 0.4	[23]
	Soy protein isolate, iron fortified	7.3 ± 0.4	[23]
Soft Tofu	Soybeans	9.6 ± 0.9	[22]
Soft Tofu	Soybeans	7.3 ± 1.0	[22]
Silken Tofu Kinugoshi	Soybeans	9.6 ± 0.7	[22]
Silken Tofu	Soybeans	4.4 ± 0.5	[22]
Silken Tofu	Soybeans	3.7 ± 0.5	[22]
Medium firm Tofu	Soybeans	14.3 ± 1.8	[22]
Organic Medium firm Tofu	Soybeans	6.7 ± 1.3	[22]
Firm Tofu	Soybeans	3.5 ± 0.2	[22]
Extra firm tofu Chinese style	Soybeans	3.7 ± 0.1	[22]
Baked tofu	Soybeans, soy sauce (wheat)	5.5 ± 0.3	[22]
Fried tofu	Soybean, soybean oil, soy sauce	0.4 ± 0.1	[22]
Dry tofu	Soybeans	2.5 ± 0.3	[22]
Soy shake	Soy and milk, chocolate flavored	1.3 ± 0.01	[23]
	Soy, milk, vanilla flavored	1.3 ± 0.01	[23]
	Soy protein isolate, milk, dark chocolate flavored	2.0 ± 0.04	[23]
	Soy, sor protein isolate, chocolate flavored	3.6 ± 0.02	[23]
Organic tempeh	Soybeans, *Rhizopus oligosporus*	n.d.	[22]
	Soybeans, brown rice, *R. oligosporus*	8.2 ± 0.4	[22]
	Soybeans, flaxseed, brown rice, *R. oligosporus*	6.1 ± 0.4	[22]
	Soybeans, brown rice, *R. oligosporus*	n.d.	[22]
Marinated bean curd	Soybeans, soy sauce	9.5 ± 1.0	[22]
Soybean curd noodle	Soybeans	n.d.	[22]
Deep fried soybean cake	Soybeans, soybean oil	1.9 ± 0.3	[22]
Baked soybean cake	Soybeans, soy sauce, sesame oil	1.1 ± 0.2	[22]

Table 1. Type, composition, and lunasin content of soybean-derived foods

In search of natural sources of lunasin besides soybean, a first screening has been carried out using different beans, grains and herbal plants. Lunasin has been found in cereal grains known for its health effects, such as barley, wheat, and rye [24-27]. Several seeds of oriental herbal and medicinal plants have been analyzed, finding that lunasin is present in all of the *Solanaceae* family, except *L. Chinensis,* but not in any of the *Phaseolus* beans [28]. These findings suggested the presence of lunasin or lunasin-like peptides in other grains and plants. This peptide has been identified in *Amaranth,* a plant well-known and used by the Aztecs for its high nutritional value and its biological properties [29]. A recent study has revealed the presence of lunasin in different *Lupinus* cultivated and wild species [30]. A more rigorous and systematic search of lunasin and lunasin homologues in different seeds should be need to carry out in order to establish a relation between the presence of this peptide and the taxonomic properties of the plants.

2.1. Bioavailability of lunasin

One of the properties of an ideal cancer preventive agent is that it can be taken orally. This means being able to survive degradation by gastrointestinal and serum proteinases and peptidases, and to reach the target organ or tissue in an active form. Simulation of gastrointestinal digestion of lunasin has demonstrated that, while synthetic pure lunasin is easily hydrolyzed by pepsin and pancreatin, lunasin in soy protein is resistant to the action of these enzymes. Bioavailability studies carried out with animals have confirmed the preliminary results obtained by *in vitro* analysis. First studies carried out in mice and rats fed lunasin-enriched soy protein found that 35% of ingested lunasin reaches the target tissues and organs in an intact and active form [17,28]. Lunasin from rye and barley have also shown stability towards pepsin and pancreatin *in vitro* digestion and the liver, kidney, and blood of rats fed with lunasin-enriched rye or barley, respectively, contained this peptide as detected by Western blot [26,27]. Naturally protease inhibitors, such as Bowman-Birk protease inhibitor and Kunitz trypsin inhibitor have been demonstrated to exert a protective effect on lunasin against digestion by gastrointestinal enzymes, playing this protection a key role in making lunasin bioavailable [31]. These authors reported that lunasin is bioavailable after its oral administration to mice, reaching different tissues, including lung, mammary gland, prostate, and brain, where this peptide might exert is chemopreventive effects. These authors also found that lunasin extracted from the blood and liver of lunasin-enriched soy flour-fed rats was bioactive and able to suppress foci formation in the same concentration as synthetic lunasin.

A clinical trial focused on evaluating lunasin's bioavailability has demonstrated that in healthy volunteer men, 4.5% of lunasin ingested in the form of soy protein reaches plasma [32]. Results from this study are relevant in supporting future clinical trials to demonstrate cancer preventive properties of lunasin.

3. Lunasin's role as chemopreventive peptide

Peptide lunasin has demonstrated to exert promising chemopreventive properties against different types of cancers by both cell culture and animal experiments (Table 2). First studies performed with mammalian cells revealed that lunasin did not affect their morphology and proliferation. However, this peptide acted preventing their transformation induced by chemical carcinogens-7,12-dimethylbenz[a]anthracene (DMBA) and 3-methylcholanthrene (MCA) [33,34], viral and ras-oncogenes [33,35,36]. These experiments made lunasin be considered a "watchdog" agent in the cell nucleus that once the transformation event occurs, it acts as a surrogate tumor suppressor that tightly binds to deacetylated core histones disrupting the balance between acetylation-deacetylation, which is perceived by the cell as abnormal and leads to cell death [37]. This first mechanism of action involving histone acetylation inhibition is considered as one of the most important epigenetic modifications acting on signal transduction pathways involved in cancer development [38,39]. When the cells are in the steady-state conditions, the core H3 and H4 histones are mostly deacetylated, as a repressed state. When cells were treated with peptide lunasin and well-known deacetylase inhibitor sodium butyrate, histone acetylation was inhibited in C3H10T1/2 fibroblasts and breast cancer MCF-7 cells [33,36]. Furthermore, lunasin has been demonstrated to compete with different histone acetyltransferase enzymes (HATs), such as yGCN5 and PCAF, inhibiting the acetylation and repressing the cell cycle progression [24,25,28]. Recently, we have reported that lunasin is a potent inhibitor of histones H3 and H4 histone acetylation [40]. Lunasin's inhibitory activity was found to be higher than that demonstrated by other compounds, such as anacardic acid and curcumin, which chemopreventive properties have been already reported [41-43]. Studies focused on elucidating lunasin's structure-activity relationship establish that lunasin's sequence is essential for inhibiting H4 acetylation whereas poly-D sequence is the main active sequence responsible for H3 acetylation inhibition [40] (Table 3).

Although first studies only established lunasin's capacity to act when transformation process happens, studies performed in the last few years have demonstrated that this peptide also acts on established cancer cells lines. This activity against different types of cancer cell lines is summarized in this chapter. Moreover, results obtained from cancer animal models are also included.

3.1. Chemopreventive properties against breast cancer

With a prevalence of about 4.4 million women and a lethality rate of more than 410,000 cases per year, breast cancer is the most common cancer disease and the leading cause of death in women worldwide [44]. Based on the prevalence of estrogen receptors (ER) within the cell, breast cancer is categorized into the ER–positive type and the ER-negative type. About 70-80% of all breast cancers are estrogen sensitive and they are treated by conventional procedures including surgery, radiation chemotherapy, and estrogen analogues. However, ER-negative tumors are more aggressive and resistant to treatments [45,46]. Therefore, searching for new preventive and/or curative strategies for this type of breast cancer has centered the interest of current investigations.

3.1.1. Lunasin against breast cancer in vitro

Up to one third on breast cancers that are initially ER-independent become resistant to endocrine therapy during tumor progression [47]. Due to this emergence of hormone-resistance, it is necessary to search for alternative therapies. Lunasin has been demonstrated to inhibit cell proliferation in ER-negative breast cancer MDA-MB-231 cells in a dose-dependent manner, showing an IC_{50} value of 181 μM [48]. Studies carried out to establish a structure/activity relationship showed an IC_{50} value of 138 μM for the 21 amino acid sequence localized at the C-terminus of lunasin, thus being the main responsible for lunasin's inhibitory effect on breast cancer cells proliferation [40].

Table 2. Biological effects of peptide lunasin demonstrated by cell culture experiments

Cell line	Cell proliferation	Cell cycle	Apoptosis	Gene expression	Protein levels	Other effects	Reference
Breast cancer MDA-MB-231	Inhibition	Arrest in S-phase	Induction	CDC25A, Caspase 8, Ets2, Myc, Erbb2, PIK3R1 and JUN genes RB gene	cyclins D1, D3, CDK4 and CDK6	Synergisms with aspirin Synergisms with anacardic acid	[48, 40, 57]
Breast cancer MCF-7 treated with Na-butyrate	n.r.	n.r.	n.r.	n.r.	n.r.	HAT activity, inhibition of H3 and H4 acetylation and RB phosphorylation	[24, 25, 28, 33]
NIH3T3 treated with Na-butyrate	n.r.	n.r.	n.r.	n.r.	n.r.	HAT activity, inhibition of H3 and H4 acetylation and RB phosphorylation	[24, 25, 28, 33]
NIH3T3 transfected with viral E1A oncogen	n.r.	n.r.	n.r.	n.r.	protein p21	foci formation	[24, 25]
NIH3T3 transfected with ras-oncogenes	n.r.	n.r.	n.r.	n.r.	n.r.	colony formation, inhibition H3 acetylation	[36]
DMBA-induced C3H10T1/2	Inhibition	n.r.	n.r.	n.r.	n.r.	foci formation	[33]
MCA-induced C3H10T1/2	Inhibition	n.r.	n.r.	n.r.	n.r.	foci formation	[33]
DMBA-induced NIH3T3	Inhibition	n.r.	n.r.	n.r.	n.r.	foci formation Synergisms with aspirin and anacardic acid	[33, 54]
MCA-induced NIH3T3	Inhibition	n.r.	n.r.	n.r.	n.r.	foci formation Synergisms with aspirin and anacardic acid	[33, 54]
Colon cancer HT-29 and KM12L4	Inhibition	Arrest in G2/M phase	Induction	Modulation of Bcl-2, Bax, nCLU, cytochrome c and caspase-3 ↓ integrin 2, matrix metalloproteinase 10, selectin E, integrin 5 and collagen type VII α1 genes	↑ proteins p21 and p27	↑ activity of caspase-9	[59, 61]
Leukemia L1210	Inhibition	Arrest in G2 phase	Induction	n.r.	n.r.	caspase-8, -9 and -3 activity	[64]
Prostate epithelial RWPE-1	n.r.	n.r.	n.r.	HIF1A, PRKAR1A, TOB1, and THBS1 genes	n.r.	H4-Lys 8 acetylation H4-Lys 16 acetylation	[66]
Macrophages RAW 264.7	n.r.	n.r.	n.r.	LPS-induced production of IL-1 , TNF- , and NO genes clusterin gene	IL-1 , NO, and TNF-	COX-2/PGE2 and inducible iNOS/nitric oxide pathways LPS-induced production of ROS generation of hydroxyl radicals	[69-71, 73, 76]

n.r.: effect not reported

Lunasin fragment	Peptide sequence	Activity	Reference
f(1-43)	SKWQHQQDSCRKQLQGVNLTPCEKHIMEKIQGRGDDDDDDDDD	Inhibition of H4 acetylation (Lys8;Lys12;Lys5,8,12,16) Inhibition of H3 acetylation (Lys9; Lys9,14) Anti-inflammation	[40, 69]
f(1-22)	SKWQHQQDSCRKQLQGVNLTPC	Unknown or null function	
f(23-35)	EKHIMEKIQGRGD EKHIMEKIQGRGDDDDDDDDD	Targets lunasin to histones Inhibition of H4 acetylation (Lys12; Lys5,8,12,16)	[40]
f(23-43)	RGD	A recognition site for integrin receptors present in the extracellular matrix, Internalize lunasin into cells	
f(34-43)	DDDDDDDDD	Directly binds to core histones Inhibition of H4 acetylation (Lys8; Lys12; Lys5,8,12,16) Inhibition of H3 acetylation (Lys9; Lys9,14)	[33, 40]

Table 3. Structure/activity relationship of lunasin and its derived fragments

A plethora of chromatin alterations appears to be responsible for the development and progression of various types of cancers, including breast cancer. Acetylation of specific lysine residues in histones is generally linked to chromatin disruption and transcriptional activation of genes [49]. In our studies, a dose-dependent inhibitory effect on H4 acetylation at positions H4-Lys8 and H4-Lys12 was observed after treatment of lunasin at 75 μM in MDA-MB-231 cells, reaching 17% and 19% inhibition, respectively, compared to control [40]. It should be needed to extensively study the relevance of these results on lunasin's chemopreventive activity to provide data about its molecular mechanism of action on epigenetic alterations. It would be useful to define new prognostic markers and therapeutic targets.

We have also demonstrated that lunasin modulates expression of different genes and proteins involved in cell cycle, apoptosis and signal transduction [48]. A pivotal regulatory pathway determining rates of cell cycle transition from G1 to S phase is the cyclin/cyclin-dependent kinases (CDK)/p16/retinoblastoma protein (RB) pathway. Over-expression of cyclins D1 and D3 is one of the most frequent alterations present in breast tumors. Cyclins D interacts with CDK4 or CDK6 to form a catalytically active complex, which phosphorylates RB to free active E2F [50]. Inhibition of deregulated cell cycle progress in cancer cells is being considered an effective strategy to delay or halt tumor growth. Lunasin up-regulates RB gene expression [48], and inhibits RB phosphorylation [28], suggesting that both transcriptional and post-translational modifications may be responsible for its inhibitory effect on cancer cell cycle progression. Moreover, lunasin has been found to inhibit cell proliferation, arrest the cell cycle in the S phase in 45%, and provokes a down-regulatory effect on the mRNA levels of CDK2, CDK4, CDC25A, Caspase 8, and Ets2, Myc, Erbb2, AKT1, PIK3R1 and Jun signaling genes in MDA-MB-231 cells [48]. Also, lunasin's down-regulatory action on levels of proteins, such as cyclin D1, cyclin D3, CDK4 and CDK6, might also contribute on its breast cancer MDA-MB-231 cells cycle arrest effect [40]. The ability of lunasin to modulate expression of genes and proteins involved in cell cycle, apoptosis and signal transduction seems to play a relevant role in its properties against breast cancer. However, further research should be needed to elucidate the complete molecular and epigenetic mechanism of action in breast cancer.

3.1.2. Lunasin against breast cancer in vivo

Lunasin's role as chemopreventive agent against breast cancer has also been demonstrated in *in vivo* mouse models. Our first findings showed a relevant inhibitory effect of a lunasin-enriched diet on mammary tumors development in DMBA-induced SENCAR mice [34]. Tumor generation and tumor incidence were reduced by 38% and 25%, respectively, in the mice fed with lunasin-enriched diet (containing 0.23% lunasin) compared with control group. Moreover, the tumor sections obtained from the lunasin-enriched group showed slight stromal invasion and degree of morphological aggressiveness due to the effect of this peptide contained in the soy protein preparation. Park and co-works have reported that isoflavone-deprived soy peptides prevent DMBA-induced rat mammary tumorigenesis, as well as inhibit the growth of human breast cancer MCF-7 cells in a dose-dependent manner, and induce cell death [51]. Lunasin might be responsible for the effects reported by these authors.

A recent study has shown that lunasin reduces tumor incidence and generation in a xenograft mouse model using human breast cancer MDA-MB-231 cells [31]. Lunasin's inhibitory effect on the tumor weight and volume was also reported by these authors. In contrast, BBI showed no effect on tumor development. The tumor histological sections obtained from the lunasin-treated group showed cell proliferative inhibition and cell apoptosis induction. These first animal models consider lunasin as a new and promising alternative to prevent and/or treat breast cancer.

3.1.3. Lunasin's combinations as a novel strategy against breast cancer

Cancer chemotherapeutic strategies commonly require multiple agents to prevent and/or treat cancer because of its ability to achieve greater inhibitory effects on cancer cells with lower toxicity potential on normal cells [3]. In the last two decades, it has been recognized the aspirin's chemopreventive role against different types of cancer. However, aspirin use has been associated with undesirable side effects, peptic ulcer complications, particularly bleeding and mucosal injury [52,53]. Studies are searching new agents to be combined to aspirin, increasing its effectiveness or decreasing its side effects. Our findings revealed that lunasin potentiates aspirin's cell proliferation inhibitory and apoptosis inducing properties in MDA-MB-231 cells [48]. This combination regulates the genes expression encoding G1 and S-phase regulatory proteins and the extrinsic-apoptosis dependent pathway, at least partially, through synergistic down-regulatory effects were observed for ERBB2, AKT1, PIK3R1, FOS and JUN signaling genes. Moreover, additional studies have demonstrated that lunasin/aspirin combination inhibits foci formation and cell proliferation in chemical carcinogens DMBA and MCA induced-NIH/3T3 cells [54]. The effect was notably higher than that observed when compounds of the combination acted as a single agent.

Anacardic acid is a natural compound found in the shell of the cashew nut. It has been linked to anti-oxidative, anti-microbial, anti-inflammatory and anti-carcinogenic activities [55,56]. Our findings revealed that lunasin/anacardic acid combination arrests cell cycle in S-phase and induces apoptosis at higher levels than that observed when each compound is used individually. This combination also promotes the inhibition of ERBB2, AKT1, JUN and RAF1 signaling genes expression. Synergistic effects have also been observed when lunasin was combined with anacardic acid to treat breast cancer cells and chemical-induced fibroblast cells [57].

The safety and efficacy of chronic use of these combinations should be further tested in animal models and human studies to establish the optimal dose and duration of treatment. Moreover, studies derived from these findings about mechanisms of action of these lunasin's combinations would open a new vision in the development of novel therapies against breast cancer.

3.2. Lunasin's chemopreventive properties against colon cancer

Colon cancer is the second leading cause of cancer death in the Western world. The high incidence, morbidity and mortality of colon cancer make necessary the effective prevention of this disease. In the last years, pathogenesis of colorectal cancer has been elucidated, giving the approach for development of new drugs to combat this malignancy. Accumulating studies have shown the capability of bioactive food components to modulate the risk of developing colon cancer [58]. Recently, lunasin's potential chemopreventive role has been also reported.

3.2.1. *Lunasin against colon cancer in vitro*

It has been demonstrated that lunasin causes cytotoxicity in four different human colon cancer cell lines, KM12L4, RKO, HCT-116, and HT-29 cell, with IC_{50} values of 13.0 μM, 21.6 μM, 26.3 μM and 61.7 μM, respectively [59]. These values suggest that lunasin is most potent killing the highly metastatic KM12L4 colon cancer cells than any other colon cell lines used in this study. Moreover, lunasin was capable to provoke cytotoxic effects on the oxaliplatin-resistant variants of these colon cancer cells [60]. Studies on mechanism of action of this peptide have revealed that lunasin causes arrest of cell cycle in G2/M phase and induction of the mitochondrial pathway of apoptosis. The cell cycle arrest was attributed with concomitant increase in the expression of the p21 protein in HT-29 colon cancer cells, while both p21 and p27 protein expressions were up-regulated by lunasin treatment in KM12L4 colon cancer cells [59,61]. Moreover, treatment with lunasin decreased the ratio of Bcl-2:Bax by up-regulating the expression of the pro-apoptotic Bax and down-regulating the expression of the anti-apoptotic Bcl-2, also increasing the activity of caspase-3 [61]. This might be attributed to the increase in the expression of the pro-apoptotic form of clusterin which is positively affected by the increase p21 expression in cell nucleus. Treatment of lunasin causes translocation of Bax into the mitochondrial membrane resulting in the release of cytochrome c and the increase of the expression of cytosolic cytochrome c in KM12L4 cells. It was also demonstrated that treatment with lunasin provokes an increase in the activity of caspase-9 and caspase-3 in both HT-29 and KM12L4 cells [59]. Furthermore, lunasin has been showed to modify the expression of human extracellular matrix and adhesion genes [59]. The Arg-Gly-Asp motif present in the lunasin structure is a recognition site for integrin receptors present in the extracellular matrix (ECM). Integrins are heterodimeric receptors associated with cell adhesion, and cancer metastasis [62]. Treatment of KM12L4 cells with lunasin resulted in the modification on the expression of 62 genes associated with ECM and cell adhesion [59]. These authors also reported that lunasin down-regulated the gene expression of collagen type VII α1, integrin β2, matrix metalloproteinase 10, selectin E and integrin α5 by 10.1-, 8.2-, 7.7-, 6.5- and 5.0-fold, respectively, compared to the untreated colorectal cancer cells. On the other hand, the expression of collagen type XIV α1 was up-regulated upon lunasin treatment by 11.6-fold. These results suggest a potential role of peptide lunasin as an agent to combat metastatic colon cancer particularly in cases where resistance to chemotherapy develops.

3.2.2. *Lunasin against colon cancer in vivo*

Colon cancer liver metastasis is a widely used model to study the effects of different markers and chemotherapy on colon cancer metastasis. Recently, Dia & de Mejia (2011b) have reported that lunasin acts as chemopreventive agent against this type of metastasis using colon cancer KM12L4 cells directly injected into the spleen of athymic mice [60]. Lunasin administered at concentration of 4 mg/kg body weight resulted in a significant inhibition of liver metastasis of colon cancer cells, potentially because of its binding to $\alpha5\beta1$ integrin and subsequent suppression of FAK/ERK/NF-κB signalling. Lunasin was also capable to potentiate the effect of oxaliplatin in preventing the outgrowth of metastasis. Moreover, lunasin potentiated the effect of oxaliplatin in modifying expression of proteins involved in apoptosis and metastasis including Bax, Bcl-2, IKK-α and P65 [60]. These results suggest that lunasin can be used as a potential integrin antagonist thereby preventing the attachment and extravasation of colon cancer cells leading to its anti-metastatic effect. These results open a new vision about the lunasin used in metastasis that might benefit to prolong the survival of mice with metastatic colon cancer.

3.3. Lunasin's chemopreventive properties against other type of cancers

Leukemia is considered to be the most common type of cancer in children. Leukemia disrupts the normal reproduction and repair processes of white blood cells causing them to divide too quickly before they mature and resulting in the arrest on the proper production of all blood cells [63]. Chemopreventive properties of peptide lunasin have also been shown in human leukaemia L1210 cells, with an IC_{50} value of 14 μM [64]. Cell cycle analysis performed by these authors showed that lunasin caused a dose-dependent G2 cell cycle arrest and induction of apoptosis. The expressions of caspases-3, -8 and -9 were significantly up-regulated by 12-, 6- and 6-fold, respectively, which resulted in the increase of percentage of L1210 leukemia cells undergoing apoptosis from 2 to 40% [64].

Prostate cancer is one of the leading causes of cancer death in worldwide men. The multistage, genetic, and epigenetic alterations nature of prostate cancer during disease progression and the response to therapy, represent fundamental challenges in our quest to understand and control this prevalent disease [65]. Recently, Galvez and co-workers have studied lunasin's effects on tumorigenic RWPE-1 and non-tumorigenic RWPE-2 human prostate epithelial cells [66]. These authors observed that HIF1A, PRKAR1A, TOB1, and THBS1 genes were up-regulated by lunasin in RWPE-1 but not in RWPE-2 cells, confirming lunasin's capacity to selectively act on cancer cells without affecting non-cancerous cells. Moreover, lunasin specifically inhibited H4-Lys8 acetylation while enhanced H4-Lys16 acetylation catalyzed by HAT enzymes p300, PCAF, and HAT1A [66]. As a dietary peptide capable of up-regulate gene expression by specific epigenetic modifications of the human genome, lunasin is suggested to represent a novel food bioactive peptide with the potential to reduce cancer risk.

4. Anti-inflammatory and antioxidant activities of lunasin

Inflammation and oxidative stress are two of the most critical factors implicated in carcinogenesis and other degenerative disorders. Accumulating evidences have revealed that chronic inflammation is involved in the development of approximately 15–20% of malignancies worldwide [67], being clearly associated with increased cancer risk and progression [68]. Lunasin has been found to exert anti-inflammatory activity that might contribute to its chemopreventive properties. First studies demonstrated that lunasin potently inhibits lipopolysaccharide (LPS)-induced production of pro-inflammatory mediators interleuquine-6 (IL-6), tumor necrosis factor-α (TNF-α), and prostaglandin (PG) E2 (PGE2) in macrophage RAW 264.7 cells [69], through modulation of cyclooxygenase-2 (COX-2)/PGE2 and inducible nitric oxide synthase/nitric oxide pathways, and suppression of NF-κB pathways [70,71]. Larkins and co-workers (2006) have demonstrated that COX-2 inhibition can decrease breast cancer cells motility, invasion and matrix metalloproteinase expression [72]. Abnormally up-regulated COX and PGs expression are features in human breast tumors, thus lunasin might have a role in treatment and prevention of this kind of cancer. Moreover, the same biological activity was observed for lunasin-like peptides purified from defatted soybean flour by combination of ion-exchange chromatography and size exclusion chromatography. These peptides showed potent anti-inflammatory activity by inhibiting LPS-induced RAW 264.7 cells through suppression of NF-κB pathways [70,71]. Interestingly, Liu and Pan (2010) used *E. coli* as a host to produce valuable bioactive lunasin that was also showed its anti-inflammatory properties. The purified recombinant lunasin form *E.coli* expressed system inhibits histone acetylation, and inhibits the production of pro-inflammatory cytokines, such as TNF-α, interleukin-1β and nitric oxide in LPS-stimulated RAW 264.7 cells [73].

Large amounts of reactive oxygen species (ROS) have been shown to participate in the etiology of several human degenerative diseases, including inflammation, cardiovascular and neurodegenerative disorders, and cancer [74]. It is believed that persistent inflammatory cells recruitment, repeated generation of ROS and pro-inflammatory mediators, as well as continued proliferation of genomically unstable cells contribute to neoplasic transformation and ultimately result in tumor invasion and metastasis [75]. Restoration/activation of improperly working or repressed antioxidant machinery or suppression of abnormally amplified inflammatory signaling can provide important strategies for chemoprevention.

Lunasin has been found to exert potent antioxidant properties, inhibiting linoleic acid oxidation and acting as a potent free radical scavenger, and reducing LPS-induced production of ROS by RAW 264.7 macrophage cells at a dose-dependent manner [69]. Recently, lunasin purified from *Solanum nigrum L.* has been found to protect DNA from oxidative damage by scavenging the generation of hydroxyl radical, as well as reducing Fe^{3+} to Fe^{2+} through blocking fenton reaction and inhibiting linoleic acid oxidation [76]. Moreover, these authors demonstrate lunasin's suppresive effects on the production of intracellular ROS and glutathione. Preliminary results indicate a similar inhibitory effect of ROS and GSH productions was also observed in Caco2 cells [77]. This activity might contribute on lunasin's chemopreventive role against cancer and other oxidative stress-related disorders.

5. Production of lunasin

Although the potential anticancer effect of lunasin has been demonstrated for over a decade, little progress has been made to test *in vivo* efficacy of purified lunasin in large-scale animal studies or human clinical trials. The main limitations of these studies have been the lack of a method for obtaining gram quantities of highly purified lunasin from plant sources needed to perform such studies. Chemical synthesis is a rapid and effective method to produce lunasin in small quantities but the high cost and difficulties of the scale-up process makes lunasin's synthesis an economically impractical alternative. In addition, the process employs chemicals that are potential environmental hazards. To date, the reported methods to isolate and purify lunasin from soybean only allowed obtaining small quantities of this peptide at 80% purity [70]. However, recently, Cavazos and co-workers (2012) have developed an improved method to isolate and purify lunasin from defatted soy flour, resulting in at least 95% purity [23]. Simultaneously, a large-scale method to generate highly purified lunasin from defatted soy flour has been developed by Seber and co-workers (2012) [78]. This method is based on the sequential application of anion-exchange chromatography, ultrafiltration, and reversed-phase chromatography, obtaining preparations of > 99% purity with a yield of 442 mg/kg of defatted soy flour. Moreover, these preparations show the same biological activity than that reported for synthetic lunasin although the sequence contains Asn as an additional C-terminal amino acid residue.

An additional alternative to increase lunasin content in soybean has been recently reported [79]. This strategy aims to exploit the potential of sourdough lactic acid bacteria to release lunasin during fermentation of cereal and non conventional flours. After fermentation, lunasin from the water soluble extracts was increased up to 2-4 times, being *Lactobacillus curvatus* SAL33 and *Lactobacillus brevis* AM7 the strains capable to release higher concentrations of this peptide. This new strategy opens new possibilities for the biological synthesis and for the formulation of functional foods containing bioactive lunasin.

The use of recombinant production by transgenic organisms is widely employed in industry owing to their ease of use, robustness and costs, and has become the most effective system for the production of long peptides and proteins. A recent study has explored efficient recombinant production of lunasin by exploiting the *Clostridium thermocellum* CipB cellulose-binding domain as a fusion partner protein [80]. This system resulted in yields of peptide of up to 210 mg/L, but the authors consider that these yields might be increased in bioreactors where oxygen and nutrients levels are tightly regulated.

6. Conclusions

Peptides are becoming a group of health-promoting food components with promising chemopreventive and chemotherapeutic properties against cancer. Among them, peptide lunasin, found in soybean and other plants, is turning into one of the most promising. This peptide has been demonstrated its bioavailability after resisting gastrointestinal and serum degradation, and reaching blood and target organs in an intact and active form. Efficacy of lunasin against breast, colon, leukemia and prostate cancer using cell culture experiments and animal models have been revealed in the last decade. These results make lunasin a good candidate for a new generation of chemopreventive/chemotherapeutical agents derived from natural seeds. However, there is still much to be learned about the effects and mechanisms of lunasin on cancer prevention/therapy. The major challenge on the use of lunasin in treating cancer would be the conversion of existing results into clinical outcomes. The next step should be to design clinical trials to confirm lunasin's chemopreventive properties against different types of cancer. Moreover, genomics, proteomics and biochemical tools should be applied to complete elucidate its molecular mechanism of action. Other aspects, such as searching for lunasin in other seeds, optimization of techniques to enrich products with this peptide and studying lunasin's interactions with other food constituents affecting its activity should also be conducted.

Author details

Blanca Hernández-Ledesma
Institute of Food Science Research (CIAL, CSIC-UAM, CEI UAM+CSIC), Madrid, Spain

Ben O. de Lumen and Chia-Chien Hsieh *
Department of Nutritional Science and Toxicology, University of California Berkeley, CA, USA

7. References

[1] Ferlay J, Shin HR, Bray F, Forman D, Mathers C, Parkin DM. GLOBOCAN 2008, Cancer incidence and mortality worldwide: IARC Cancer Base No. 10 [Internet]. Lyon, France: International Agency for Research on Cancer; 2010. Available from: http://globocan.iarc.fr.

[2] Manson M. Cancer prevention – the potential for diet to modulate molecular signaling. Trends in Molecular Medicine 2003;9:11-8.

[3] de Kok TM, van Breda SG, Manson MM. Mechanisms of combined action of different chemopreventive dietary compounds. European Journal Nutrition 2008;47:51-9.

[4] de Mejia EG, Dia VP. The role of nutraceutical proteins and peptides in apoptosis, angiogenesis, and metastasis of cancer cells. Cancer Metastasis Review 2010; 29:511-28.

[5] Bhutia SK, Maiti TK. Targeting tumors with peptides from natural sources. Trends in Biotechnology 2008;26:210-7.

[6] Losso JN. The biochemical and functional food properties of the Bowman-Birk Inhibitor. Critical Reviews in Food Science & Nutrition 2008;48:94-118.

[7] Armstrong WB, Kennedy AR, Wan XS, Atiba J, McLaren E, Meyskens FL. Single-dose administration of Bowman-Birk inhibitor concentrate in patients with oral leukoplakia. Cancer Epidemiology Biomarkers & Prevention 2000;9:43-7.

[8] Armstrong WB, Wan XS, Kennedy AR, Taylor TH, Meyskens FL. Development of the Bowman-Birk inhibitor for oral cancer chemoprevention and analysis of neu immunohistochemical staining intensity with Bowman-Birk inhibitor concentrate treatment. Laryngoscope 2003;113:1687-702.

[9] Meyskens FL. Development of Bowman-Birk inhibitor for chemoprevention of oral head and neck cancer. Cancer Prevention 2001;952:116-23.

---

* Corresponding Author

[10] Rodrigues L, Teixeira J, Schmitt F, Paulsson M, Lindmark Mansson H. Lactoferrin and cancer disease prevention. Critical Reviews in Food Science and Nutrition 2009;49:203–17.

[11] Lizzi AR, Carnicelli V, Clarkson MM, Di Giulio A, Oratore A. Lactoferrin derived peptides: mechanisms of action and their perspectives as antimicrobial and antitumoral agents. Mini-reviews in Medicinal Chemistry 2009;9:687-95.

[12] Picot L, Bordenave S, Didelot S, Fruitier-Arnaudin I, Sannie F, Thorkelsson G, Berge JP, Guerard F, Chabeaud A, Piot JM. Antiproliferative activity of fish protein hydrolysates on human breast cancer cell lines. Process Biochemistry 2006;41:1217-22.

[13] Jang A, Jo C, Kang KS, Lee M. Antimicrobial and human cancer cell cytotoxic effect of synthetic angiotensin-converting enzyme (ACE) inhibitory peptides. Food Chemistry 2007;107:327-36.

[14] Kannan A, Hettiarachchy NS, Lay JO, Liyanage R. Human cancer cell proliferation inhibition by a pentapeptide isolated and characterized from rice bran. Peptides 2010;31:1629-34.

[15] Galvez AF, de Lumen BO. A soybean cDNA encoding a chromatin binding peptide inhibits mitosis of mammalian cells. Nature Biotechnology 1999;17:495-500.

[16] de Mejia EG, Vasconez M, de Lumen BO, Nelson R. Lunasin concentration in different soybean genotypes, commercial soy protein, and isoflavone products. Journal of Agricultural and Food Chemistry 2004;52:5882-87.

[17] Jeong HJ, Jeong JB, Kim DS, de Lumen BO. Inhibition of core histone acetylation by the cancer preventive peptide lunasin. Journal of Agricultural and Food Chemistry 2007;55:632-7.

[18] Park JH, Jeong HJ, de Lumen BO. Contents and bioactivities of lunasin, Bowman-Birk inhibitor, and isoflavones in soybean seed. Journal of Agricultural and Food Chemistry 2005;53:7686-90.

[19] Wang WY, Dia VP, Vasconez M, de Mejia EG, Nelson RL. Analysis of soybean protein-derived peptides and the effect of cultivar, environmental conditions, and processing on lunasin concentration in soybean and soy products. Journal of AOAC International 2008;91:936-46.

[20] Paucar-Menacho LM, Berhow MA, Mandarino JMG, de Mejia EG, Chang YK. Optimisation of germination time and temperature on the concentration of bioactive compounds in Brazilian soybean cultivar BRS 133 using response surface methodology. Food Chemistry 2010;119:636-42.

[21] Paucar-Menacho LM, Berhow MA, Mandarino JMG, Chang YK, de Mejia EG. Effect of time and temperature on bioactive compounds in germinated Brazilian soybean cultivar BRS 258. Food Research International 2010; 43:1856-65.

[22] Hernandez-Ledesma B, Hsieh C-C, de Lumen BO. Lunasin and Bowman-Birk protease inhibitor (BBI) in US commercial soy foods. Food Chemistry 2009;115:574-80.

[23] Cavazos A, Morales EI, Dia VP, González de Mejia E. Analysis of lunasin in comercial and pilot plant produced soybean products and an improved method of lunasin purification. Journal of Food Science 2012;77:C539-45.

[24] Jeong HJ, Lam Y, de Lumen BO. Barley lunasin suppresses ras-induced colony formation and inhibits core histone acetylation in mammalian cells. Journal of Agricultural and Food Chemistry 2002;50:5903-8.

[25] Jeong HJ, Jeong JB, Kim DS, Park JH, Lee JB, Kweon DH, Chung GY, Seo EW, de Lumen BO. The cancer preventive peptide lunasin from wheat inhibits core histone acetylation. Cancer Letters 2007;255:42-8.

[26] Jeong HJ, Lee JR, Jeong JB, Park JH, Cheong YK, de Lumen BO. The cancer preventive seed peptide lunasin from rye is bioavailable and bioactive. Nutrition and Cancer 2009;61:680-6.

[27] Jeong HJ, Jeong JB, Hsieh C-C, Hernández-Ledesma B, de Lumen BO. Lunasin is prevalent in barley and is bioavailable and bioactive in *in vivo* and *in vitro* studies. Nutrition and Cancer 2010;62:1113-9.

[28] Jeong JB, Jeong HJ, Park JH, Lee SH, Lee JR, Lee HK, Chung GY, Choi JD, de Lumen BO. Cancer-preventive peptide lunasin from *Solanum nigrum* L. inhibits acetylation of core histones H3 and H4 and phosphorylation of retinoblastoma protein (Rb). Journal of Agricultural and Food Chemistry 2007;55:10707-13.

[29] Silva-Sanchez C, de la Rosa APB, Leon-Galvan MF, de Lumen BO, de Leon-Rodriguez A, de Mejia EG. Bioactive peptides in amaranth (*Amaranthus hypochondriacus*) seed. Journal of Agricultural and Food Chemistry 2008;56:1233-40.

[30] Ramos Herrera OJ, Sepúlveda Jiménez G, López Laredo AR, Hernández-Ledesma B, Hsieh C-C, de Lumen BO, Bermúdez Torres K. Identification of chemopreventive peptide lunasin in some *Lupinus* species. Proceeding 13th International Lupin Conference, June 6-10, 2011, Poznan, Poland.

[31] Hsieh C-C, Hernández-Ledesma B, de Lumen BO. Complementary roles in cancer prevention: Protease inhibitor makes the cancer preventive peptide lunasin bioavailable. PLoS ONE 2010;5:e8890.

[32] Dia VP, Torres S, de Lumen BO, Erdman JW, de Mejia EG. Presence of lunasin in plasma of men after soy protein consumption. Journal of Agricultural and Food Chemistry 2009;57:1260-6.

[33] Galvez AF, Chen N, Macasieb J, de Lumen BO. Chemopreventive property of a soybean peptide (Lunasin) that binds to deacetylated histones and inhibit acetylation. Cancer Research 2001;61:7473-8.

[34] Hsieh C-C, Hernández-Ledesma B, de Lumen BO. Soybean peptide lunasin suppresses in vitro and in vivo 7,12-dimethylbenz[a]anthracene-induced tumorigenesis. Journal of Food Science 2010;75:H311-6.

[35] Lam Y, Galvez AF, de Lumen BO. Lunasin suppresses E1A-mediated transformation of mammalian cells but does not inhibit growth of immortalized and established cancer cell lines. Nutrition and Cancer 2003;47:88-94.

[36] Jeong HJ, Park JH, Lam Y, de Lumen BO. Characterization of lunasin isolated from soybean. Journal of Agricultural and Food Chemistry 2003;51:7901-6.

[37] de Lumen BO. Lunasin: A cancer preventive soy peptide. Nutrition Reviews 2005;63:16-21.

[38] Dwarakanath BS, Verma A, Bhatt AN, Parmar VS, Raj HG. Targeting protein acetylation for improving cancer therapy. Indian Journal of Medicinal Research 2008;128:13-21.

[39] Dalvai M, Bystricky K. The role of histone modifications and variants in regulating gene expression in breast cancer. Journal of Mammary Gland Biology and Neoplasia 2010;15:19-33.

[40] Hernández-Ledesma, B, Hsieh C-C, de Lumen BO. Relationship between lunasin's sequence and its inhibitory activity of histones H3 and H4 acetylation. Molecular Nutrition and Food Research 2011;55:989-98.
[41] Balasubramanyam K, Swaminathan V, Ranganathan A, Kundu TK. Small molecule modulators of histone acetyltransferase p300. Journal of Biological Chemistry 2003;278:19134-40.
[42] Balasubramanyam K, Altaf M, Varier RA, Swaminathan V, Ravindran A, Sadhale PP, Kundu TK. Polyisoprenylated benzophenone, garcinol, a natural histone acetyltransferase inhibitor, represses chromatin transcription and alters global gene expression. Journal of Biological Chemistry 2004;279:33716-26.
[43] Balasubramanyam K, Varier RA, Altaf M, Swaminathan V, Siddappa NB, Ranga U, Kundu TK. Curcumin, a novel p300/CREB-binding protein-specific inhibitor of acetyltransferase, represses the acetylation of histone/nonhistone proteins and histone acetyltransferasedependent chromatin transcription. Journal of Biological Chemistry 2004;279:51163-71.
[44] Mangiapane S, Blettner M, Schlattmann P. Aspirin use and breast cancer risk: a meta-analysis and meta-regression of observational studies from 2001 to 2005. Pharmacoepidemiology and Drug Safety 2008;12:115-24.
[45] Li Y, Brown PH. Translational approaches for the prevention of estrogen receptor-negative breast cancer. European Journal of Cancer Prevention 2007;16:203-15.
[46] Cuzick J. Chemoprevention of breast cancer. Breast Cancer 2008;15:10-6.
[47] Im JY, Park H, Kang KW, Choi WS, Kim HS. Modulation of cell cycles and apoptosis by apicidin in estrogen receptor (ER)-positive and-negative human breast cancer cells. Chemico-Biological Interactions 2008;172:235-44.
[48] Hsieh C-C, Hernández-Ledesma B, de Lumen BO. Lunasin, a novel seed peptide, sensitizes human breast cancer MDA-MB-231 cells to aspirin-arrested cell cycle and induced-apoptosis. Chemico-Biological Interactions 2010;186:127-34.
[49] Strahl BD, Allis CD. The language of covalent histone modifications. Nature 2000;403:41-5.
[50] Sutherland RL, Musgrove EA. Cyclins and breast cancer. Journal of Mammary Gland Biology and Neoplasia 2004;9:95-104.
[51] Park K, Choi K, Kim H, Kim K, Lee MH, Lee JH, Rim JCK. Isoflavone-deprived soy peptide suppresses mammary tumorigenesis by inducing apoptosis. Experimental and Molecular Medicine 2009;41:371-80.
[52] Lanas A, Bajador E, Serrano P, Fuentes J, Carreño S, Guardia J. Nitrovasodilators, low-dose aspirin, other nonsteroidal antiinflammatory drugs, and the risk of upper gastrointestinal bleeding. New England Journal of Medicine 2000;343:834-9.
[53] Laine L. Review article: Gastrointestinal bleeding with low-dose aspirin: What's the risk? Alimentation and Pharmacology Therapy 2006;24:897-908.
[54] Hsieh C-C, Hernández-Ledesma B, de Lumen BO. Lunasin-aspirin combination against NIH/3T3 cells transformation induced by chemical carcinogens. Plant Foods for Human Nutrition 2011;66:107-13.
[55] Kubo I, Ochi M, Vieira PC, Komatsu S. Antitumor agents from the cashew (*Anacardium occidentale*) apple juice. Journal of Agricultural and Food Chemistry 1993;41:1012-5.

[56] Sung B, Pandey MK, Ahn KS, Yi TF, Chaturvedi MM, Liu MY, Aggarwal BB. Anacardic acid (6-nonadecyl salicylic acid), an inhibitor of histone acetyltransferase, suppresses expression of nuclear factor-κB-regulated gene products involved in cell survival, proliferation, invasion, and inflammation through inhibition of the inhibitory subunit of nuclear factor- κ Bα kinase, leading to potentiation of apoptosis. Blood 2008;111:4880-91.

[57] Hsieh C-C, Hernández-Ledesma B, de Lumen BO. Cell proliferation inhibitory and apoptosis inducing properties of anacardic acid and lunasin in human breast cancer MDA-MB-231 cells. Food Chemistry 2010;125:630-6.

[58] Kim YS, Milner JA. Dietary modulation of colon cancer risk. Journal of Nutrition 2007;173:2576S–9S.

[59] Dia VP, de Mejia EG. Lunasin induces apoptosis and modifies the expression of genes associated with extracellular matrix and cell adhesion in human metastatic colon cancer cells. Molecular Nutrition and Food Research 2011;55:623-34.

[60] Dia VP, de Mejia EG. Lunasin potentiates the effect of oxaliplatin preventing outgrowth of colon cancer metastasis, binds to a5b1 integrin and suppresses FAK/ERK/NF-κB signaling. Cancer Letters 2011;313:167–80.

[61] Dia VP, de Mejia EG. Lunasin promotes apoptosis in human colon cancer cells by mitochondrial pathway activation and induction of nuclear clusterin expression. Cancer Letters 2010;295:44-53.

[62] Dittmar T, Heyder C, Gloria-Maercker E, Hatzmann W, Zanker KS. Adhesion molecules and chemokines: the navigation system for circulating tumor (stem) cells to metastasize in an organ-specific manner. Clinical and Experimental Metastasis 2008; 25:11-32.

[63] Kasteng F, Sobocki P, Svedman C, Lundkvist J. Economic evaluations of leukemia: a review of the literature. International Journal of Technology Assessment in Health Care 2007;23:43-53.

[64] de Mejia EG, Wang W, Dia VP. Lunasin, with an arginine-glycine-aspartic acid motif, causes apoptosis to L1210 leukemia cells by activation of caspase-3. Molecular Nutrition and Food Research 2010;54:406-14.

[65] Strope SA, Andriole GL. Update on chemoprevention for prostate cancer. Current Opinion in Urology 2010;20:194-7.

[66] Galvez AF, Huang L, Magbanua MMJ, Dawson K, Rodriguez R. Differential expression of thrombospondin (THBS1) in tumorigenic and nontumorigenic prostate epithelial cells in response to a chromatin-binding soy peptide. Nutrition Cancer 2011;63:623-6.

[67] Kuper H, Adami HO, Trichopoulos D. Infections as a major preventable cause of human cancer. Journal of International Medicine 2000;248:171-83.

[68] Allavena P, Garlanda C, Borrello MG, Sica A, Mantovani A. Pathways connecting inflammation and cancer. Current Opinion in Genetics Development 2008;18:3-10.

[69] Hernández-Ledesma B, Hsieh C-C, de Lumen BO. Anti-inflammatory and antioxidant properties of peptide lunasin in RAW 264.7 macrophages. Biochemical and Biophysical Research Communications 2009;390:803-8.

[70] Dia VP, Wang W, Oh VL, de Lumen BO, de Mejia EG. Isolation, purification and characterization of lunasin from defatted soybean flour and in vitro evaluation of its anti-inflammatory activity. Food Chemistry 2009;114:108-15.

[71] de Mejia EG., Dia VP. Lunasin and lunasin-like peptides inhibit inflammation through suppression of NF-kB pathway in the macrophage. Peptides 2009;30:2388-98.

[72] Larkins TL, Nowell M, Singh S, Sanford GL. Inhibition of cyclooxygenase-2 decreases breast cancer cell motility, invasion and matrix metalloproteinase expression. BMC Cancer 2006;6:181.

[73] Liu CF, Pan TM. Recombinant expression of bioactive peptide lunasin in *Escherichia coli*. Applied Microbiology and Biotechnology 2010;88:177-86.

[74] Ames BN, Shigenaga MK, Hagen TM. Oxidants, antioxidants, and the degenerative diseases of aging. Proceedings of the National Academy of Sciences of USA 1993;90:7915-22.

[75] Khan N, Afaq F, Mukhtar H. Cancer chemoprevention through dietary antioxidants: progress and promise. Antioxidant & Redox Signaling 2008;10:475-510.

[76] Jeong JB, de Lumen BO, Jeong HJ. Lunasin peptide purified from *Solanum nigrum* L. protects DNA from oxidative damage by suppressing the generation of hydroxyl radical via blocking fenton reaction. Cancer Letters 2010;293:58-64.

[77] Hernández-Ledesma B, Espartosa DM, Hsieh C-C, de Lumen BO, Recio I Role of peptide lunasin's antioxidant activity on its chemopreventive properties. Annals of Nutrition and Metabolism 2011;58:354.

[78] Seber LE, Barnett BW, McConnell EJ, Hume SD, Cai J, Boles K, Davis KR. Scalable purification and characterization of the anticancer lunasin peptide from soybean. PLoS ONE 2012;7:e35409.

[79] Rizzello CG, Nionelli L, Coda R, Gobbetti M. Synthesis of the cancer preventive peptide lunasin by lactic acid bacteria during sourdough fermentation. Nutrition and Cancer 2011;64:111-20.

[80] Kyle S, James KAR, McPherson MJ. Recombinant production of the therapeutic peptide lunasin. Microbial Cell Factories 2012; 11:28.

Chapter 2

Bowman-Birk Inhibitors from Legumes: Utilisation in Disease Prevention and Therapy

Alfonso Clemente, Maria del Carmen Marín-Manzano,
Maria del Carmen Arques and Claire Domoney

Additional information is available at the end of the chapter

http://dx.doi.org/10.5772/51262

1. Introduction

Serine proteases have long been recognized as major players in a wide range of biological processes including cell signaling, cell cycle progression, digestion, immune responses, blood coagulation and wound healing. Their role in the physiology of many human diseases, ranging from cancer and inflammatory disorders to degenerative diseases, now represents an increasingly important feature of this family of enzymes. Proteases are tightly controlled through a number of different mechanisms, including regulation of gene expression, recognition of the substrate by the active site, activity regulation by small molecules, changes in cellular location, post-translational modifications, interaction with other proteins and/or through inhibition of proteolysis by protease inhibitors (PI) [1-3]. This last mechanism usually involves competition with substrates for access to the active site of the enzyme and the formation of tight inhibitory complexes. An understanding of the role played by serine proteases and their specific inhibitors in human diseases offers novel and challenging opportunities for preventive and/or therapeutic intervention [4].

Within this framework, there is a growing interest in naturally-occurring serine protease inhibitors of the Bowman-Birk family due to their potential chemopreventive and/or therapeutic properties which can impact on several human diseases, including cancer, neurodegenerative diseases and inflammatory disorders. In light of the Food and Drug Administration (FDA) approval for trials of Bowman-Birk inhibitors (BBI) concentrate (BBIC), a protein extract of soybean (*Glycine max*) enriched in BBI, as an 'Investigational New Drug', human trials have been completed in patients with benign prostatic hyperplasia [5], oral leukoplakia [6-8] and ulcerative colitis [9] (**Table 1**). Although, in most of these cases, the intrinsic ability of BBI to inhibit serine proteases has been related to beneficial health properties, the mechanisms of action and the identity of their therapeutic targets are largely unknown. In this chapter, we describe the emerging evidence for the positive contribution of BBI from legumes to disease prevention and therapy.

Disease	Type of trial	Experimental design	Main results	Ref.
Benign prostatic hyperplasia	Phase I trial	Duration: 6 months. 19 patients. Daily doses up to 800 CIU [a]	Significant decrease (up to 43 %) in PSA[b] levels after treatment. Decrease of prostate volume in patients. No dose-limiting toxicity	[5]
Oral leukoplakia	Phase I trial	Duration: 1 month. 24 patients. Single daily dose: 800 CIU	BBI was well tolerated. No clinical evidence of toxicity or any adverse reaction	[6]
	Plase II trial	Duration: 1 month. 32 patients. Administration: twice daily, up to 1066 CIU	31 % of patients achieved clinical response and lesion area decreased after treatment. Dose-dependent effect. BBI was non-toxic. The positive clinical effect of BBIC could be due to the inhibition of serine proteases involved in the cleavage of neu-oncogen protein on the cell surface, preventing the release of the extracellular domain of the protein into the bloodstream	[7]
	Double-blind randomized, Placebo-controlled phase IIb trial	148 patients. Daily dose: 600 CIU	Although this study has not been completed yet, preliminary results suggest that BBIC is not fully effective in patients	[8]
Ulcerative colitis	A randomized, double blind, placebo-controlled trial	12 weeks of therapy. 28 patients. Daily dose: 800 CIU	BBIC might be associated with the regression of disease without apparent toxicity or adverse side effects	[9]

[a]CIU: chymotrypsin inhibitory units; [b]PSA: prostate specific antigen

Table 1. Clinical trials utilizing a protein extract of soybean enriched in Bowman-Birk inhibitors (BBIC)

2. The Bowman-Birk family

2.1. Sources and occurrence

Plant PI can be categorized into at least 12 different families with 10 of these targeting serine proteases and adopting the standard mechanism of inhibition [10]. Members of the

Bowman-Birk family are canonical serine PI of low molecular weight, being particularly abundant in legume seeds. Soybean BBI represent the most extensively studied members of the Bowman-Birk family, but related BBI from other dicotyledonous legumes [including chickpea (*Cicer arietinum*), common bean (*Phaseolus vulgaris*), lentil (*Lens culinaris*) and pea (*Pisum sativum*)] and from monocotyledonous grasses (*Poaceae*) [including wheat (*Triticum aestivum*), rice (*Oryza sativa*) and barley (*Hordeum vulgare*) species], have been identified and characterized.

The BBI that are expressed in seeds are the products of multi-gene families. Several isoinhibitors have been identified in seeds of individual species [11, 12]. The expression of distinct genes, together with the post-translational modifications of primary gene products, combine to give rise to the array of BBI-like variants described for many legume species. Variants in overall and active site sequences, size, functional properties and spatial pattern of expression have been described [13]. As a result, qualitative and quantitative differences in protease inhibitory activities have been shown in comparisons of pea genotypes [14, 15]. The close linkage of the genes encoding BBI, demonstrated for a number of legume species [16], allows the development of near-isolines having distinct haplotypes. In pea, the co-segregation of quantitative and qualitative variation has been used to develop a series of near-isolines, which have allowed the biological significance of a five-fold variation in seed protease inhibitory activity to be investigated at the level of ileal digestibility [17, 18]. These lines now facilitate related studies on the positive contribution of seed BBI to the prevention of disease states.

The occurrence of BBI in soy foods (soymilk, soy infant formula, defatted soy meal, oilcake, tofu, soybean protein isolate and soybean protein concentrate, among others) is noteworthy, where BBI may be present in different amounts. The soy varieties used, the products themselves and the technological processes used in their preparations all contribute to variation in BBI concentration. In order to quantify BBI in soy foods, enzymatic and immunological methods have been developed; however, no comprehensive information on the concentration of BBI in soy foods is available currently. Recently, Hernández-Ledesma *et al.* [19] reported BBI concentrations in 12 soymilk samples ranging from 7.2 to 55.9 mg BBI/100 mL of soymilk, while BBI was not detected in a soy-based infant formula. BBI was also reported in tofu samples, with concentrations ranging between 2.9 and 12.4 mg/100 g product. Since BBI could not be detected in natto and miso samples, it may be assumed that BBI were degraded during the fermentation process.

2.2. Inhibitory properties

The inhibitory activity of BBI is due to the formation of stable complexes between the inhibitor and target proteases. The conformation of the reactive site loop is complementary to the active site of the protease inhibited and allows BBI to bind tightly to proteases in a substrate-like manner [20, 21]; the resulting non-covalent complex renders the target protease inactive. Upon complex formation, BBI may be cleaved very slowly (low K_{cat}). In legumes, the enzyme inhibitory activity is associated with two subdomains of the BBI, located at opposite sides of the molecule; each canonical loop is contained within a nonapeptide region joined via a disulphide bond between flanking cysteine residues. The double-headed BBI have a characteristic highly conserved array of intra-molecular disulphide bridges occurring among 14 Cys residues [22]. The two binding loops can each inhibit independently and simultaneously an enzyme molecule, which may be the same or distinct types of enzyme [13]. The specificity of each reactive site is determined by the identity of the amino acid residue in position P_1 [23]. In double-headed BBI, the first active site is usually involved in trypsin inhibition, with the P_1 position being occupied by Arg or Lys. The presence of an Ala residue at the P_1 position has been reported to be associated with inhibition of elastase [24]. Chymotrypsin is often the target of the second inhibitory domain as it has a hydrophobic amino acid in its P_1 position [13, 25]. Additional residues adjacent to the reactive site peptide bond (P_1-P_1') can have a significant effect on the affinity of BBI for particular protease targets [13]. BBI from legumes include potent inhibitors of both trypsin and chymotrypsin, with K_i values within the nanomolar range reported for different members, including soybean [26], pea [27, 28], lentil [29, 30] and lupin (*Lupinus albus*) BBI [31].

3. Bioavailability and metabolism of BBI

In order to exert any local or systemic health benefits, dietary BBI must resist degradation and maintain biological activity, at least to some extent, after food processing and further passage through the gastrointestinal tract (GIT) [22]. BBI from several legume sources have been shown to resist thermal treatment (up to 100 °C), under either neutral or acidic conditions [32]. Most of the heat-resistant trypsin inhibitory activity in processed legumes is attributable to BBI. At temperatures of 80 °C or lower, chickpea BBI were found to be stable and their inhibitory activities to be unaffected by thermal treatment [33]. Soybean BBI do not lose activity at pH values as low as 1.5 in the presence of pepsin at 37 °C for 2 h [34]; these proteins are also stable to both the acidic conditions and the action of digestive enzymes under simulated gastric and intestinal digestion [35]. Such stability is associated with the rigid structure provided by the seven intra-molecular disulphide bridges that maintain the structural and functional features of the binding sites by adding covalent attachment to the inhibitor core [10, 36]. BBI are fully inactivated by autoclaving or reduction of their disulphide bridges followed by alkylation of the cysteinyl sulfhydryl groups [26].

The resistance of BBI to extreme conditions within the GIT may favour the transport of biologically active BBI across the gut epithelium and could allow their distribution to target organs or tissues in order to exert their beneficial effects locally. The uptake and distribution of soybean BBI, following oral administration, has been examined in rodents. By using [^{125}I] BBI, it was demonstrated that BBI becomes widely distributed in mice 3 h after oral administration [37]. Labelled BBI was detected in the luminal contents of the small and large intestine; analysis of tissue homogenates revealed also the presence of active BBI in internal organs where soybean BBI have been shown to exert anti-carcinogenic effects (see next section). By using inverted sacs from different sections of the small intestine, it was demonstrated that active BBI could be transported effectively across the gut epithelium. It has been shown that soybean BBI have a serum half-life of 10 h in rats and hamsters, and are excreted in urine and faeces [38]. In humans, BBI are taken up rapidly and can be detected in the urine within 24-48 h [6]. These findings suggest that BBI are absorbed after oral administration and can reach several tissues and organs.

BBI have potential health-promoting properties within the GIT [22]. *In vivo* studies have demonstrated the presence of active BBI in the small intestine. Hajós *et al.* [39] reported the survival (~ 5 % of total ingested) of soybean BBI in an immunological reactive form in the small intestine of rats; unfortunately, the inhibitory activities of BBI were not evaluated in these experiments. More recently, it has been demonstrated that BBI from chickpea seeds can resist both acidic conditions and the action of digestive enzymes, and transit through the stomach and small intestine of pigs, generally held as a suitable model for human digestive physiology [40]. The presence of active BBI (5-8 % of the total ingested BBI) at the terminal ileum revealed the resistance of at least some, or a significant proportion, of these proteins to the extreme conditions of the GIT *in vivo*. Chromatographic, electrophoretic and enzymatic data obtained from ileal samples suggested that most of the BBI activity is derived from a protein core containing the two binding domains, and resistant to proteolysis. *In vitro* incubation studies of soybean BBI with mixed fecal samples of pigs showed that BBI remained active and their intrinsic ability to inhibit serine proteases was not significantly affected by the enzymatic or metabolic activity of fecal microbiota [41]. All of these results are significant to investigations of the potential uses of BBI in preventive or therapeutic medicine.

4. Chemopreventive properties of Bowman-Birk inhibitors

Chemoprevention is the use of natural agents or synthetic drugs to halt or reverse the carcinogenesis process before the emergence of invasive cancer. The fact that certain dietary constituents can exert chemopreventive properties has major public health implications and the widespread, long-term use of such compounds should be promoted in populations at normal risk, based on understanding the scientific basis of their beneficial effects. In particular, BBI have been linked to a possible protective effect against both inflammatory disorders and cancer development (**Table 2**).

4.1. Colorectal cancer

Nutritional intervention and/or dietary manipulation have been suggested as key strategies to prevent and/or control colorectal carcinogenesis [42, 43], one of the major causes of cancer-related mortality worldwide in both men and women [44]. There is now robust preclinical evidence to suggest that dietary BBI from several legume sources can prevent or suppress cancer development and associated inflammatory disorders within the GIT [22]. Soybean BBI have been reported to be effective at concentrations as low as 10 mg/100 g diet, in reducing the incidence and frequency of colorectal tumors, in studies based on the dimethylhydrazine (DMH) rat model, where no adverse effect of BBI was documented for animal growth or organ physiology [45]. When the inhibitory activity of BBI is abolished, any suppressive effect on colorectal tumor development disappears, suggesting that the inhibitory properties of BBI against serine proteases may be required for their reported chemopreventive properties. Proteases play a critical role in tumorigenesis, where their activities become dysregulated in colorectal cancer and neoplastic polyps [46]. In particular, serine proteases are key players in several biological functions linked to tumor development, including cell growth (dys)regulation and cell invasion as well as angiogenesis and inflammatory disorders. Some of these proteases have been reported as promising cancer biomarkers [47-49] (**Table 3**). An understanding of the role played by specific serine proteases in the biological processes associated with disease may suggest modes of therapeutic intervention [1, 50]. Successful examples of therapeutic intervention using PI include ubiquitin-proteasome inhibitors in the treatment of multiple myeloma [51]. The ubiquitin-proteasome pathway is essential for most cellular processes, including protein quality control, cell cycle, transcription, signalling pathways, protein transport, DNA repair and stress responses [52]. Inhibition of proteasome activity leads to accumulation of poly-ubiquitinylated and misfolded proteins, endoplastic reticulum stress and eventually apoptosis. Although soybean BBI has been demonstrated to inhibit the proteasomal activity of MCF7 breast cancer cells (see section 4.4), the proteasomal inhibition in colon cancer cells need to be unambiguously demonstrated. Another potential therapeutic target of BBI is matriptase (also known as MT-SP1 or epithin), an epithelial-specific member of the type II transmembrane serine protease family, which plays a critical role in differentiation and function of the epidermis, gastrointestinal epithelium and other epithelial tissues. Several studies suggest that matriptase is over-expressed in a wide variety of malignant tumors including prostate, ovarian, uterine, colon, epithelial-type mesothelioma and cervical cell carcinoma [53]. It has been proposed to have multiple functions, acting as a potential activator of critical molecules associated with tumor invasion and metastasis. MT-SP1 contributes to the upstream activation of tumor growth and its progression through the selective degradation of extracellular matrix proteins and activation of cellular regulatory proteins, such as urokinase-type plasminogen activator, hepatocyte-growth factor/scatter factor and protease-activated receptor [54]. Although the ability of soybean BBI to inhibit a secreted form of recombinant MT-SP1 has been demonstrated [55], the clinical relevance of such inhibition has not been proven yet. The validation of specific serine proteases as BBI targets, together with the identification of natural BBI variants, and the design of specific potent inhibitors of these proteases, will contribute to the assessment of BBI as colorectal chemopreventive agents for preventive and/or therapeutic medicine [22].

Cancer type	BBI source	Carcinogen	Model system	Effect and/or mechanisms of action	Refs.
Colorectal	Soybean	DMH[a]	Colon carcinogenesis in rodents	Reduction of incidence and frequency of tumors likely via protease inhibition	[45]
	Soybean	DMH	Mouse colon and anal inflammation	Suppression of adenomatous tumors of the GIT	[56]
	Soybean	DSS[b]	Mouse colon inflammation	Suppression of histological inflammation parameters, lower mortality rate and delayed onset of mortality	[57]
	Horsegram	DMH	Colorectal carcinogenesis	Protective role against inflammation and pre-neoplastic lesions	[58]
	Lentil	-	Colon cancer cells	Proliferation of HT29 colon cancer cells was decreased (IC_{50} = 32 μM), whereas the non- malignant fibroblastic CCD18Co cells were unaffected	[30]
	Pea	-	Colon cancer cells	The anti-proliferative effect of BBI in colon cancer cells are demonstrated	[15]
	Soybean	-	Colon cancer cells	Time- and concentration-dependent anti-proliferative effect on HT29 cells, arrest at G0-G1 phase; trypsin- and chymotrypsin-like proteases are potential targets	[26]
	Recombi-nant proteins	-	Colon cancer cells	rTI1B, a major BBI isoinhibitor from pea, having trypsin and chymotrypsin inhibitory activity, affected the proliferation of colon cancer cells; however, a derived inactive mutant did not show any anti-proliferative effect	[68]
Gastric	Field bean	Benzo-pyrene	Mouse stomach carcinogenesis	BBI was more effective in prevention than in therapeutic treatment, with activity related to its protease inhibitory ability	[100]

Cancer type	BBI source	Carcinogen	Model system	Effect and/or mechanisms of action	Refs.
Breast	Black-eyed pea	-	Breast cancer cells	BBI induced apoptosis, cell death and lysosome membrane permeabilization	[79]
	Soybean	-	Breast cancer cells	Proteasome was reported as potential therapeutic target in MCF-7 cells	[78]
Prostate	Soybean	-	Prostate cancer cells and rat prostate carcinogenesis	BBI exerted chemopreventive activity associated with induction of connexin-43 expression and apoptosis	[76,77]
	Soybean	-	Prostate cancer xeno-grafts in nude mice	BBI and BBIC inhibited the growth of LNCaP cells	[72]
	Soybean	-	Prostate cancer cells	BBI prevented the generation of activated oxygen species and activated DNA repair through a p53-dependent mechanism	[74, 75]
Oral	Soybean	-	Oral leukoplakia	BBIC exerted a dose-dependent reduction in oral lesion size in 31% of patients without any adverse effects; modulation of protease activity and *neu* oncogene levels was observed	[6,7,69]

[a]DMH: dimethylhydrazine; [b]DSS: dextran sulphate sodium

Table 2. *In vitro* and *in vivo* studies showing chemopreventive properties of Bowman-Birk inhibitors

A strong interest exists in investigating the potential of BBI as anti-inflammatory agents within the GIT. In rodents, soybean BBI treatment appears to have a potent suppressive effect on colon and anal gland inflammation, following exposure to carcinogenic agents [56], or when assessed in an acute injury/colitis model [57]. The protective effect of BBI from soybean or those from perennial horsegram (*Macrotymola axillare*) against inflammation and development of pre-neoplastic lesions induced in the DMH mouse model was reported recently [58]. Given the lack of toxicity as well as the reported anti-inflammatory properties in animals, the potential for BBIC to benefit patients with ulcerative colitis has been evaluated. In a randomized double-blind placebo-controlled trial, a dose of 800 chymotrypsin inhibitor units (CIU) per day over a three-month treatment period was associated with a clinical response and induction of remission, as assessed by the Sutherland Disease Activity Index score [59], in patients with ulcerative colitis, without apparent toxicity [9]. Approximately 50 % of patients responded clinically and 36 % showed remission of disease. Several mechanisms have been proposed to explain the anti-inflammatory properties of BBI. The ability of BBI to decrease the production and release of superoxide anion radicals, mediators of inflammatory processes, in purified human polymorphonuclear leukocytes [60] and in differentiated HL60 cells [59] has been reported. The decrease in superoxide radical levels may reduce free radical-induced DNA damage and transformation to malignant phenotypes. In addition, superoxide radicals can initiate a wide range of toxic oxidative reactions, including lipid peroxidation. In this regard, it has been demonstrated that BBI can reduce the content of lipid peroxides in irradiated cells *in vitro* [61], a reduction that is presumed to be linked to the anti-inflammatory activity of BBI. The role that certain serine proteases play during proteolysis in acute and chronic inflammatory processes is well recognized. The ability of soybean BBI to inhibit serine proteases involved in inflammatory processes, such as cathepsin G [62, 63], elastase [64] and mast cell chymase [64] has been reported. The last enzyme acts as a chemo-attractant and may play a role in the accumulation of inflammatory cells during development of allergic and non-allergic diseases [65]. The interaction of chymase and BBI may impact on other processes involved in anti-inflammatory responses, such as the regulation of collagenase [66] and interleukin 1β (IL-1β) [67]. Nevertheless, a clear correlation between the inhibition of these serine proteases and the anti-inflammatory properties associated with soybean BBI has not been demonstrated clearly [22].

Serine protease	Function	Pathological process	Refs.
Tryptase	Phagocytosis, degradation of ECM[a] compounds, regulation of inflammatory responses, blood coagulation	Atherosclerosis, asthma, inflammatory disorders	[97, 98]
Cathepsin G	ECM degradation, migration, regulation of inflammatory disorders	Inflammation, metastasis	[62]
Matriptase	Matrix degradation, regulation of intestinal barrier, iron metabolism	Pathogenesis of epithelial tissues, tumor growth and progression	[55]
Human elastase	Pathogen killing, ECM degradation, inflammatory disorders	Pulmonary disease, inflammation	[62, 99]
Chymase	Degradation of ECM compounds, regulation of inflammatory responses	Inflammation, asthma, gastric cancer	[64]
Proteasome	Protein degradation, cell proliferation, differentiation, angiogenesis and apoptosis	Carcinogenesis, inflammation, neurodeg-eneration	[58, 78]

[a]ECM: extracellular matrix

Table 3. Serine proteases involved in pathological processes as potential therapeutic targets of soybean BBI and related proteins (adapted from Clemente et al., 2011 [22]).

In previous studies, a significant concentration- and time-dependent decrease in the growth of an array of colon cancer cells (HT29, Caco2, LoVo) has been demonstrated *in vitro*, following treatment with BBI variants from several legume sources, including pea, lentil and soybean; in contrast, the growth of non-malignant colonic fibroblastic CCD18-Co cells was unaffected by BBI [15, 26, 30]. Recently, the anti-proliferative effect of rTI1B, a major pea isoinhibitor expressed heterologously in *Pichia pastoris*, has been evaluated using colon cancer cells grown *in vitro*. Comparisons of the effects of rTI1B with those observed using a related synthetic mutant derivative, showed that the proliferation of HT29 colon cancer cells was inhibited significantly by rTI1B in a dose-dependent manner, whereas the mutant which lacked trypsin and chymotrypsin inhibitory activity did not show any significant effect on colon cancer cell growth (**Figure 1**) [68]. Although the molecular mechanism(s) of this chemopreventive activity remains unknown, the reported data indicate that both trypsin- and chymotrypsin-like serine proteases involved in carcinogenesis are likely primary targets for BBI.

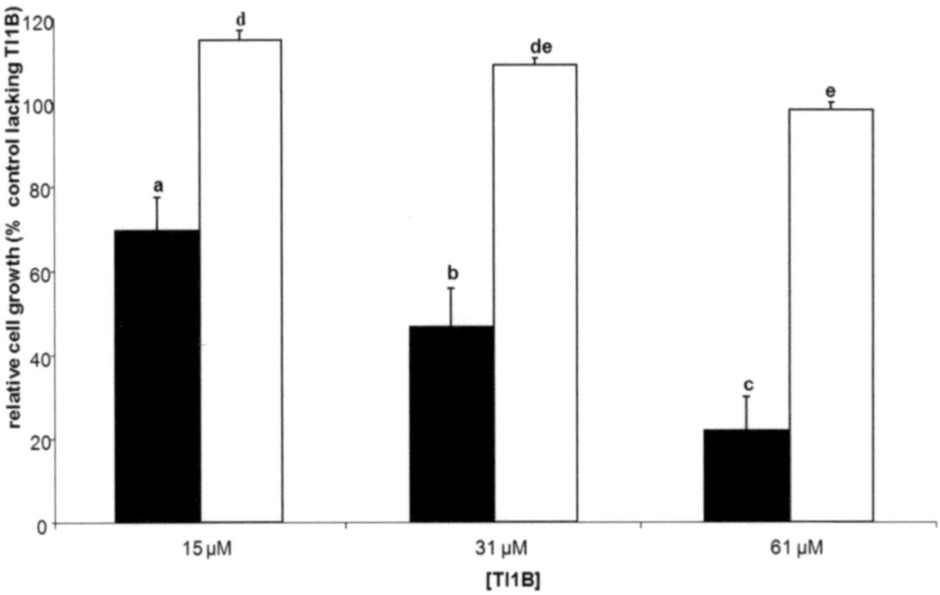

Figure 1. Dose–response effects of rTI1B (closed bars), a major pea isoinhibitor expressed in *Pichia pastoris*, and the corresponding inactive mutant (open bars), having amino acid substitutions at the P_1 positions in the two inhibitory domains, on the *in vitro* growth of HT29 human colorectal adenocarcinoma cells. Growth media were supplemented with rTI1B in the concentration range 15–61 µM and cells harvested after a period of 96 h. Data are means of at least three independent experiments, each having four technical replicates; bars represent standard deviations. Means not sharing superscript letters differ significantly ($p < 0.05$; Bonferroni's test) (Adapted from Clemente et al., (2012) [68]).

4.2. Oral leukoplakia

Leukoplakia in the oral cavity is considered a suitable model for the study of chemoprevention because the precancerous lesions are readily accessible to visual examination, diagnostic sampling and evaluation of response to treatment. In a Phase I clinical trial, no clinical evidence of toxicity or any adverse effect was apparent when BBIC was administered as a single oral dose of up to 800 CIU to twenty-four patients with oral leukoplakia over one month-period [6]. The study revealed that BBIC was well-tolerated and no allergic reactions, gastrointestinal side-effects or other clinical symptoms were elicited. In a non-randomized phase IIa clinical trial, treatment with BBIC for one month resulted in a dose-dependent reduction in oral lesion size in 31% of patients [7]. The positive clinical effect of BBIC was associated with modulation of protease activity and *neu* oncogene levels (as the surrogate endpoint biomarker for the trial) in exfoliated oral mucosal cells [7, 69]; however, this evidence is indirect and the specific target proteases inhibited were not reported. A recent phase IIb randomized, double-blind, placebo-controlled trial involving patient treatment with BBIC for six months was performed. Even though this multi-institutional study has not been completed yet, BBIC does not seem to be fully effective as chemopreventive agent for the management of oral leukoplakia [8].

4.3. Prostate cancer

Prostate cancer is the second most frequently diagnosed cancer in men although the incidence of cancer varies greatly throughout the world. Dietary habits and lifestyle have been identified as major risk factors in prostate cancer growth and progression, suggesting that prostate cancer might be preventable [70]. Epidemiological studies have shown an inverse association between soy intake and the risk of developing prostate cancer [71]. Preclinical and clinical studies have shown the potential chemopreventive properties of BBI in prostate cancer. Purified soybean BBI and BBIC have been shown to inhibit the growth of LNCaP human prostate cancer xenografts in nude mice [72], and to decrease the growth, invasion and clonogenic survival of several human prostate cancer cells [73]. The effectiveness of soybean BBI in preventing the generation of activated oxygen species in prostate cancer cells [74] and in activating DNA repair through a p53-dependent mechanism has been reported [75]. More recently, BBIC has prevented the growth of prostate tumors in transgenic rats developing adenocarcinoma, most likely as a consequence of its anti-proliferative activity via induction of connexin 43 expression [76, 77]. In humans, a double-blind, randomized, phase I trial was carried out in nineteen male subjects with benign prostatic hyperplasia, which is a precursor condition for prostate cancer, and lower urinary tract symptoms [5]. In this study, the authors demonstrated that BBIC treatment for six months reduced levels of prostate-specific antigen (PSA), a clinical marker for prostate cancer, and prostate volume in patients. Additional clinical studies will be necessary to determine the potential of BBIC as prostate cancer chemopreventive agent.

4.4. Breast cancer

Breast cancer is one of the most frequent cancer types and is responsible for the highest mortality rate among women. Novel complementary strategies, including chemoprevention, have been suggested. As ^{125}I-BBI, when orally administrated in rodents, has been demonstrated in the bloodstream and distributed through the body [37-38], its chemopreventive properties could occur in breast tissue. *In vitro* studies have reported the potential of BBI as chemopreventive agents in breast cancer. Soybean BBI has been shown to inhibit, specifically and potently, the chymotrypsin-like proteasomal activity in MCF7 breast cancer cells *in vitro* and *in vivo* [78]. The proteasomal inhibition was associated with an accumulation of ubiquitinylated proteins and the proteasome substrates, $p21^{Cip1/WAF1}$ and $p27^{Kip1}$; a down-regulation of cyclin D1 and E was also observed. These authors suggested that soybean BBI abates proteasome function and, dependent on dose and time, up-regulates MAP kinase phosphatase-1 (MKP-1), which in turn suppresses phosphorylation coupled to extracellular signal-related kinase activity in MCF7 treated cells. The ability of soybean BBI to inhibit the proteasomal chymotrypsin-like activity in intact MCF7 cells suggests that the protease inhibitor can penetrate cells and facilitate the inhibition of intracellular target proteases. Recent findings have demonstrated that BBI from black-eye pea (*Vigna unguiculata*) induced apoptotic cell death in MCF7 breast cancer cells associated with severe cell morphological alterations, including the alteration of the nuclear morphology, plasma membrane fragmentation, cytoplasm disorganization, presence of double-membrane vesicles, mitochondrial swelling and lysosome membrane permeabilization [79].

4.5. Radioprotection

Radiotherapy is used in the treatment of a broad range of malignant tumors with the aim to inflict maximal damage on the tumor tissue. Exposure of surrounding normal tissue to therapeutic radiation should be minimized to avoid side effects that can have a significant impact on general status and quality of life of patients. The use of radioprotective agents to reduce the damage in normal tissue may improve the therapeutic benefit of radiotherapy. The radioprotective properties of BBI have been tested on cell cultures; so far, no data regarding efficacy in humans are available. Soybean BBI have shown potent and selective radioprotection of normal tissue *in vitro* [80] and *in vivo* [81], without protecting tumor tissue. Dittmann *et al.* [81] showed that soybean BBI increased clonogenic survival after irradiation only in cells, either normal or transformed, having a wild-type *p53* tumor suppressor gene. In a cell line with inducible expression of mutated *p53*, the radioprotective effect of BBI was only detected when the expression of the mutated *p53* was switched off [75]. The activation of the DNA–repair machinery, induced by pre-treatment of fibroblastic cells with soybean BBI, suggests a possible BBI-mediated *p53*-dependent mechanism for radioprotection [82, 83]. Since a high number of tumors have lost *p53* function during their development, the clinical application of BBI to protect normal tissue from radiation damage would effectively improve the therapeutic outcome of radiotherapy.

The involvement of BBI in radiation-induced signaling cascades, and their role in stabilizing a specific tyrosine phosphatase that interferes with the activation of an epidermal growth factor receptor in response to radiation exposure, could be responsible for such protection [84]. Experiments carried out with linear forms of BBI demonstrated that the secondary structure of BBI, required for the protease inhibitory activity, was not necessary for its radioprotective properties [85]. The radioprotective effect of soybean BBI was mainly associated with the chymotrypsin inhibitory site [86] and could be mimicked using a synthetic linearized nonapeptide (CALSYPAQC), corresponding to the active site for chymotrypsin inhibition, but lacking protease inhibitor activity [85]. These observations provide opportunities for the use of synthetic peptides for protecting against ionizing radiation. BBI, when applied topically, once a day for 5 days, to SKH-1 hairless mice with a high risk of developing UV-induced skin tumors, inhibited the formation and growth of skin tumors [85]. In addition, topical application of nondenatured soymilk, once a day for a period of five days prior to UV irradiation, to mini-swine skin reduced or completely eliminated UV-induced formation of thymine dimers and apoptotic cells. Finally, BBIC appears to play a radioprotective role in radiation-induced cataract formation reducing the prevalence and severity of the lens opacifications in mice exposed to high-energy protons [88].

5. Beneficial properties of Bowman-Birk inhibitors in non-related cancer diseases

The loss of muscle protein due to inactivity, disease or aging is a process known as muscular atrophy or wasting. Skeletal muscular atrophy in response to disuse involves both a decrease in protein synthesis and increased protein degradation, predisposing humans to undergo a substantial loss of muscle mass. In connection with this, complex proteolytic cascades may provide a mechanism for the initiation of protein degradation during atrophy. Dietary intervention suggests possible therapeutic strategies *via* protease inhibition to diminish muscular atrophy and loss of strength following unloading. Skeletal muscular atrophy can be reproduced experimentally in rodents by hind-limb unloading. Dietary supplementation containing 1% BBIC has been reported to inhibit unloading-induced weakness in mice [89], promoting redox homeostasis in muscle fibers and blunting atrophy-induced weakness [90]. Morris *et al.* [89] suggested that inhibiting muscle degrading proteases may provide a new pharmacological strategy in treating skeletal muscular atrophy; however, such proteases remain uncharacterized. More recently, oral administration of BBIC was reported to improve muscle mass and function and to modulate pathological processes in the mouse model of Duchenne muscular dystrophy (mdx mouse) [91]. Chymase, a serine protease that is released from mast cells, is involved in inflammatory processes and is susceptible to inhibition by BBI, and hence has been suggested as a BBI target in mdx mice. Additional studies are necessary to elucidate the potential therapeutic role of BBI in muscular dystrophy.

Multiple sclerosis (MS) is a chronic inflammatory disease of the central nervous system characterized by progressive demyelination of the brain and spinal cord. Available therapeutic treatments have only limited efficacy and show significant side effects. The search for novel therapeutic agents that can be administered orally, and act synergistically with existing therapies, would be useful for patients with MS. Purified soybean BBI and BBIC have been shown to be effective in the suppression of experimental autoimmune encephalomyelitis in rodents, a model to study the pathogenic mechanisms of MS and to test potential therapies [92]. The oral administration of BBI in mice caused an improvement of several disease parameters (onset, severity, weight loss, inflammation, neuronal loss and demyelination), with no apparent adverse effects [93, 94]. Interestingly, BBI ameliorated disease, even when treatment was initiated after disease onset, *via* an IL-10-dependent mechanism [94]; the molecular basis for the induction of IL-10 production by BBI remains to be elucidated. Recent studies have demonstrated that BBI is responsible for delayed onset of disease but did not stop disease development, which became similarly severe in treated mice as in control animals [95, 96]. The ability of soybean BBI to delay both inflammatory and neurodegenerative aspects of autoimmune encephalomyelitis suggests that it may be useful for treating acute MS exacerbations and neurological dysfunction.

6. Concluding remarks

In recent years, much effort has focused on clarifying the potential chemopreventive properties of BBI. Preclinical and clinical studies have clearly demonstrated that BBI uptake is well-tolerated and no side-effects were elicited. This is particularly relevant and lack of toxicity is a major consideration, given the necessity for prolonged duration of administration. Consistently, several studies have shown that serine proteases are potential BBI targets in prevention and therapy; however, these targets have not been proven thus far. The validation of specific serine proteases as BBI targets will contribute to the assessment of BBI as chemopreventive agents that may be used in preventive and/or therapeutic medicine.

Author details

Alfonso Clemente*, María del Carmen Marín-Manzano and María del Carmen Arques
Department of Physiology and Biochemistry of Nutrition, Estación Experimental del Zaidín, Spanish Council for Scientific Research (CSIC), Granada, Spain

Claire Domoney
Department of Metabolic Biology, John Innes Centre, Norwich Research Park, Norwich, UK

Acknowledgement

A.C. acknowledges support by ERDF-co-financed grants from the Spanish CICYT (AGL2010-15877AGL and AGL2011-26353). A.C. is involved in COST Action FA1005 INFOGEST on Food Digestion. C.D. acknowledges support from the European Union (Grain Legumes Integrated Project, a Framework Programme 6 project, grant no. FOOD-CT-2004-506223) and from Defra, United Kingdom (grant nos. AR0105 and AR0711).

7. References

[1] Turk B. Targeting Proteases: Successes, Failures and Future Prospects. Nature Reviews Drug Discovery 2006; 5: 785-799.

---

* Corresponding Autor

[2] Shen A. Allosteric Regulation of Protease Activity by Small Molecules. Molecular Biosystems 2010; 6: 1431-1443.

[3] Drag M, Salvesen GS. Emerging Principles in Protease-based Drug Discovery. Nature Reviews Drug Discovery 2010; 9: 690-701.

[4] Deu E, Verdoes M, Bogyo M. New Approaches for Dissecting Protease Functions to Improve Probe Development and Drug Discovery. Nature Structural & Molecular Biology 2012; 19: 9-16.

[5] Malkowicz SB, McKenna WG, Vaughn DJ, Wan XS, Propert KJ, Rockwell K, Marks SHF, Wein AJ, Kennedy AR. Effects of Bowman-Birk Inhibitor Concentrate (BBIC) in Patients with Benign Prostatic Hyperplasia. Prostate 2001; 48: 16-28.

[6] Armstrong WB, Kennedy AR, Wan XS, Atiba J, McLaren CE, Meyskens FL. Single-dose Administration of Bowman-Birk Inhibitor Concentrate in Patients with Oral Leukoplakia. Cancer Epidemiology, Biomarkers & Prevention 2000; 9: 43-47.

[7] Armstrong WB, Kennedy AR, Wan XS, Taylor TH, Nguyen QA, Jensen J, Thompson W. Clinical Modulation of Oral Leukoplakia and Protease Activity by Bowman-Birk Inhibitor Concentrate in a Phase IIa Chemoprevention Trial. Clinical Cancer Research 2000; 6: 4684-4691.

[8] Meyskens FL, Taylor T, Armstrong W, Kong L, Gu M, Gonzalez R, Villa M, Wong V, Garcia A, Perloff M, Kennedy A, Wan S, Ware JH, Messadi D, Lorch J, Wirth L, Jaffe Z, Goodwin J, Civantos F, Sullivan M, Reid M, Merciznu M, Jayaprakash V, Kerr AR, Le A. Phase IIb Randomized Clinical Chemoprevention Trial of a Soybean-derived Compound (Bowman-Birk Inhibitor Concentrate) for Oral Leukoplakia. Cancer Prevention Research 2010; 3: CN02-05.

[9] Lichtenstein GR, Deren J, Katz S, Lewis JD, Kennedy AR, Ware JH. Bowman-Birk Inhibitor Concentrate: A Novel Therapeutic Agent for Patients with Active Ulcerative Colitis. Digestive Diseases and Sciences 2008; 53: 175-180.

[10] Bateman KS, James MNG. Plant Proteinase Inhibitors: Structure and Mechanism of Inhibition. Current Protein & Peptide Science 2011; 12: 341-347.

[11] Domoney C, Welham T, Ellis N, Mozzanega P, Turner L. Three Classes of Proteinase Inhibitor Gene Have Distinct but Overlapping Patterns of Expression in *Pisum sativum* Plants. Plant Molecular Biology 2002; 48: 319–329.

[12] De Almeida B, Garcia da Silva W, Alves M, Gonzalves E. *In silico* Characterization and Expression Analysis of the Multigene Family Encoding the Bowman-Birk Protease Inhibitor in Soybean. Molecular Biology Reports 2012; 39: 327-334.

[13] Clemente A, Domoney C. Biological Significance of Polymorphism in Legume Protease Inhibitors from the Bowman-Birk Family. Current Protein & Peptide Science 2006; 7: 201-216.

[14] Domoney C, Welham T. Trypsin Inhibitors in *Pisum*: Variation in Amount and Pattern of Accumulation in Developing Seed. Seed Science Research 1992; 2: 147–154.

[15] Clemente A, Gee JM, Johnson IT, Domoney C. Pea (*Pisum sativum* L.) Protease Inhibitors from the Bowman-Birk Class Influence the Growth of Human Colorectal Adenocarcinoma HT29 Cells *in vitro*. Journal of Agricultural and Food Chemistry 2005; 53: 8979–8986.

[16] Page D, Aubert G, Duc G, Welham T, Domoney C. Combinatorial Variation in Coding and Promoter Sequences of Genes at the *Tri* Locus in *Pisum sativum* Accounts for Variation in Trypsin Inhibitor Activity in Seeds. Molecular Genetics and Genomics 2002; 267: 359-369.

[17] Wiseman J, Al-Mazooqi W, Welham T, Domoney C. The Apparent Ileal Digestibility, Determined with Young Broilers, of Amino Acids in Near-isogenic Lines of Peas (*Pisum sativum* L.) Differing in Trypsin Inhibitor Activity. Journal of the Science of Food and Agriculture 2003; 83: 644-651.

[18] Wiseman J, Al-Marzooqi W, Hedley C, Wang TL, Welham T, Domoney C. The Effects of Genetic Variation at *r, rb* and *Tri* Loci in *Pisum sativum* L. on Apparent Ileal Digestibility of Amino Acids in Young Broilers. Journal of the Science of Food and Agriculture 2006; 86: 436-444.

[19] Hernández-Ledesma B, Hsieh CC, de Lumen BO. Lunasin and Bowman-Birk Protease Inhibitor (BBI) in US Commercial Soy Foods. Food Chemistry 2009; 115: 574-580.

[20] Bode W, Huber R. Natural Protein Proteinase-Inhibitors and their Interaction with Proteinases. European Journal of Biochemistry 1992; 204: 433-451.

[21] Chen P, Rose J, Love R, Wei CH, Wang BC. Reactive Sites of an Anticarcinogenic Bowman-Birk Proteinase Inhibitor are Similar to Other Trypsin Inhibitors. The Journal of Biological Chemistry 1992; 267: 1990-1994.

[22] Clemente A, Sonnante G, Domoney C. Bowman-Birk Inhibitors from Legumes on Human Gastrointestinal Health: Current Status and Perspectives. Current Protein & Peptide Science 2011; 12: 358-373.

[23] Schechter I, Berger A. On the Size of the Active Site in Proteases. I. Papain. Biochemical and Biophysical Research Communications 1967; 27: 157-162.

[24] Rocco M, Marloni L, Chambery A, Poerio E, Parente A, Di Maro A. A Bowman-Birk Inhibitor with Anti-elastase Activity from *Lathyrus sativus* L. Seeds. Molecular Biosystems 2011; 7: 2500-2507.

[25] Piergiovanni AR, Galasso I. Polymorphism of Trypsin and Chymotrypsin Binding Loops in Bowman-Birk Inhibitors from Common Bean (*Phaseolus vulgaris* L.). Plant Science 2004; 166: 1525-1531.

[26] Clemente A, Moreno J, Marín-Manzano MC, Jiménez E, Domoney C. The Cytotoxic Effect of Bowman-Birk Isoinhibitors, IBB1 and IBBD2, from Soybean (*Glycine max*) on HT29 Human Colorectal Cancer Cells is Related to their Intrinsic Ability to Inhibit Serine Proteases. Molecular Nutrition & Food Research 2010; 54: 396-405.

[27] Ferrasson E, Quillien L, Gueguen J. Proteinase Inhibitors from Pea Seeds: Purification and Characterization. Journal of Agricultural and Food Chemistry 1997; 45: 127-131.

[28] Clemente A, MacKenzie DA, Jeenes DJ, Domoney C. The Effect of Variation within Inhibitory Domains on the Activity of Pea Protease Inhibitors from the Bowman-Birk Class. Protein Expression and Purification 2004; 36: 106-114.

[29] Ragg EM, Galbusera V, Scarafoni A, Negri A, Tedeschi G, Consoni A, Sessa F, Duranti M. Inhibitory Properties and Solution Structure of a Potent Bowman-Birk Protease Inhibitor from Lentil (*Lens culinaris,* L.) Seeds. FEBS Journal 2006; 273: 4024-4039.

[30] Caccialupi P, Ceci LR, Siciliano RA, Pignone D, Clemente A, Sonnante G. Bowman-Birk Inhibitors in Lentil: Heterologous Expression, Functional Characterisation and Anti-proliferative Properties in Human Colon Cancer Cells. Food Chemistry 2010; 120: 1058-1066.

[31] Scarafoni A, Consonni A, Galbusera V, Negri A, Tedeshi G, Rasmussen P, Magni C, Duranti M. Identification and Characterization of a Bowman-Birk Inhibitor Active Towards Trypsin but not Chymotrypsin in *Lupinus albus* Seeds. Phytochemistry 2008; 69: 1820-1825.

[32] Osman MA, Reid PM, Weber CW. Thermal Inactivation of Tepary Bean (*Phaseolus acutifolius*), Soybean and Lima Bean Protease Inhibitors: Effect of Acidic and Basic pH. Food Chemistry 2002; 78: 419-423.

[33] Clemente A, Vioque J, Sánchez-Vioque R, Pedroche J, Bautista J, Millán F. Factors Affecting the *in vitro* Protein Digestibility of Chickpea Albumins. Journal of the Science of Food and Agriculture 2000; 80: 79-84.

[34] Weder JK. Inhibition of Human Proteinases by Grain Legumes. Advances in Experimental Medicine and Biology 1986; 199: 239-279.

[35] Park JH, Jeong HJ, Lumen BOD. *In Vitro* Digestibility of the Cancer-Preventive Soy Peptides Lunasin and BBI. Journal of Agricultural and Food Chemistry 2007; 55: 10703-10706.

[36] Trivedi MV, Laurence JS, Siahann TJ. The Role of Thiols and Disulfides on Protein Stability. Current Protein & Peptide Science 2009; 10: 614-625.

[37] Billings PC, St Clair WH, Maki PA, Kennedy AR. Distribution of the Bowman-Birk Protease Inhibitor in Mice Following Oral Administration. Cancer Letters 1992; 62: 191-197.

[38] Kennedy AR. Chemopreventive Agents: Protease Inhibitors. Pharmacology & Therapeutics 1998; 78: 167-209.

[39] Hajós G, Gelencser E, Pustzai A, Grant G, Sakhri M, Bardocz S. Biological Effects and Survival of Trypsin Inhibitors and the Aglutinin from Soybean in the Small Intestine of the Rat. Journal of Agricultural and Food Chemistry 1995; 43: 165-170.

[40] Clemente A, Jiménez E, Marín-Manzano MC, Rubio LA. Active Bowman-Birk Inhibitors Survive Gastrointestinal Digestion at the Terminal Ileum of Pigs fed Chickpea-Based Diets. Journal of the Science of Food and Agriculture 2008; 88: 513-521.

[41] Marín-Manzano MC, Ruiz R, Jiménez E, Rubio LA, Clemente A. Anti-carcinogenic Soyabean Bowman-Birk Inhibitors Survive Fermentation in their Active Form and do not Affect the Microbiota Composition *In Vitro*. The British Journal of Nutrition 2009; 101: 967-971.

[42] Reddy BS. Novel Approaches in the Prevention of Colon Cancer by Nutritional Manipulation and Chemoprevention. Cancer Epidemiology, Biomarkers & Prevention 2000; 9: 239-247.

[43] Pan MH, Lai CS, Wu JC, Ho CT. Molecular Mechanisms for Chemoprevention of Colorectal Cancer by Natural Dietary Compounds. Molecular Nutrition & Food Research 2011; 55: 32-45.

[44] Jemal A, Siegel R, Ward E, Hao YP, Xu JQ, Thun MJ. Cancer Statistics, 2009. Cancer Journal for Clinicians 2009; 59: 225-249.

[45] Kennedy AR, Billings OC, Wan XS, Newberne PM. Effects of Bowman-Birk Inhibitor on Rat Colon Carcinogenesis. Nutrition and Cancer 2002; 43: 174-186.

[46] Chan AT, Baba Y, Sima K, Nosho K, Chung DC, Hung KE, Mahmood U, Madden K, Poss K, Ranieri A, Shue D, Kucherlapati R, Fuch CS, Ogino S. Cathepsin B Expression and Survival in Colon Cancer: Implications for Molecular Detection of Neoplasia. Cancer Epidemiology, Biomarkers & Prevention 2010; 19: 2777-2785.

[47] Weldon S, McNally P, McElvaney NG, Elborn JS, McAuley DF, Wartelle J, Belaaouaj A, Levine RJ, Taggart CC. Decrease Levels of Secretory Leukoprotease Inhibitor in the *Pseudomonas*-Infected Cystic Fibrosis Lung are Due to Neutrophil Elastase Degradation. Journal of Immunology 2009; 183: 8148-8156.

[48] Inoue Y, Yokobori T, Yokoe T, Toiyama Y, Miki C, Mimori K, Mori M, Kusunoki M. Clinical Significance of Human Kallikrein7 Gene Expression in Colorectal Cancer. The Annals of Surgical Oncology 2010; 17: 3037-3042.

[49] Petraki C, Dubinski W, Scorilas A, Saleh C, Pasic MD, Komborozo V, Khalil B, Gabril MY, Streutker C, Diamandis EP, Yousef GM. Evaluation and Prognostic Significance of Human Tissue Kallikrein-related Peptidase 6 (KLK6) in Colorectal Cancer. Pathology Research and Practice 2012; 208: 104-108.

[50] Scott CJ, Taggart CC. Biologic Protease Inhibitors as Novel Therapeutic Agents. Biochimie 2010; 92: 1681-1688.

[51] Wu WKK, Cho CH, Lee CW, Wu K, Fan D, Yu J, Sung JJY. Proteasome Inhibition: a New Therapeutic Strategy to Cancer Treatment. Cancer Letters 2010; 293: 15-22.

[52] Latonen L, Moore HM, Bai B, Jaamaa S, Laiho M. Proteasome Inhibitors Induce Nucleolar Aggregation of Proteasome Target Proteins and Polyadenylated RNA by Altering Ubiquitin Availability. Oncogene 2011; 30: 790-805.

[53] Bugge TH, Antalis TM, Wu Q. Type II Transmembrane Serine Proteases. Journal of Biological Chemistry 2009; 284: 23177-23181.

[54] Lee SL, Dickson RB, Lin CY. Activation of Hepatocyte Growth Factor and Urokinase/Plasminogen Activator by Matriptase, an Epithelial Membrane Serine Protease. Journal of Biological Chemistry 2000; 275: 36720-36725.

[55] Yamasaki Y, Satomi S, Murai N, Tsuzuki S, Fushiki T. Inhibition of Membrane-Type Serine Protease 1/Matriptase by Natural and Synthetic Protease Inhibitors. Journal of Nutritional Science and Vitaminology 2003; 49: 27-32.

[56] Billings PC, Newberne P, Kennedy AR. Protease Inhibitor Suppression of Colon and Anal Gland Carcinogenesis Induced by Dimethylhydrazine. Carcinogenesis 1990; 11: 1083-1086.

[57] Ware HW, Wan S, Newberne P, Kennedy AR. Bowman-Birk Concentrate Reduces Colon Inflammation in Mice with Dextran Sulphate Sodium-Induced Ulcerative Colitis. Digestive Diseases and Sciences 1999; 44: 986-990.

[58] de Paula A, de Abreu P, Santos KT, Guerra R, Martins C, Castro-Borges W, Guerra MH. Bowman-Birk Inhibitors, Proteasome Peptidase Activities and Colorectal Pre-neoplasias

Induced by 1,2-dimethylhydrazine in Swiss Mice. Food and Chemical Toxicology 2012; 50: 1405-1412.

[59] Sutherland LR, Martin F, Greer S, Robinson M, Greenberger N, Saibil F, Martin T, Sparr J, Prokipchuck E, Borgen L. 5-aminosalicylic Acid Enema in the Treatment of Distal Ulcerative Colitis, Proctosigmoiditis and Proctitis. Gastroenterology 1987; 92: 1894-1898.

[60] Frenkel K, Chranzan K, Ryan CA, Wiesner R, Troll W. Chymotrypsin-specific Protease Inhibitors Decrease H_2O_2 Formation by Activated Human Polymorphonuclear Leukocytes. Carcinogenesis 1987; 8: 1207-1212.

[61] Baturay NZ, Roque H. *In vitro* Reduction of Peroxidation in UVC Irradiated Cell Cultures by Concurrent Exposure with Bowman-Birk Protease Inhibitor. Teratogenesis, Carcinogenesis and Mutagenesis 1991; 11: 195-202.

[62] Larionova NI, Gladysheva IP, Tikhonova TV, Kazanskaya NF. Inhibition of Cathepsin G and Human Granulocyte Elastase by Multiple Forms of Bowman-Birk Type Soybean Inhibitor. Biochemistry-Moscow 1993; 58: 1046-1052.

[63] Gladysheva IP, Zamolodchikova TS, Sokolova EA, Larionova NI. Interaction Between Duodenase, a Proteinase with Dual Specificity, and Soybean Inhibitors of Bowman-Birk and Kunitz Type. Biochemistry-Moscow 1999; 64: 1244-1249.

[64] Ware JH, Wan XS, Rubin H, Schechter NM, Kennedy AR. Soybean Bowman-Birk Protease Inhibitor is a Highly Effective Inhibitor of Human Mast Cell Chymase. Archives of Biochemistry and Biophysics 1997; 344: 133-138.

[65] Tani K, Ogushi K, Kido H, Kawano T, Kumori Y, Kamikura T, Cui P, Sone S. Chymase is a Potent Chemoattractant for Human Monocytes and Neutrophils. Journal of Leukocyte Biology 2000; 67: 585-589.

[66] Saarinen J, Kalkkinen N, Welgus HG, Kovanen PT. Activation of Human Interstitial Procollagenase through Direct Cleavage of the Leu83- Thr84 Bond by Mast Cell Chymase. Journal of Biological Chemistry 1994; 269: 18134-18140.

[67] Mizutani H, Schechter NM, Lazarus G, Black RA, Kupper TS. Rapid and Specific Conversion of Precursor Interleukin 1beta (1L-beta) to an Active IL-1 Species by Human Mast Cell Chymase. Journal of Experimental Medicine 1991; 174: 821-825.

[68] Clemente A, Marín-Manzano MC, Jiménez E, Arques MC, Domoney C. The Anti-proliferative Effects of TI1B, a Major Bowman-Birk isoinhibitor from Pea (*Pisum sativum* L), on HT29 Colon Cancer Cells are Mediated Through Protease Inhibition. The British Journal of Nutrition 2012 (doi.10.1017/S000711451200075X).

[69] Wan XS, Meyskens FL, Armstrong WB, Taylor TH, Kennedy AR. Relationship Between Protease Activity and *neu* Oncogene Expression in Patients with Oral Leukoplakia Treated with the Bowman-Birk Inhibitor. Cancer Epidemiology, Biomarkers & Prevention 1999; 8: 601-608.

[70] Shirai T. Significance of Chemoprevention for Prostate Cancer Development: Experimental *in vivo* Approaches to Chemoprevention. Pathology International 2008; 58: 1-6.

[71] Yan L, Spitznagel EL. Meta-analysis of Soy Food and Risk of Prostate Cancer in Men. International Journal of Cancer 2005; 117: 667-669.

[72] Wan XS, Ware JH, Zhang L, Newberne PM, Evans SM, Clark CL, Kennedy AR. Treatment with Soybean-derived Bowman Birk Inhibitor Increases Serum Prostate-specific Antigen Concentration while Suppressing Growth of Human Prostate Cancer Xenografts in Nude Mice. Prostate 1999b; 41: 243-252.

[73] Kennedy AR, Wan XS. Effects of the Bowman-Birk Inhibitor on Growth, Invasion, and Clonogenic Survival of Human Prostate Epithelial Cells and Prostate Cancer Cells. Prostate 2002; 50: 125-133.

[74] Sun XY, Donald SP, Phang JM. Testosterone and Prostate Specific Antigen Stimulate Generation of Reactive Oxygen Species in Prostate Cancer Cells. Carcinogenesis 2001; 22: 1775-1780.

[75] Dittmann K, Virsik-Kopp P, Mayer C, Rave-Frank M, Rodemann HP. The Radioprotective Effect of BBI Is Associated with the Activation of DNA Repair-Relevant Genes. International Journal of Radiation Oncology 2003; 79: 801-808.

[76] McCormick DL, Johnson WD, Bosland MC, Lubet RA, Steele VE. Chemoprevention of Rat Prostate Carcinogenesis by Soy Isoflavones and Bowman-Birk Inhibitor. Nutrition and Cancer 2007; 57: 184-193.

[77] Tang MX, Asamoto M, Ogawa K, Naiki-Ito A, Sato S, Takahashi S, Shirai T. Induction of Apoptosis in the LNCaP Human Prostate Carcinoma Cell Line and Prostate Adenocarcinomas of SV40T Antigen Transgenic Rats by the Bowman-Birk. Pathology International 2009; 59: 790-796.

[78] Chen YW, Huang SC, Lin-Shiau SY, Lin JK. Bowman-Birk Inhibitor Abates Proteasome Function and Suppresses the Proliferation of MCF7 Breast Cancer Cells Through Accumulation of MAP Kinase Phosphatase-1. Carcinogenesis 2005; 26: 1296-1305.

[79] Joanitti GA, Azevedo RB, Freitas SM. Apoptosis and Lysosome Membrane Permeabilization Induction on breast Cancer Cells by an Anticarcinogenic Bowman-Birk Inhibitor from *Vigna unguiculata* Seeds. Cancer Letters 2010; 293: 73-81.

[80] Dittmann K, Löffler H, Bamberg M, Rodemann HP. Bowman-Birk Proteinase Inhibitor (BBI) Modulates Radiosensitivity and Radiation-Induced Differentiation of Human Fibroblasts in Culture. Radiotherapy and Oncology 1995; 34: 137-143.

[81] Dittmann K, Toulany M, Classen J, Heinrich V, Milas L. Selective Radioprotection of Normal Tissues by Bowman-Birk Proteinase Inhibitor (BBI) in Mice. Strahlentherapie Und Onkologie 2005; 181: 191-196.

[82] Dittmann KH, Gueven N, Mayer C, Ohneseit P, Zell P, Begg AC, Rodemann HP. The Presence of Wild-Type TP53 is Necessary for the Radioprotective Effect of the Bowman-Birk Proteinase Inhibitor in Normal Fibroblasts. Radiation Research 1998; 150: 648-655.

[83] Dittmann K, Mayer C, Kehlbach R, Rodemann HP. The Radioprotector Bowman-Birk Proteinase Inhibitor Stimulates DNA Repair via Epidermal Growth Factor Receptor Phosphorylation and Nuclear Transport. Radiotherapy and Oncology 2008; 86: 375-382.

[84] Gueven N, Dittmann K, Mayer C, Rodemann HP. Bowman-Birk Protease Inhibitor Reduces the Radiation-Induced Activation of the EGF Receptor and Induces Tyrosine Phosphatase Activity. International Journal of Radiation Oncology 1998; 73: 157-162.

[85] Gueven N, Dittmann K, Mayer C, Rodemann HP. The Radioprotective Potential of the Bowman-Birk Protease Inhibitor is Independent of its Secondary Structure. Cancer Letters 1998; 125: 77-82.

[86] Yavelow J, Collins M, Birk Y, Troll W, Kennedy AR. Nanomolar Concentrations of Bowman-Birk Soybean Protease Inhibitor Suppress X-ray Induced Transformation *In Vitro*. Proceedings of the National Academy of Sciences of the United States of America 1985; 82: 5395-5399.

[87] Huang MT, Xie JG, Lin CB, Kizoulis M, Seiberg M, Shapiro S, Conney A. Inhibitory Effect of Topical Applications of Non-denatured Soymilk on the Formation and Growth of UVB-Induced Skin Tumors. Oncology Research 2004; 14: 387-397.

[88] Davis JG, Wan XS, Ware JH, Kennedy AR Dietary Supplements Reduce the Cataractogenic Potential of Proton and HZE-Particle Radiation in Mice. Radiation Research 2010; 173: 353-361.

[89] Morris CA, Morris LD, Kennedy AR, Sweeney HL. Attenuation of Skeletal Muscle Atrophy via Protease Inhibition. Journal of Applied Physiology 2005; 99: 1719-1727.

[90] Arbogast S, Smith J, Matuszczak Y, Hardin BJ, Moylan JS, Smith JD, Ware J, Kennedy AR, Reid MB. Bowman-Birk Inhibitor Concentrate Prevents Atrophy, Weakness, and Oxidative Stress in Soleus Muscle of Hindlimb-Unloaded Mice. Journal of Applied Physiology 2007; 102: 956-964.

[91] Morris CA, Selsby JT, Morris LD, Pendrak K, Sweeney HL. Bowman-Birk Inhibitor Attenuates Dystrophic Pathology in mdx Mice. Journal of Applied Physiology 2010; 109: 1492-1499.

[92] Cruz-Orengo L, Holman DW, Dorsey D, Zhou L, Zhang P, Wright M, McCandless EE, Patel JR, Luker GD, Littmann DR, Rusell JH, Klein RS. CXCR7 Influences Leukocyte Entry into the CNS Parenchyma by Controlling Abluminal CXCL12 Abundance During Autoimmunity. Journal of Experimental Medicine 2011; 208: 327-339.

[93] Gran B, Tabibzadeh N, Martin A, Ventura ES, Ware JH, Zhang GX, Parr JL, Kennedy AR, Rostami AM. The Protease Inhibitor, Bowman-Birk inhibitor, Suppresses Experimental Autoimmune Encephalomyelitis: a Potential Oral Therapy for Multiple Sclerosis. Multiple Sclerosis 2006; 12: 688-697.

[94] Touil T, Ciric B, Ventura E, Shindler KS, Gran B, Tostami A. Bowman-Birk Inhibitor Suppresses Inflammation and Neuronal Loss in a Mouse Model of Multiple Sclerosis. Journal of the Neurological Sciences 2008; 271: 191-202.

[95] Dai H, Ciric B, Zhang GX, Rostami A. Bowman-Birk Inhibitor Attenuates Experimental Autoimmune Encephalomyelitis by Delaying Infiltration of Inflammatory Cells into the CNS. Immunologic Research 2011; 51: 145-152.

[96] Dai H, Ciric B, Zhang GX, Rostami A. Interleukin-10 Plays a Crucial Role in Suppression of Experimental Autoimmune Encephalomyelitis by Bowman-Birk Inhibitor. Journal of Neuroimmunology 2012; 245: 1-7.

[97] Scarpi D, McBride JD, Leatherbarrow RJ. Inhibition of Human Beta-Tryptase by Bowman-Birk Inhibitor Derived Peptides: Creation of a New Tri-Functional Inhibitor. Bioorganic & Medicinal Chemistry 2004; 12: 6045-6052.

[98] Muricken DG, Gowda LR Molecular Engineering of a Small Trypsin Inhibitor Based on the Binding Loop of Horsegram Seed Bowman-Birk Inhibitor. Journal of Enzyme Inhibition and Medicinal Chemistry 2011; 26: 553-560.

[99] Rocco M, Malorni L, Chambery A, Poerio E, Parente A, Di Maro A. A Bowman-Birk Inhibitor with Anti-Elastase Activity from Lathyrus sativus L. Seeds. Molecular Biosystems 2011; 7: 2500-2507.

[100] Fernandes AO, Banerji AP. Inhibition of Benzopyrene-Induced Forestomach Tumors by Field Bean Protease Inhibitor. Carcinogenesis 1995; 16: 1843-1846.

Chapter 3

Antihypertensive Peptides from Food Proteins

Roseanne Norris and Richard J. FitzGerald

Additional information is available at the end of the chapter

http://dx.doi.org/10.5772/51710

1. Introduction

Hypertension or elevated blood pressure (BP) is a global health concern, thought to affect up to 30 % of the adult population in developed and developing countries. It is defined by a BP measurement of 140/90 mmHg or above. Hypertension is a major risk factor concomitant with cardiovascular disease (CVD) states such as coronary heart disease, peripheral artery disease and stroke, and kidney disease. Essential hypertension, the most common type of hypertension and to which 90-95% of cases belong, is manifested as an increase in an individual's BP due to an unknown cause. This class of hypertension can be improved with lifestyle choices such as regular exercise, heart-healthy eating, non smoking, reducing sodium intake and reducing the level of stress [1]. For these reasons it is defined as a controllable risk factor of CVD. At present there is a range of synthetic drugs on the market for treatment of hypertension including diuretics, adrenergic inhibitors such as α- and β-blockers, direct vasodilators, calcium channel blockers, angiotensin II (Ang II) receptor blockers and angiotensin converting enzyme (ACE) inhibitors. However, although hypertension can be controlled by pharmacological agents, it represents a major burden on annual global healthcare costs. According to the Centre for Disease Control and Prevention (CDC) [2], it was estimated that hypertension-related costs reached $76.6 billion in the USA in 2010. It is thought that prevention through lifestyle choices and early treatment for individuals with mild hypertension can significantly reduce global health-care costs.

Food proteins contain numerous biologically-active peptides (BAPs). These BAPs can exert positive physiological responses in the body beyond their basic nutritional roles in the provision of nitrogen and essential amino acids. Many bioactivities have been found including peptides with antihypertensive capabilities. This has led to significant research on the discovery and generation of peptides with antihypertensive properties *in vivo.* Food proteins such as the casein and whey protein components of milk, meat, egg, marine and meat proteins have all been found to contain peptides with potential antihypertensive properties within their primary sequences. These peptides may become active when released through enzymatic/bacterial hydrolysis [3]. The food industry has recognised the potential of these *natural* antihypertensive agents as possible future functional ingredients, aiding in the primary prevention and/or management of hypertension.

2. Hypotensive mechanisms of action

The regulation of BP is complex, involving a variety of intertwining metabolic pathways. By far, the most studied BP control pathways with regard to food-derived peptides involve those shown to inhibit ACE *in vitro*. This enzyme is one of the main regulators of BP and is involved in two main systems, the renin-angiotensin system (RAS) and the kinin-nitric oxide system (KNOS). Inhibition of ACE in these systems leads to dilation of the artery walls or vasodilation and subsequent lowering of BP. However, it is not yet known whether this is the main mechanism followed *in vivo* or whether there are a number of other BP control mechanisms involved [4].

2.1. ACE inhibition

ACE inhibition is an excellent physiological target for clinical hypertensive treatment due to its involvement in two BP related systems, the RAS and the KNOS. The RAS is thought to be one of the predominant pressor systems in BP control. In the RAS the N-terminus of the prohormone angiotensinogen, which is derived from the liver, is cleaved by renal renin to produce the decapeptide angiotensin I (Ang I). ACE then removes the C-terminal dipeptide HL to form Ang II, a potent vasoconstrictory peptide which acts directly on vascular smooth muscle cells. Thus, inhibition of ACE consequentially leads to BP reduction. Ang II binds to AT_1 and AT_2 receptors which are located in peripheral tissues around the body and in the brain. The vasocontriction produced by Ang II is mediated by the AT_1 receptor. [5-7]. In the KNOS, ACE inactivates the vasodilatory peptides bradykinin and kallidin. Kallidin is synthesised from kininogen by kallikrein, and its further action on kallidin leads to the formation of bradykinin among other vasoactive peptides. Bradykinin binds to β-receptors which lead to an eventual increase in intracellular Ca^{2+} level. The binding of bradykinin to β-receptors and the increase in Ca^{2+} stimulates nitric oxide synthase (NOS) to convert L-arginine to nitric oxide (NO), a potent vasodilator. ACE can therefore, indirectly inhibit the production of NO as it hydrolyses bradykinin into inactive fragments [7].

There are a number of widely-used synthetic ACE inhibitors currently on the market that serve as the first line of approach for the treatment of hypertension. Such inhibitors include Captopril, Enalapril and Lisinopril. However, their use is associated with a range of side-effects including cough, skin rashes, hypotension, loss of taste, angiodema reduced renal function and fetal abnormalities [8]. Natural ACE inhibitory peptides from food are not associated with the side-effects brought about by the synthetic drugs. They are not as potent inhibitors of ACE as the synthetic inhibitors which can have IC_{50} values in the nM region. As they inhibit ACE to a lesser extent, this potentially allows for safer levels of bradykinin in the body. Thus, for this reason, ACE-inhibitory peptides have gained interest as potential preventative agents for hypertension control.

ACE-inhibitory peptides have been identified in a range of food proteins including casein, whey, ovalbumin, red algae, wakame, soy, gelatin, chicken muscle, dried bonito, corn, sardines, rapeseed, potato, chick pea, tuna muscle, pea albumin, garlic, wheat germ, sake, porcine haemoglobin and squid. The ACE inhibitory peptides found in different food proteins has been extensively reviewed (for review see [9-12; 3; 131]. Examples of recently reported food protein ACE inhibitory peptide sources include loach (*Misgurnus anguillicaudatus*) [13], pork meat [14], lima bean (*Phaseolus lunatus)* [15], skate skin [16] and boneless chicken leg meat [17]. ACE inhibitory peptides have been generated in a number of different ways. They can be produced naturally during gastrointestinal (GI) digestion by the hydrolytic action of the proteinases pepsin, trypsin, chymotrypsin and by brush border peptidases [18]. Simulated GI digestion has been carried out on a range of protein sources to assess the effect of GI digestion on ACE-inhibitory peptides [19-24]. More commonly, ACE-inhibitory peptides are produced through enzymatic hydrolysis with GI enzymes such as pepsin and trypsin or with enzyme combinations such as Alcalase™ [25]. ACE-inhibitory peptides have also been produced during the fermentation of milk during cheese production. *Lactobacillus* and *Lactococcus lactis* strains have been shown to produce ACE inhibitory peptides. Furthermore, fermented soy products such as soy paste, soy sauce, natto and tempeh have been found to produce ACE-inhibitory peptides [26-29].

ACE inhibitory peptides can work in three ways and are classed as inhibitor-type, substrate-type or prodrug-type based on changes in ACE inhibitory activity after hydrolysis of peptides by ACE [30]. Inhibitor-type peptides are ACE inhibitory peptides whose activity is not significantly altered as the peptides are resistant to cleavage by ACE. Substrate-type ACE inhibitors show a decrease in ACE activity due to cleavage by ACE. Prodrug type refers to the conversion to potent ACE inhibitors following hydrolysis of larger peptide fragments by ACE itself. The resulting peptides tend to produce long-lasting hypotensive effects *in vivo* [30]. A prodrug type ACE inhibitor was isolated from a thermolysin-digest of Katsuo-bushi, a Japanese traditional food processed from dried bonito. The study reported an 8-fold increase in ACE-inhibitory activity when the peptide Leu-Lys-Pro-Asn-Met (IC_{50}=2.4 μM) was hydrolyzed by ACE to produce Leu-Lys-Pro [IC_{50}=0.32 μM; 30]. When Leu-Lys-Pro-Asn-Met and Leu-Lys-Pro were orally administered to spontaneously hypertensive rats (SHR), Leu-Lys-Pro-Asn-Met showed a maximal decrease of BP after 4 and 6 h, results which are comparable to that of Captopril inhibition. However, the maximal hypotensive effect of Leu-Lys-Pro was seen at 2 h [30].

Inhibition of ACE is by far the most studied mechanism of BP control with regard to food-derived biologically-active peptides. Most peptides have been found to inhibit ACE to some degree. However, in most cases, it has yet to be answered whether this is the BP mechanism being employed *in vivo*. There are other regulatory pathways of BP control, independent of ACE, that are also potential targets for the action of antihypertensive peptides (see Figure 1 for vasorelaxative peptides and molecules).

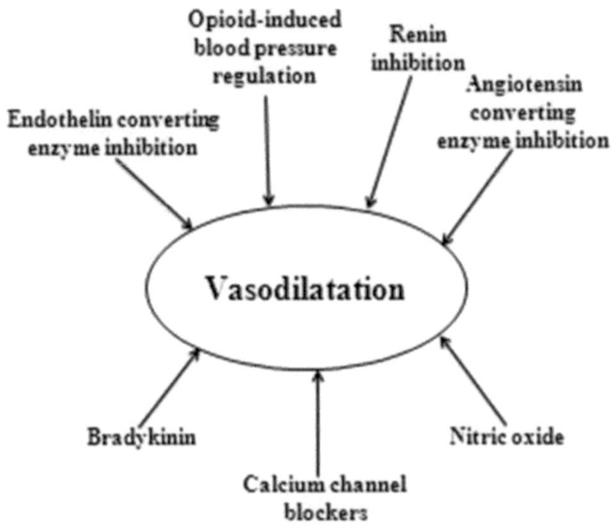

Figure 1. Vasorelaxative peptides and molecules in blood pressure control systems.

2.2. Renin Inhibition

Renin inhibition is another potential target for BP control. It is thought that inhibition of renin could provide a more effective treatment for hypertension it prevents the formation of Ang-I, which can be converted to Ang-II in some cells independent of ACE, by the enzyme chymase [31]. In addition, unlike ACE which acts on a number of substrates in various biochemical pathways, angiotensinogen is the only known substrate of renin. Therefore, renin inhibitors could ensure a higher specificity in antihypertensive treatment compared to ACE inhibitors [31-32]. Food peptides have recently been found to be inhibitors of renin. Peptides from enzymatic flaxseed fractions were found to inhibit both human recombinant renin and ACE. The study concluded that such peptides with the ability to inhibit both ACE and renin may potentially provide better antihypertensive effects *in vivo* in comparison to peptides that only inhibit ACE [33]. A similar outcome was seen in a study carried out by Li & Aluko [34] where fractions of pea protein isolates inhibited both ACE and renin to a high degree with IC_{50} values <25 mM.

2.3. Calcium channel blocking effects

Calcium channel blockers interact with voltage-gated calcium channels (VGCCs) in cardiac muscle and blood vessel walls, reducing intracellular calcium and consequently lowering vasoconstriction. It has been shown in various studies that peptides can have the ability to act as calcium channel blockers. Fifteen synthetic peptides based on Trp-His skeleton analogues were tested for their vasodilatory effects in 1.0 μM phenylephrine-contracted thoracic aortic rings from Sprague-Dawley rats. It was previously reported that Trp-His induced the most potent vasodilation among 67 synthetic di-and tripeptides. The study demonstrated that His-Arg-Trp had an endothelium-independent vasorelaxative effect in the phenylephrine-contracted thoracic aorta. It was also shown that His-Arg-Trp, at a concentration of 100 μM, caused a significant reduction in intracellular Ca^{2+} concentration. The increase intracellular [Ca^{2+}], brought about by the action of Bay K8644 or Ang II, was significantly inhibited by His-Arg-Trp (>30%). It was proposed that His-Arg-Trp may have supressed extracellular Ca^{2+} influx through voltage-gated L-type Ca^{2+} channels [35]. Another recent study reported a similar result with Trp-His which was also found to block L-type Ca^{2+} channels. Trp-His at 300 μM elicited an intracellular Ca^{2+} reduction of 23 % in 8 week-old male Wistar rat thoracic aortae smooth muscle cells. In addition, the reduction in [Ca^{2+}] brought about by Trp-His was eliminated by verapamil indicating that Trp-His specifically works on L-type Ca^{2+} channels [36].

2.4. Opioid peptide vasorelaxive effects

Food-derived peptides have also been found to be sources of opioid like-activities. These peptides bind to opioid receptors to produce morphine-like effects. Natural opioid peptides include endorphins, enkephalins and dynorphins. In humans opioid receptors are found in the nervous, endocrine and immune systems, and in the intestinal tract. These receptors may be involved in various regulatory processes in the body including the regulation of circulation which can affect BP [37; 38]. Nurminen *et al* [39] found an antihypertensive effect on oral administration of the tetrapeptide, α-lactorphin (Tyr-Gly-Leu-Phe), to SHR and to normotensive Wistar Kyoto rats (WKY). Maximum BP reductions were found in SHR, with a decrease of 23 ± 4 and 17 ± 4 mm Hg in systolic BP (SBP) and diastolic BP (DBP), respectively. However, the α-lactophin-induced reduction in BP was not found after administration of the specific opioid receptor antagonist, Naloxone. Therefore, the antihypertensive effect was considered to be a result of interaction with opioid receptors. A follow-up study looked at the effects of α-lactophin along with a second milk-derived peptide β-lactorphin (Tyr-Leu-Leu-Phe) on mesenteric arterial function to demonstrate the regulatory mechanisms of action. It was shown with the NOS inhibitor N^G-nitro-L-arginine methyl ester (L-NAME) that α-lactophin produced an endothelium-dependant vasorelaxation, whereas, β-lactorphin also enhanced endothelium-independent vasorelaxation. The study concluded that α-lactophin may stimulate opioid receptors which in turn releases NO causing the vasorelaxative effect [40]. The casein-derived peptide casoxin D (Tyr-Val-Pro-Phe-Pro-Pro-Phe) has also been reported to have an hypotensive effect via opioid receptors. The peptide was found to have an endothelium-dependent relaxation in canine mesenteric artery strips. Anti-opioid and vasorelaxing effects were mediated by the opioid μ-receptor and BK B1-receptor, respectively [41-42]. Furthermore, it has been suggested that opioid-induced BP regulation by such peptides may act upon receptors in the intestinal tract. Interestingly, this would mean that the peptide would not need to be absorbed into the blood stream at the brush border membrane [43]. It could very well be that opioid-mediated reduction in BP may be the principal mechanism for antihypertensive peptides.

2.5. Endothelin-1 and endothelin converting enzyme (ECE) inhibition

The vasoconstrictory peptide endothelin-1 (ET-1) is released from big endothelin-1 (big ET-1) by the action of endothelin-converting enzyme (ECE). ET-1 mediates vasoconstriction via 2 receptors, ETa and. ETb. Both receptors mediate contractions on smooth muscle, but ETb also induces relaxation of endothelial cells by the production of nitric oxide. ET-1 is known to have a greater vasocontrictive effect than Ang II [44; 7]. Endothelial-dependent release of NOS was found to be the mechanism of action for the antihypertensive egg protein derived ovokinin (f2-7) peptide (Arg-Ala-Asp-His-Pro-Phe). Dilation of isolated SHR mesenteric arteries was found to be inhibited by L-NAME but not by indomethacin, demonstrating NO release from the endothelial cells [45]. A later study showed that ovokinin (2–7) modulates a hypotensive effect through interaction via B_2 bradykinin receptors [46].

It has been found that food proteins have the ability to act as inhibitors of ECE. Okitsu *et al* [47] found ECE inhibitory peptides in pepsin digests of beef and bonito pyrolic appendix. Up to 45 and 40 % of ECE activity could be inhibited with the beef and bonito peptides, respectively. A second study showed that the ACE-inhibitory peptide Ala-Leu-Pro-Met-His-Ile-Arg, released through tryptic digestion of bovine β-lactoglobulin, can inhibit the release of ET-1 in cultured porcine aortic endothelial cells (PAECs). At a concentration of 1 mM Ala-Leu-Pro-Met-His-Ile-Arg, ET-1 release was reduced by 29 %. The study concluded that the ET-1 reduction may be due to indirect reduction of ET-release by ACE inhibition through the BK pathway, rather than direct action on ET-1 by the peptide [48]. ACE breaks down BK into inactive fragments in the KNOS. Subsequent accumulation of BK (vasodilator) due to ACE inhibition leads to increased release of the vasodilator NO, and antagonises the release of the ET-1 by endothelial cells.

3. Structure activity relationships

An understanding of the relationship between a peptide and its bioactivity allows for the targeted release of potentially potent peptide sequences. This would eliminate the need for the time-consuming conventional peptide discovery strategy. There is limited knowledge on the structure-activity relationship of hypotensive peptides. To date, the main focus with regard to bioactive peptide research has been on the generation and characterisation of these peptides. ACE inhibition is by far the most widely studied biomarker with regard to antihypertensive effects of bioactive food peptides. ACE can work on a wide range of peptide substrates, and appears to have a broad specificity. Some structural features that influence the binding of a peptide to the ACE active site have been recognised (Table 1). However, potent inhibitory peptides of ACE are generally short sequences, i.e., 2-12 amino acids in length. However, some larger inhibitory sequences have been identified. Studies have indicated that binding to ACE is strongly influenced by the substrate's C-terminal tripeptide sequence. Hydrophobic amino acid residues with aromatic or branched side chains at each of the C-terminal tripeptide positions are common features among potent inhibitors. The presence of hydrophobic Pro residues at one or more positions in the C-terminal tripeptide region seems to positively influence a peptide's ACE inhibitory activity.

Tyr, Phe and Trp residues are also present at the C-terminus of many potent ACE inhibitors, especially with di- and tripeptide inhibitors [9]. It has been suggested that Leu residues may also contribute to ACE inhibition [49]. Furthermore, the positive charge on the side chains of Arg and Lys residues at the C-terminus have been noted to contribute to the ACE inhibitory potential of a peptide [50-51; 9]. An L-configured amino acid at position three at the C-terminus of the inhibitory peptide may be a requirement for potent inhibition. A study showed that the IC_{50} for the tripeptide D-Val-Ala-Pro (2 μM) increased to 550 μM with L-Val-Ala-Pro, yet only a slight increase in IC_{50} was seen for the peptide L-Phe-Val-Ala-Pro (17 μM; Maruyama *et al.*, 1987). It is thought that conformation contributes to the ACE inhibitory potential of long-chain peptide inhibitors [3].

N-Terminus-------------------------	--	-----------------------C-Terminus
Hydrophobic residues	2-12 amino acids in length Peptide conformation important for longer peptides	C-terminal tripeptide Bulky hydrophobic residues Aromatic or branched side chains Proline at one or more positions Positively charged residues in position two, Arg, Lys Tyr, Phe, Trp, Leu L-configured residue in position three

Table 1. Some structural features of potent angiotensin converting enzyme (ACE) inhibitory peptides.

Both domains of ACE (C- and N-domains) contain an active site containing the sequence His-Glu-XX-His. These active sites are located within the cleft of the two domains, and are protected by an N-terminal 'lid'. This 'lid' blocks access of large polypeptides to the active site. This is thought to explain why small peptides are more effective in inhibiting ACE. In addition, ACE inhibition may include inhibitor interaction with subsites on the enzyme that are not generally occupied by substrates or with an anionic inhibitor binding site that is different for the catalytic site of the enzyme. With the catalytic sites of ACE having different conformational requirements, this could indicate that for a more complete inhibition of ACE, there may be a need to use a variety of peptide inhibitors each with slightly different conformational features [52-53].

Quantitative computational tools are increasingly been applied in medicinal and pharmaceutical drug discovery. Recently it has been acknowledged that such models could be adapted to food-derived bioactive peptide sequences. Quantitative structure-activity relationship modelling (QSAR) and substrate docking can be used as an effective tool to assess *in silico* numerous peptide structures for their bioactivity potential. Thus, this work allows for a molecular understanding of peptide structure and bioactivity. QSAR studies are based on the relationship between chemical structure of ligands and receptors, and biological activity. Physicochemical variables or descriptor variables of a ligand such as steric properties, hydrophobicity and electronic properties, molecular mass and shape are used to quantitatively correlate the ligand's chemical structure with bioactivity [54]. A small number of QSAR studies have been carried out on ACE-inhibitory peptides. The structure-activity relationship of di-and tri-peptides using partial least square analysis (PLS) QSAR was assessed by constructing a database of known ACE-inhibitory peptides. Using a 3-z scale descriptor approach, two models were developed for the amino acid components of the peptide datasets. The dipeptide model had a predictive power of 71.1 % while the tripeptide model had a predictive power of 43.4 %. The dipeptide model indicated that amino acids with bulky and hydrophobic side chains were favoured by ACE while the tripeptide model suggested that C-terminal aromatic residues, positively charged residues in position two and hydrophobic residues at the amino terminus were preferred [55]. Another study by the same authors used a 5-z scale model to assess peptides of 4-10 amino acids in length. The study concluded that the tetrapeptide residue at the C-terminus has a large influence on the potency of peptide's 4-10 amino acids in length [56].

Substrate docking involves the docking of molecules (ligands) to a receptor or into a protein target such as an enzyme. All possible docking or binding conformations are assessed for their binding affinity to a molecule, and their potential as high affinity binding ligands is estimated by use of a scoring function. An integrated QSAR and Artificial Neural Network (ANN) approach was used to assess the ACE-inhibitory potential of 58 dipeptides present in the sequence of defatted wheat germ protein. The model was used to investigate preferred structural characteristics of ACE-inhibitory dipeptides and following this, appropriate proteases were successfully selected to produce the dipeptides predicted to be potent inhibitors by the QSAR-ANN model. The QSAR model predicted that the C-terminal of the peptide had principal importance on ACE inhibitory activity, with hydrophobic C-terminal residues being essential for high potency. Furthermore, proteins with a high abundance of hydrophobic residues were considered to be good substrates for the production of potent ACE inhibitory peptides [57]. Recently, the ability of docking to predict ACE inhibitory dipeptide sequences was assessed using the molecular docking program AutoDock Vina. All potential dipeptides and phospho-dipeptides were docked and scored. Phospho-dipeptides were predicted by the program to be good inhibitors of ACE. However, the experimentally determined IC_{50} results for selected phospho-dipeptides did not correlate and the study concluded that phospho-dipeptides may not be potent inhibitors of ACE *in vivo*. Furthermore, LIGPLOT analysis, a program to plot schematic diagrams of protein-ligand interactions, carried out on two newly identified ACE inhibitory dipeptides Asp-Trp and Trp-Pro (ACE IC_{50} values of 258 and 217 μM, respectively) interestingly showed no zinc interaction with the ACE active site [58].

4. Peptide bioavailability

The potential antihypertensive effect of a peptide depends on the peptides ability to reach their target organ intact and in an active form. However, there are several barriers which lie in the way of this outcome. Antihypertensive peptides must be resistant to digestive proteinases and peptidases; they must be able to be transported through the bush border membrane intact and must be resistant to serum peptidases. With regard to ACE inhibition, while there have been many studies focusing on the production, isolation and characterisation of ACE-inhibitory peptides, to date little attention has been placed on their bioavailability. It is therefore difficult to determine the relationship between *in vitro* ACE-inhibitory activity and an *in vivo* hypotensive effect. This is made even more difficult with the utilisation of several different *in vitro* assays and assay conditions for the determination of ACE-inhibition [59]. Furthermore, variations in *in vivo* experimental design such as administration by intravenous subcutaneous or oral administration, and the use of animal models or hypertensive patients, all hinder the ability to compare results among different studies [60].

Bioactive peptides when taken orally may be inactived by several digestive proteinases and peptidases including pepsin in the stomach, and the pancreatic enzymes trypsin, elastase, α-chymotrypsin and carboxypeptidase A and B in the small intestine. A number of studies have been carried out investigating ACE-inhibitory peptides and their ability to resist gastrointestinal digestion by these enzymes. These studies involve simulating the gastrointestinal process by sequential hydrolysis of ACE inhibitory peptides with pepsin and Pancreatin™, each concluding the importance of gastrointestinal digestion analysis in the ACE inhibitory activity of the peptide [61-66; 49; 21; 20] . It has been noted that certain protein/peptide structures are resistant to gastrointestinal digestion due to the composition and position of amino acids in their primary chains. The rate of hydrolysis of a peptide is also dependent on the peptide's amino acid composition. Peptides containing Pro and hydroxy Pro residues have been found to be resistant to hydrolysis. Furthermore, glycosylated peptides and peptides which have undergone changes during food processing such as during the formation of Maillard reaction products have been shown to be resistant to GI tract enzyme cleavage [67]. Once the peptides reach the brush border membrane of the large intestine, they may also be subjected to further cleavage by a variety of membrane anchored epithelial cell intestinal peptidases. These include a number of aminotripeptidases and several dipeptidases, each with varying specificities [60]. However, it has been found that certain free amino acids released during gastrointestinal breakdown may in turn serve as inhibitors of the brush border membrane dipeptidases, Moreover, it has been reported that during gastrointestinal proteolysis at the brush border membrane, the large variety and high concentration of peptides present would exceed the apparent *V*max for hydrolysis, allowing for safe passage of many di- and tripeptides through the membrane wall. Absorption through the membrane is possible for both di- and tripeptides with the help of a peptide transporter termed PepT1. PepT1 operates as an electrogenic proton/peptide symporter having wide substrate specificity [67]. There is an increasing body of research that shows the presence of the lactotripeptides (LTPs) Ile-Pro-Pro and Val-Pro-Pro in human and animal circulatory systems after oral administration, suggesting the resistance of these peptides to gastrointestinal degradation and their absorption intact across the brush border membrane [68-72]. However, it has been suggested that intestinal absorption of Val-Pro-Pro may operate via paracellular transport, rather than with the help of PepT1 [73]. Larger Pro-rich peptides have also been found to be transported intact across the brush border membrane. A study found that the ACE-inhibitory and antihypertensive peptide Leu-His-Leu-Pro-Leu-Pro, β-casein (f133-138), was resistant to gastrointestinal digestion. However, this peptide was hydrolysed to the pentapeptide His-Leu-Pro Leu-Pro by cellular peptidases before transportation across the intestinal epithelium. The study concluded by use of a Caco-2 monolayer model that the likely mechanism of transport was via paracellular passive diffusion [74]. An earlier study quantifying ACE-inhibitory peptides in human plasma found the pentapeptide to be present in human plasma after oral administration which demonstrates the ability of the peptide to be absorbed through the human brush border membrane [75].

Absorption of peptides across the brush border membrane can be studied by Caco-2 cell monolayers, the representative model for human intestinal epithelial cell barrier. The intestinal transport of pea and whey ACE inhibitory peptides was also studied using a Caco-2 monolayer. It was found that only minor ACE inhibitory activity crossed the Caco-2 cell monolayer in 1 h. However, it was concluded that the extent of ACE inhibitory peptides that may be transported *in vivo* would be higher, as the Caco-2 model is tighter than intestinal mammalian tissue [76]. The transepithelial transport of oligopeptides across the intestinal wall was assessed using a Caco-2 cell monolayer [133]. The study showed that the hydrolysis of peptides by brush-border peptidases is the rate-limiting step for the transepithelial transport of oligopeptides (≥4 residues in length). Bradykinin and Gly-Gly-Tyr-Arg, which were found to be resistant to cellular peptidases, were investigated for their apical-to-basolateral transport mechanism. Bradykinin and its analogues were found to be transported by the intracellular pathway, probably the adsorptive transcytosis. The transport rate was found to be dependent on the hydrophobic properties of the peptides. Gly-Gly-Tyr-Arg was suggested to be transported mainly via the paracellular pathway [133]. Foltz *et al.*, [77] devised a predictive *in silico* amino acid clustering model for dipeptides which can predict a dipeptide's ability to withstand small intestinal digestion. Dipeptides (220 in total) were tested for small intestinal stability by simulated digestion and their relative stability (% of initial dipeptide concentration) was plotted against time. Using the area under the curve (AUC) approach, the contribution of N- and C-terminal amino acids were calculated, based on the average AUC of all peptides containing the amino acid of interest. Data clustering allowed for ranking of the N- and C-terminal amino acid residues and they were grouped by their average AUC values. Correlations with experimentally measured stability allowed for classification of dipeptides as intestinally 'stable', 'neutral' or 'instable' using the clustering model.

Following absorption of a peptide into the blood stream, it may undergo hydrolysis by serum peptidases. The ACE inhibitory peptide may need to be able to with-stand hydrolysis in order to reach their target organs intact and yield their antihypertensive effect. It has been suggested that potent ACE inhibitors may be produced in circulation by the action of serum peptidases on less potent inhibitors of ACE and by the action of ACE itself. These peptides have been referred to pro-drug type inhibitors of ACE [78; 60].

Thus, the bioavailability of ACE-inhibitory peptides is essential for their activity. Several approaches to aid in peptide delivery are been considered. Peptides may be chemically modified in order to reduce the rate of enzymatic degradation and to increase bioavailability while also in some cases enhancing bioactivity. The half-life of unmodified peptides in the blood is in most cases very short. They also generally have poor bioavailability in tissues and organs, limiting their ability as preventative therapeutic agents [79; 80]. Modifications such as end changes, glycosylation, alkylation, and conformational changes to amino acids within the peptide may therefore have potential for ACE inhibitory peptides [80; 81]. These approaches have already been adapted to opioid peptides [81]. There is significant scope for these modifications to also be applied to ACE inhibitory and antihypertensive peptides. Encapsulation via nanoparticles and liposomes is also a strategy previously employed for opioid peptides that has possibility for adaption to for ACE inhibitory peptides. These approaches may aid in the passage of a peptide through the GI tract and may enhance the plasma half-life of the peptides. Furthermore, there is potential for bioactive peptides to be produced by microorganisms through genetic engineering to be delivered to target organs *in situ* [60]. Lastly, there is also the possibility to cross-link BAP to protein transduction domains that have been found to be able to cross biological membranes thus promoting peptide and protein delivery into cells [60; 82]. Morris *et al* [82] also devised a similar strategy using the peptide transporter PepT-1 to carry target peptides into cells.

5. *Ex-vivo* and *in vivo* animal studies

The first step employed to determine if a peptide is hypotensive is to conduct trials with small animals such as SHRs, the accepted model for human essential hypertension. A bioactive peptide can only be referred to as 'antihypertensive' after a significant decrease in BP is observed in trials with SHR. There have been many studies carried out in animals to elucidate whether food-derived ACE-inhibitory peptides can lead to an antihypertensive effect *in vivo*. Antihypertensive peptides from milk, egg, animal (including meat and marine animals), plant and macroalgae have recently been reviewed [4; 25; 83-86].

Ile-Pro-Pro (β-casein f74-76; κ-casein f108-110) and Val-Pro-Pro (β-casein f84-86) were among the first dietary peptides found to have a hypotensive effect in SHR. The peptides were first isolated from milk fermented with *Lactobacillus helveticus* and *Saccharomyces cerevisiae* (Ameal S) and their ACE-inhibitory IC_{50} values were obtained (Val-Pro-Pro and Ile-Pro-Pro having IC_{50}s of 9 and 5 μM, respectively [87]). The antihypertensive effect was first demonstrated when SHR were administered with a single oral dose of the LTPs which resulted in a significant decrease SBP between 6 to 8 h after administration [88]. Thereafter, several studies have been conducted to further characterise the *in vivo* effect of the LTPs. Their long-term effects (12-20 weeks) of administration have been assessed [89-93]. Administration of the peptides via a peptide supplement and via a sour milk drink to SHR resulted in a decrease in SBP of 12 and 17 mm Hg, respectively, compared to the control (water) after 12 weeks [90]. Endothelial function protective effects of Ile-Pro-Pro and Val-Pro-Pro were investigated using isolated SHR mesenteric arteries stored in solutions of

Krebs containing Ile-Pro-Pro and Val-Pro-Pro (1 mM), when mounted in an organ bath. Vascular reactivity measurements demonstrated better preservation of endothelium-dependent relaxation in arteries stored with the LTPs compared to controls [94]. Their bioactive effect in double transgenic rats (dTGR) harbouring human renin and angiotensinogen genes was also assessed. These transgenic rats develop malignant hypertension, cardiac hypertrophy, renal damage, and endothelial dysfunction due to increased Ang II formation. A decrease of 19 mm Hg in SBP was seen in rats administered with fermented milk supplemented with the peptides (Ile-Pro-Pro (1.8 mg/100 ml) and Val-Pro-Pro (1.8 mg/100 ml) compared to the control group Thus, it was concluded that the supplemented fermented milk product can aid in preventing the development of malignant hypertension. There was no effect on BP reported from a group receiving the peptides dissolved in water, despite the higher intake level of peptides. The authors concluded that the reported antihypertensive effect of the fermented milk product can not be explained solely by the Ile-Pro-Pro and Val-Pro-Pro supplements and suggested that a combination of factors such as calcium and potassium content, and less sodium may have contributed to the observed hypotensive effect [95].

Other rat models, such as the normotensive WKY rat, have been used to evaluate the effect of food peptides on arterial BP. However, a significant hypotensive effect is not always observed in WKY. Single oral administration of Ameal S containing the LTPs decreased BP from 6 to 8 h after administration in SHR. However, no change in SBP was observed in normotensive WKYs [88]. Similarly, the ACE inhibitory peptide Leu-Arg-Pro-Val-Ala-Ala from bovine lactoferrin was found to have a significant antihypertensive effect in SHR but no change in BP was found when the peptide was administered via intravenous injection to WKY rat [96]. Thus, the hypotensive effect of some food-derived peptides may be specific to the hypertensive state of the animal. The effect of fermented milk with LTPs on BP and vascular function in salt-loaded type II diabetic Goto–Kakizaki rats has also been assessed. GK rats are characterized by impaired glucose-induced insulin secretion, abnormal glucose regulation, insulin resistance and polyuria. They are normotensive but when on a high-salt diet can develop hypertension. The study showed a significant decrease in BP and enhanced endothelium-dependent relaxation of mesenteric arteries [97].

There are wide variations in BP responses from different food proteins. These variations may be due to the different food sources themselves but also may be due to differences in experimental models such as the type of animal used, the dosage of peptide required for a significant decrease in BP, duration of administration and administration route, i.e., oral versus intravenous administration. In general, it has been found that peptides administered intravenously have a higher decrease in BP than peptides administered orally. This may be due to lower bioavailability of these peptides in the blood stream as transport of the peptides across the brush border membrane in an intact state may not be possible. Hydrolysis or partial hydrolysis of the peptides by GI and serum enzymes may lead to inactive or less active hypotensive peptide forms. Thus, bioavailability studies are essential to assess the antihypertensive potential of a peptide [3].

Furthermore, it must be noted that although dietary peptides have lower ACE IC_{50} *in vitro* in comparison to the synthetic ACE drug inhibitor Captopril (IC_{50} in nM range), in most cases they display higher *in vivo* hypotensive effects than are expected with respect to their *in vitro* results. It has been suggested that this may be due to a higher affinity of dietary peptides to the tissues and a slower elimination in comparison to Captopril. Moreover, it is possible that several BP mechanisms of action are being employed [30; 98]. It was demonstrated that neither the egg-protein derived peptide ovokinin (2-7) (Arg-Ala-Asp-His-Pro-Phe) or Arg-Pro-Leu-Lys-Pro-Trp, the most potent derivative obtained from the structural modification of ovokinin,, inhibit ACE *in vitro*. IC_{50} values obtained were >1000 µmol/L, despite having a significant effect on BP when orally administered to SHR [99]. Thus, it must be acknowledged that *in vitro* ACE inhibitory determination may not be the best approach to assess the potential of a peptide as an antihypertensive agent. In a study by da Costa *et al* [100] it was found that the most potent *in vitro* ACE inhibitory peptides from whey did not have a significant effect on BP when orally administered to SHR. However, whey peptides with relatively low *in vitro* ACE-inhibitory activity in comparison achieved significant reductions in BP.

6. Human studies

The majority of the clinical trials regarding the antihypertensive effects of milk-derived peptides to date have been carried out on the LTPs, Ile-Pro-Pro and Val-Pro-Pro. Although some conflicting results exist, the majority of these trials have reported a significant decrease in BP. Their effect on office BP has been well documented (for reviews see 101; 3; 25). It is essential that the BP of test subjects is evaluated in comparison to placebo values and not to baseline values of the test product. As with test products, placebo groups have been found to often decrease in BP over the test period [101]. Furthermore, the 24-h ambulatory BP monitoring (ABPM), of patients BP is thought to be a more reliable method for evaluation of BP as it reduces the 'white coat effect'. Recent clinical trials with LTPs have employed ABPM [102-104; 91]. ABPM was used to assess the effects of LTP administration on dipper (where BP decreases at night time) and non-dipper (where BP does not decrease at night time) hypertensive subjects. Non-dippers are thought to have a higher cardiac vascular risk and BP monitoring in the morning and night can help predict cardiac events such as stroke and myocardial infarction. Twelve patients received a fermented milk product containing Ile-Pro-Pro (1.52 mg) and Val-Pro-Pro (2.53 mg) daily for 4 weeks. The study reported a significant reduction in night-time and early-morning SBP for nondipper subjects but not for dipper subjects [105]. A range of hemodynamic parameters was recently evaluated for 52 human subjects with high-normal BP or first-degree hypertension. These included office BP and ABPM, stress-induced BP increase and cardiac output-related parameters. Subjects were treated with LTPs (3 mg/day) for 6 weeks. The study reported a reduction in office SBP as well as an improvement in pulse wave velocity (an instrumental biomarker for vascular rigidity), stroke volume and stroke volume index (markers of cardiac flow) and acceleration and velocity index (markers of cardiac contractility). No effect on ABPM and BP reaction to stress was observed [106]. LTPs have also been reported to reduce arterial stiffness in humans. In a double-blind parallel group intervention study, 89 hypertensive subjects received daily milk containing a low dose of 5 mg/day of Ile-Pro-Pro and Val-Pro-Pro for 12 weeks and a dose of 50 mg/day for the following 12 weeks. Arterial stiffness, measured by the augmentation index (AI), decreased in the peptide group by -1.53% compared to 1.20% in the placebo group at the end of the second intervention period [107]. A similar result was seen in a study by Nakamura *et al* [108]. Twelve hypertensive subjects were administered four tablets containing Val-Pro-Pro (2.05 mg) and Ile-Pro-Pro (1.13 mg) daily for 9 weeks and were monitored for various hemodynamic parameters. A significant reduction in AI as well as peripheral SBP and DBP along with central SBP (cSBP) was observed. Furthermore, it has been suggested that LTPs may also have a positive effect on vascular endothelial function in subjects with stage-I hypertension [109] and may improve arterial compliance in postmenopausal women [110].

Other hypotensive peptides and food preparations used in human trials include peptides from casein [111], whey [112], dried bonito [113], fermented milk containing gamma-aminobutyric acid (GABA; 114) sardine muscle [115] and wakame (*Undaria pinnatifida;* 116). Recently, a number of meta-analyses on antihypertensive peptides have been carried out. Pripp *et al* [117] performed a meta-analysis on antihypertensive peptides from milk and fish proteins which included 15 human trials. A pooled decrease in SBP of -5.13 mm Hg and a decrease of -2.42 mm Hg for DBP were found. A similar result was found with a meta-analysis of 12 trials with LTPs, (623 participants in total) when pooled data in forest plots found a decrease of -4.8 mm Hg and 2.2 mm Hg in SBP and DBP, respectively. The observed hypotensive effects also seemed to be greater in hypertensive patients than in patients with pre-hypertension [118]. Another meta-analysis carried out on data from LTPs trials interestingly found that the effect of LTPs on BP was more evident in Asian subjects (SBP = -6.93 mm Hg; DBP=-3.98 mm Hg) than in Caucasians (SBP=-1.17 mm Hg; DBP = -0.52 mm Hg). The study also found that the LTP-induced hypotensive effects were not related to subject age, baseline BP value, administered dose or length of treatment [119]. Conflicting results however, were reported in a recent meta-analysis by Usinger *et al* [132]. Data from 15 controlled trials (1232 subjects in total) that observed the effect of fermented milk or similar products produced by *Lactobacilli* fermentation of milk proteins were used in the meta-analysis. The study reported a pooled decrease in SBP of just -2.45 mm Hg and found no significant decrease for pooled DBP data. Furthermore, the authors stated that the included studies were of variable quality and when excluding the studies with a high risk of bias no significant decrease in SBP or DBP were found.

7. Hypotensive peptides as functional food ingredients

The main considerations which need to be taken into account in the utilisation of antihypertensive peptides as functional ingredients in food products include characterisation of their organoleptic and physicochemical properties. In the first instance, establishment of the optimal method for peptide release via protein hydrolysis is required. Industrial scale processing of BAPs requires that peptides be able to withstand

and retain their bioactivity during multi-step processing including pasteurisation, homogenisation, pressure-driven membrane-based processing such as ultrafiltration and nanofiltration, dehydration by spray-drying or freeze-drying, and peptides must also be stable during long-term storage. Little data exists on the effects of different processing techniques on BAPs in food products. Dehydration via spray drying has been found to produce changes in peptide confirmation, a reduction in amino acid content and may also lead to non-enzymatic browning reactions [43]. The industrial-scale production of a casein hydrolysate containing the antihypertensive peptides Arg-Tyr-Leu-Gly-Tyr (α_{s1}-CN f90-94) and Ala-Tyr-Phe-Tyr-Pro-Glu-Leu (α_{s1}-CN f143-149) and the stability of the hydrolysate incorporated in a yoghurt to processing conditions, i.e. drying, homogenisation and pasteurisation, and to storage at 4°C was recently investigated. The study showed the hydrolysate to be stable after processing as both *in vitro* ACE-inhibitory activity and the *in vivo* antihypertensive properties in SHR were maintained. Analysis by reverse phase-high pressure liquid chromatography-mass spectrometry (RP-HPLC-MS) showed that the integrity of the antihypertensive peptides was also maintained during storage at 4 °C for 1 month [120]. Similar studies are required for other antihypertensive peptides in order to evaluate the optimal processing conditions required for retention of bioactivity.

It has been shown that heat treatments and mechanical damage can reduce peptides bioactivity. As a result of changes in protein structure, the profile of peptides released may differ as digestive enzymes may be capable of digesting these regions of the protein that were previously inaccessible to the enzyme. This has been previously shown to be the case for whey protein [121-122]. Furthermore, ACE inhibitory peptides from whey protein isolates (WPI) pretreated at 65 °C were shown to have greater inhibitory activity than peptides from WPI pretreated at 95 °C. This result can be explained by the formation of whey aggregates [100].

Optimisation of the hydrolytic process should also be considered when planning to up-scale the production of BAPs. Bioactive peptides may be produced by enzymatic or microbial hydrolysis. However, it is thought that enzymatic hydrolysis is more suited for food-grade BAP production over microbial fermentation [123]. Enzyme immobilisation offers several advantages over the addition of soluble enzymes directly to the product. They can be recycled and the use of immobilized enzymes potentially avoids the generation of interfering metabolite products due to autolysis of the enzymes. Furthermore, protein hydrolysis using immobilized enzymes can also be carried out in milder more controlled conditions and does not need to be inactivated by heat or acidification, which may be damaging for the product [124; 123]. The use of membrane bioreactors may be a substitute for the development of functional materials from food proteins. This system integrates a reaction vessel with a membrane separation system allowing for the recycling of the enzyme, separation, fractionation and/or concentration of the bioactive compound. The use of membrane bioreactors for the development of functional materials from sea-food processing wastes has been recently reviewed [125].

BAP fractionation and enrichment steps include membrane processing incorporating ultrafiltration and liquid chromatography, ion exchange, gel filtration and reverse phase matrices. Electro-membrane filtration (EMF), a combination of conventional membrane filtration and electrophoresis, may be a consideration for industrial scale isolation of BAPs. EMF is more selective than conventional membrane filtration (ultrafiltration) and is less costly than chromatography [123].

As mentioned earlier, a large portion of antihypertensive peptides are of low molecular weight and many contain hydrophobic residues, attributes which have been classically associated with bitterness in foods. Hence, this is notably an obstacle that must be resolved during the processing of antihypertensive food products. A number of strategies have been applied with the aim of debittering protein hydrolysates including absorption of bitter peptides on activated carbon, selective extraction with alcohols and chromatographic removal using different matrices. Peptidase-mediated debittering has also been applied. This involves the concomitant or sequential incubation of the protein hydrolysates with exopeptidases, with priority cleavage at hydrophobic residues [126]. However, these debittering strategies may lead to the loss of some amino acid residues from hydrolysates. As bioactivity relies greatly on peptide sequence, these debittering methods may not therefore be suitable for debittering of BAPs including antihypertensive peptides, as hydrolysis may result in loss of activity. Changes in peptide structure may also have implications for absorption at the brush border membrane. Therefore, enzymatic debittering strategies need to be approached on a case-by-case basis.

The widespread commercialisation of antihypertensive food products is dependent on the availability of scientific data from *in vivo* animal and human models that positively demonstrates their contribution in reducing BP. Furthermore, legislation which governs health claims in relation to functional foods needs to be taken into account. In Japan, the FOSHU (food for specified health use) licensing system was put in place whereby foods claiming health benefits must first be approved by the system before been allowed to be put on the market [127]. Since then a number of antihypertensive products currently on the market in Japan have been granted FOSHU approval. 'Ameal-S' which is manufactured by Calpis Co., Ltd. is a fermented sour milk containing the LTPs Ile-Pro-Pro and Val-Pro-Pro. The soft drink Casein DP 'Peptio' manufactured by Kanebo Co., Ltd. contains the antihypertensive peptide Phe-Phe-Val-Ala-Pro-Phe-Pro-Gln-Val-Phe-Gly-Phe (α_{s1}-casein f23–34) and is also FOSHU approved. There has been a new European Regulation on nutrition and health claims in the EU since 2007 (Regulation 1924/2006). Advised by the European Commission (EC), the European Food Safety Authority (EFSA) reviews evidence of health claims made by food companies. Interestingly, EFSA has not allowed/approved any peptide related hypotensive claim to date [128-129]. For both the C12-peptide and the bonito protein-derived peptide Leu-Lys-Pro-Asn-Met, it was concluded that a cause and effect relationship has not been established between the consumption of the peptides and maintenance of normal BP [128-129]. In the US, the Food and Drug Administration (FDA) assesses the scientific evidence for health claims under the 1990 Nutrition Labelling and Education Act [130] and the 1994 Dietary Supplement Health and Education Act [131].

Therefore, industrial manufacturers of functional food products need to provide a significant amount of scientific evidence that satisfies the legislative governing body in the specific market region before any new food products can be put on the market claiming to have hypotensive effects.

8. Conclusion

Antihypertensive peptides have major potential as functional ingredients aiding in the prevention and management of hypertension. Although these peptides have been found to be less potent than antihypertensive synthetic drugs, as part of the daily diet they could play an important part as natural and safe BP control agents. Further detailed mechanistic studies on food protein-derived antihypertensive peptides must be carried out to elucidate the BP mechanism(s) involved. With regard to ACE-inhibitory peptides, a better understanding of the interactions involved in the binding of peptides to the active site of ACE is required such that more effective food peptide-based inhibitors of ACE can be discovered. The use of bioinformatics and *in silico* methods for identification of potential bioactive sequences may allow for more substrates to be assessed in a shorter time scale. Cost effective production methods including enrichment, isolation and purification procedures must first be considered and ease of scalability must be achieved. Before advancement of functional hypotensive products onto the market, moreover, the physicochemical, technofunctional and sensory properties must be considered prior to production of new antihypertensive food products.

Author details

Roseanne Norris and Richard J. FitzGerald*
Department of Life Sciences, University of Limerick, Limerick, Ireland

Acknowledgement

Financial support for this work was provided by the Irish Research Council for Science, Engineering and Technology (IRCSET) in the form of a studentship to author Norris.

9. References

[1] Kearney PM, Whelton M, Reynolds K, Whelton PK, He J. Worldwide prevalence of hypertension: a systematic review. Journal of Hypertension 2004;22 11-19.
[2] Centre for Disease Control and Prevention: High Blood Pressure Frequently Asked Questions. http://webcache.googleusercontent.com/search?q=cache:jGRR07v0WXIJ:www.cc.gov/bloodpressure/faqs.htm (Accessed 22 September 2011).

* Corresponding Author

[3] Murray BA, FitzGerald RJ. Angiotensin converting enzyme inhibitory peptides derived from proteins: biochemistry, bioactivity, and production. Current Pharmaceutical Design 2007;13(8) 773-91.

[4] Martínez-Maqueda D, Miralles B, Recio I, Hernández-Ledesma B. Antihypertensive peptides from food proteins: a review. Food & Function 2012;3 350-361.

[5] Inagami T. The renin-angiotensin system. Essays in Biochemistry 1994;28 147-164.

[6] Turner AJ, Hooper NM. The angiotensin converting enzyme gene family, genomics and pharmacology. Trends in Pharmacological Science 2002;23 177-183.

[7] FitzGerald RJ, Murray BA, Walsh DJ. Hypotensive peptides from milk proteins. The Journal of Nutrition 2004;134(4) 980S-988S.

[8] Libby P, Bonow RO, Mann DL, Zipes DP. Braunwald's Heart Disease: A textbook of cardiovascular Medicine (8th ed.). Philadelphia: Saunders; 2008.

[9] Meisel H, Walsh DJ, Murray BA, FitzGerald RJ. ACE Inhibitory Peptides. In: Mine Y, Shahidi F. (ed.) Nutraceutical proteins and peptides in health and disease. New York: CRC Press, Taylor and Francis Group; 2006. p269-315.

[10] Guang C, Phillips RD. Plant food-derived angiotensin I converting enzyme inhibitory peptides. Journal of Agricultural Food Chemistry 2009;57(12) 5113–5120.

[11] Wijesekara I, Kim SK. Angiotensin-I-converting enzyme (ACE) inhibitors from marine resources: prospects in the pharmaceutical industry. Marine Drugs 2010;8(4) 1080–1093.

[12] Ahhmed AM, Muguruma M. A review of meat protein hydrolysates and hypertension. Meat Science 2010;86 110-118.

[13] Li Y, Zhou J, Huang K, Sun Y, Zeng X. Purification of a Novel Angiotensin I-Converting Enzyme (ACE) Inhibitory Peptide with an Antihypertensive Effect from Loach (*Misgurnus anguillicaudatus*). Journal of Agricultural and Food Chemistry 2012;60(5) 1320–1325.

[14] Escudero E, Sentandreu MA, Arihara K, Toldrá F. Angiotensin I-converting enzyme inhibitory peptides generated from *in vitro* gastrointestinal digestion of pork meat. Journal of Agricultural and Food Chemistry 2010;58 2895–2901.

[15] Chel-Guerrero L, Domínguez-Magaña M, Martínez-Ayala A, Dávila-Ortiz G, Betancur-Ancona D. Lima bean (*Phaseolus lunatus*) protein hydrolysates with ACE-I inhibitory activity. Food and Nutrition Science 2012;3 511-521.

[16] Lee JK, Jeon JK, Byun HG. Effect of angiotensin I converting enzyme inhibitory peptide purified from skate skin hydrolysate. Food Chemistry 2011;125(2) 495-499.

[17] Terashima M, Bara T, Ikemoto N, Katayama M, Morimoto T, Matsumura S. Novel angiotensin converting enzyme (ACE) inhibitory peptides derived from boneless chicken leg meat. Agricultural and Food Chemistry 2010;58 7432-7436.

[18] Chabance B, Marteau P, Rambaud JC, Migliore-Samour D, Bynard M, Perrotin P, Guillet R, Jollés P, & Fiat AM. Casein peptide release and passage to the blood in humans during digestion of milk and yoghurt. Biochimie 1998;80 155-165.

[19] Vermeirssen V, Van Camp J, Devos L, Verstraete W. Release of angiotensin I converting enzyme inhibitory activity during *in vitro* gastrointestinal digestion: from batch

experiment to semi-continuous model. Journal of Agricultural Food Chemistry 2003;51 5680-5687.

[20] Hernandez-Ledesma B, Amigo L, Ramos M, Recio I. Angiotensin converting enzyme inhibitory activity in commercial fermented products. Formation of peptides under simulated gastrointestinal digestion. Journal of Agricultural Food Chemistry 2004;52(6) 1504-1510.

[21] Walsh DJ, Bernard H, Murray BA, MacDonald J, Pentzien AK, Wright GA, Wal JM, Struthers AD, Meisel H, FitzGerald RJ. *In vitro* generation and stability of the lactokinin β-lactoglobulin fragment (142-148). Journal of Dairy Science 2004;87 3845–3857.

[22] Jang A, Jo C, Lee M. Storage stability of the synthetic angiotensin converting enzyme (ACE) inhibitory peptides separated from beef sarcoplasmic protein extracts at different pH, temperature, and gastric digestion. Food Science and Biotechnology 2007;16(4) 572-575.

[23] Cinq-Mars CD, Hu C, Kitts DD, Li-Chan ECY. Investigations into inhibitor type and mode, simulated gastrointestinal digestion, and cell transport of angiotensin I-converting enzyme-inhibitory peptides in pacific hake (*Meriuccius productus*) fillet hydrolysate. Journal of Agricultural Food Chemistry 2008;56(2) 410-419.

[24] Akillioglu HG, Karakaya S. Effects of heat treatment and *in vitro* digestion on the angiotensin converting enzyme inhibitory activity of some legume species. European Food Research and Technology 2009;229(6) 915-921.

[25] Jäkälä P, Vapaatalo H. Antihypertensive peptides from milk proteins. Pharmaceuticals 2010;3 251-272.

[26] Shin AI, Yu R, Park SA, Chung DK, Ahn CW, Nam HS, Kim HS, Lee HJ. His-His-Leu, an angiotensin converting enzyme inhibitory peptide derived from Korean soybean paste, exerts antihypertensive activity *in vivo.* Journal of Agricultural Food Chemistry 2001;49(6) 3004-3009.

[27] Okamoto A, Hanagata H, Matsumoto E, Kawamura Y, Koizumi Y, Yanagiga F. Angiotensin I converting enzyme inhibitory activities of various fermented foods. Bioscience, Biotechnology, & Biochemistry 1995;59(6) 1147-1149.

[28] Nakahara T, Sano T, Yamaguchi H, Sugimoto KRI, Chikata H, Kinoshita E, Uchida R. Antihypertensive effect of peptide-enriched soy sauce-like seasoning and identification of its angiotensin I-converting enzyme inhibitory substances. Journal of Agricultural Food Chemistry 2010;58(2) 821-827.

[29] Gibbs BF, Zoygman A, Masse R, Mulligan C. Production and characterisation of bioactive peptides from soy hydrolysates and soy-fermented food. Food Research International 2004;37(2) 123-131.

[30] Fujita H, Yoshikawa M. LKPNM: a prodrug-type ACE-inhibitory peptide derived from fish protein. Immunopharmacology 1999;44(1-2) 123-127.

[31] Staessen JA, Li Y, Richart T. Oral renin inhibitors. Lancet 2006;368 1449-1456.

[32] Udenigwe CC, Li H, Aluko RE. Quantitative structure-activity relationship modelling of renin-inhibiting dipeptides. Amino Acids 2011;42(4) 1379-1386.

[33] Udenigwe CC, Lin YS, Hou WC, Aluko RE. Kinetics of the inhibition of renin and angiotensin I-converting enzyme by flaxseed protein hydrolysate fractions. Journal of Functional Foods 2009;1(2) 199–207.

[34] Li H, Aluko RE. Identification and inhibitory properties of multifunctional peptides from pea protein hydrolysate. Journal of Agricultural and Food Chemistry 2010;58 11471–11476.

[35] Tanaka M, Watanabe S, Wang Z, Matsumoto K, Matsui T. His-Arg-Trp potently attenuates contracted tension of thoracic aorta of Sprague-Dawley rats through the suppression of extracellular Ca^{2+} influx. Peptides 2009;30(8) 1502-1507.

[36] Wang Z, Watanabe S, Kobayashi Y, Tanaka M, Matsui T. Trp-His, a vasorelaxant dipeptide, can inhibit extracellular Ca^{2+} entry to rat vascular smooth muscle cells through blockade of dihydropyridine-like L-type Ca^{2+} channels. Peptides 2010;31(11) 2060-2066.

[37] Meisel H, FitzGerald RJ. Opioid peptides encrypted in intact milk protein sequences. British Journal of Nutrition 2000;1 S27-S31.

[38] Jauhiainen T, Korpela R. Milk peptides and blood pressure. The Journal of Nutrition 2007;137 825S-829S.

[39] Nurminen ML, Sipola M, Kaarto H, Pihlanto-Leppala A, Piilola K, Korpela R, Tossavainen O, Korhonen H, Vapaatalo H. Alpha-lactorphin lowers blood pressure measured by radiotelemetry in normotensive and spontaneously hypertensive rats. Life Sciences 2000;66 1535-1543.

[40] Sipola M, Finckenberg P, Vapaatalo H, Pihlanto-Leppälä A, Korhonen H, Korpela R, Nurminen ML. Alpha-lactorphin and beta-lactorphin improve arterial function in spontaneously hypertensive rats. Life Sciences 2002;71(11) 1245-1253.

[41] Yoshikawa M., Tani F., Shiota A., Suganuma H., Usui H., Kurahashi K., Chiba H. Casoxin D, an opioid antagonist/ileum-contracting/vasorelaxing peptide derived from human α_{s1}-casein. In: Brantl V, Teschemacher H. (eds.) β-Casomorphins and related peptides: Recent developments. Weinheim: VCH; 1994. p43–48.

[42] Kato M, Fujiwara Y, Okamoto A, Yoshikawa M, Chiba H, Shigezo U. Efficient production of Casokin D, a bradykinin agonist peptide derived from human casein, by *Bacillus brevis*. Bioscience, Biotechnology, & Biochemistry 1995;59(11) 2056-2059.

[43] Hernández-Ledesma B, del Mar Contreras M, Recio I. Antihypertensive peptides: production, bioavailability and incorporation into foods. Advances in Colloid and Interface Science 2011;165 23-35.

[44] Okitsu M, Morita A, Kakitani M, Okada M, Yokogoshi H. Inhibition of the endothelin-converting enzyme by pepsin digests of food proteins. Bioscience Biotechnology Biochemistry 1995;59 325–326.

[45] Matoba N, Usui H, Fujita H, Yoshikawa M. A novel anti-hypertensive peptide derived from ovalbumin induces nitric oxide-mediated vasorelaxation in an isolated SHR mesenteric artery. *FEBS Letters* 1999;452(3) 181-184.

[46] Scruggs P, Filipeanu CM, Yang J, Chang JK, Dun NJ. Interaction of ovokinin(2–7) with vascular bradykinin 2 receptors. *Regulatory Peptides* 2004;120(1-3) 85-91.

[47] Okitsu M, Morita A, Kakitani M, Okada M, Yokogoshi H. Inhibition of the endothelin-converting enzyme by pepsin digests of food proteins. Bioscience, Biotechnology, & Biochemistry 1995;59(2) 325-326.

[48] Maes W, Van Camp J, Vermeirssen V, Hemeryck M, Ketelslegers JM, Schrezenmeir J, Van Oostveldt P, Huyghebaert A. Influence of the lactokinin Ala-Leu-Pro-Met-His-Ile-Arg (ALPMHIR) on the release of endothelin-1 by endothelial cells. Regulatory peptides 2004;118(1-2) 105-109.

[49] Gomez-Ruiz JA, Ramos M, Recio I. Angiotensin converting enzyme inhibitory activity of peptides isolated from Manchego cheese. Stability under simulated gastrointestinal digestion. International Dairy Journal 2004;14 1075-1080.

[50] Ondetti MA, Rubin B, Cushman DW. Design of specific inhibitors of angiotensin-converting enzyme: new class of orally active antihypertensive agents. Science 1977;196(4288) 441-444.

[51] Cheung HS, Wang FL, Ondetti MA, Sabo EF, Cushman DW. Binding of peptide substrates and inhibtiors of angiotensin-converting enzyme: importance of the COOH-terminal dipeptide sequence. Journal of Biological Chemistry 1980;255(2) 401-407.

[52] Gobbetti M, Stepaniak L, De Angelis M, Corsetti A, Cagno RD. Latent bioactive peptides in milk proteins: proteolytic activation and significance in dairy processing. Critical Reviews in Food Science and Nutrition 2002;42 223-239.

[53] López-Fandiño R, Otte J, van Camp J. Physiological, chemical and technological aspects of milk-protein-derived peptides with antihypertensive and ACE-inhibitory activity. International Dairy Journal 2006;16 1277-1293.

[54] Pripp, AH, Isakasson T, Stepaniak L, Sorhaug T, Ardo Y. Quantitative structure activity relationship modelling of peptides and proteins as a tool in food science. Trends in Food Science and Technology 2005;16 484-494.

[55] Wu J, Aluko RE, Nakai S. Structural requirements of angiotensin 1-converting enzyme inhibitory peptides: quantitative structure-activity relationship study of di- and tripeptides. Journal of Agricultural & Food Chemistry 2006;54 732-738.

[56] Wu J, Aluko RE, Nakai S. Structural requirements of angiotensin 1-converting enzyme inhibitory peptides: quantitative structure-activity relationship modelling of peptides containing 4-10 amino acid residues. QSAR and Combinatorial Science 2006;25 873-880.

[57] He R, Ma H, Zhao W, Qu W, Zhao J, Luo L, Zhu W. Modelling the QSAR of ACE-inhibitory peptides with ANN and its applied illustration. International Journal of Peptides 2011;2012 1-9.

[58] Norris R, Casey F, FitzGerald RJ, Shields D, Mooney C. Predictive modelling of angiotensin converting enzyme inhibitory dipeptides. Food Chemistry 2012;133(4), 1349–1354.

[59] Murray BA, Walsh DJ, FitzGerald RJ. Modification of the furanacryloyl-L-phenylalanylglycylglycine assay for determination of angiotensin-I-converting enzyme inhibitory activity. Journal of Biochemical and Biophysical Methods 2004;59(2) 127-137.

[60] Vermeirssen V, Van Camp J, Verstraete W. Bioavailability of angiotensin I converting enzyme inhibitory peptide. British Journal of Nutrition 2004;92 357-366.

[61] Vermeirssen V, Van Camp J, Decroos K, Van Wijmelbeke V, Verstraete W. The impact of fermentation and the *in vitro* digestion on the formation of angiotensin-I-converting enzyme inhibitory activity from pea and whey protein. Journal of Dairy Science 2003;86 429-438.

[62] Vermeirssen V, Van Camp J, Devos J, Verstraete W. Release of Angiotensin I Converting Enzyme (ACE) inhibitory activity during in vitro gastrointestinal digestion: from batch experiment to semi-continuous model. Journal of Agricultural and Food Chemistry 2003;51(19) 5680-5687.

[63] Matsui T, Li CH, Osajima Y. Preparation and characterisation of novel bioactive peptides responsible for angiotensin I-converting enzyme inhibition from wheat germ. Journal of Peptide Science 1999;5 289-297.

[64] Miguel M, Aleixandre MA, Ramos M, López-Fandiño R. Effect of simulated gastrointestinal digestion on the antihypertensive properties of ACE-inhibitory peptides derived from ovalbumin. Journal of Agricultural Food Chemistry 2006;54 726-731.

[65] Wu J, Ding X. Characterization of inhibition and stability of soy-protein-derived angiotensin I-converting enzyme inhibitory peptides. Food Research International 2002;35 367-375.

[66] Tavares T, del Mar Contreras M, Amorim M, Pintado M, Recio I, Malcata FX. Novel whey-derived peptides with inhibitory effect against angiotensin-converting enzyme: In vitro effect and stability to gastrointestinal enzymes. Peptides 2011;32 1013-1019.

[67] Hannelore D. Molecular and integrative physiology of intestinal peptide transport. Annual Reviews in Physiology 2004;66 361–384.

[68] Nakamura Y, Masuda O, Takano T. Decrease in tissue angiotensin I-converting enzyme activity upon feeding sour milk in spontaneously hypertensive rats. Bioscience, Biotechnology and Biochemistry 1996;60 488–489.

[69] Masuda O, Nakamura Y, Takano T. Antihypertensive peptides are present in aorta after oral administration of sour milk containing these peptides to spontaneously hypertensive rats, Journal of Nutrition 1996;126 3063-3068.

[70] Foltz M, Meynen EM, Bianco V, van Platerink C, Koning TM, Kloek J. Angiotensin converting enzyme inhibitory peptides from a lactotripeptide-enriched milk beverage are absorbed intact into the circulation. Journal of Nutrition 2007;137(4) 953-958.

[71] Foltz M, Cerstiaens A, van Meensel A, Mols R, van der Pijl PC, Duchateau GS, Augustijns P. The angiotensin converting enzyme inhibitory tripeptides Ile-Pro-Pro and Val-Pro-Pro show increasing permeabilities with increasing physiological relevance of absorption models. Peptides 2008;29(8) 1312-1320.

[72] Ohsawa K, Satsu H, Ohki K, Enjoh M, Takano T, Shimizu, M. Producibility and digestibility of antihypertensive β-casein tripeptides, Val-Pro-Pro and Ile-Pro-Pro, in the

gastrointestinal tract: analyses using an in vitro model of mammalian gastrointestinal digestion. Journal of Agricultural Food Chemistry 2008;56(3) 854-858.

[73] Satake M, Enjoh M, Nakamura Y, Takano T, Kawamura Y, Arai S, Shimizu M. Transepithelial transport of the bioactive tripeptide, Val-Pro-Pro, in human intestinal Caco-2 cell monolayers. Bioscience, Biotechnology & Biochemistry 2002;66 378-384.

[74] Quirós A, Dávalos A, Lasunción MA, Ramos M, Recio I. Bioavailability of the antihypertensive peptide LHLPLP: Transepithelial flux of HLPLP. International Dairy Journal 2008;18 279-286.

[75] van Platerink CJ, Janssen HGM, Horsten R, Haverkamp J. Quantification of ACE inhibiting peptides in human plasma using high performance liquid chromatography–mass spectrometry. Journal of Chromatography B 2006;830 151-157.

[76] Vermeirssen V, Augustijns P, Morel N, van Camp J, Opsomer A, Verstraete W. *In vitro* intestinal transport and antihypertensive activity of ACE inhibitory pea and whey digests. International Journal of Food Sciences and Nutrition 2005;56(6) 415-430.

[77] Foltz M, van Buren L, Klaffke W, Duchateau GS. Modeling of the relationship between dipeptide structure and dipeptide stability, permeability, and ACE inhibitory actvitiy. Journal of Food Science 2009;74 H243-H251.

[78] Fujita H, Yokoyama K, Yoshikawa M. Classification and antihypertensive activity of angiotensin I-converting enzyme inhibitory peptides derived from food proteins. Journal of Food Science 2000;65 564-569.

[79] Gardener MLG. Transmucosal Passage of Intact Peptides. In: Grimble GK, Backwell FRC. (eds.) Peptides in Mammalian Metabolism. Tissue Utilisation and Clinical Targeting. London: Portland Press Ltd; 1998.

[80] Adessi C, Soto C. Converting a peptide into a drug: Strategies to improve stability and bioavailability.,Current Medicinal Chemistry, 2002;9 963-978.

[81] Witt KA, Davis TP. CNS drug delivery: opioid peptides and the blood-brain barrier. In: Rapaka RS, Sadée W. (eds.) Drug Addiction. New York: Springer Science and Business Media; 2008.

[82] Morris MC, Depollier J, Mery J, Heitz F, Divita G. A peptide carrier for the delivery of biologically active proteins into mammalian cells. Nature Biotechnology 2001;19 1173-1176.

[83] Miguel M, Aleixandre MA. Antihypertensive peptides derived from egg proteins. The Journal of Nutrition 2006;136(6) 1457-1460.

[84] Harnedy PA, FitzGerald RJ. Bioactive proteins, peptides, and amino acids from macroalgae, Journal of Phycology 2011;47 218–232.

[85] Harnedy PA, FitzGerald RJ. Bioactive peptides from marine processing waste and shellfish: a review. Journal of Functional Foods 2011;4 6-24.

[86] Ryan JT, Ross RP, Bolton D, Fitzgerald GF, Stanton C. Bioactive peptides from muscle sources: meat and fish. *Nutrients* 2011;74(7) H243-H251.

[87] Nakamura Y, Yamamoto N, Sakai K, Okubo A, Yamazaki S, Takano T. Purificiation and characterisation of angiotensin I-converting enzyme inhibitors from sour milk. Journal of Dairy Science 1995;78 777-783.

[88] Nakamura Y, Yamamoto N, Sakai K, Takano T. Antihypertensive effect of sour milk and peptides isolated from it that are inhibitors of angiotensin I- converting enzyme. Journal of Dairy Science 1995;78 1253-1257.

[89] Sipola M, Finckenberg P, Korpela R, Vapaatalo H, Nurinen ML. Effect of long-term intake of milk products on blood pressure in hypertensive rats. Journal of Dairy Research 2002;69 103-111.

[90] Sipola M, Finckenberg P, Santisteban J, Korpela R, Vapaatalo H, Nurminen ML. Long-term intake of milk peptides attenuates development of hypertension in spontaneously hypertensive rats. Journal of Physiology and Pharmacology 2001;52 745-754.

[91] Seppo L, Jauhiainen T, Poussa P, Korpela R. A fermented milk high in bioactive peptides has a blood pressure–lowering effect in hypertensive subjects. The American Journal of Clinical Nutrition 2003;77 326-330.

[92] Tuomilehto J, Lindström J, Hyyrynen J, Korpela R, Karhunen ML, Mikkola L, Jauhiainen T, Seppo L, Nissinen A. Effect of ingesting sour milk fermented using *Lactobacillus helveticus* bacteria producing tripeptides on blood pressure in subjects with mild hypertension. Journal of Human Hypertension 2004;18 795-802.

[93] Jauhiainen T, Collin M, Narva M, Cheng ZJ, Poussa T, Vapaatalo H, Korpela R. Effect of long-term intake of milk peptides and minerals on blood pressure and arterial function in spontaneously hypertensive rats. Milchwissenschaft 2005;60 358-362.

[94] Jäkälä P, Jauhiainen T, Korpela R, Vapaatalo H. Milk protein-derived bioactive tripeptides Ile-Pro-Pro and Val-Pro-Pro protect endothelial function *in vitro* in hypertensive rats. Journal of Functional Foods 2009;1(3) 266-273.

[95] Jauhiainen T, Pilvi T, Cheng ZJ, Kautuaunen H, Müller DN, Vapaatalo H, Korpela R, Mervaala E. Milk products containing bioactive tripeptides have an antihypertensive effect in double transgenic rats (dTGR) habouring human renin and human angiotensinogen genes. Journal of Nutrition and Metabolism, 2009;2010 1-6.

[96] Lee NY, Cheng JT, Enomoto T, Nakamura I. The antihypertensive activity of angiotensin-converting enzyme inhibitory peptide containing in bovine lactoferrin. *Chinese Journal of Physiology* 2006;49(2) 67-73.

[97] Jäkälä P, Hakal A, Turpeinen AM, Korpela R, Vapaatalo H. Casein-derived bioactive tripeptides Ile-Pro-Pro and Val-Pro-Pro attenuate the development of hypertension and improve endothelial function in salt-loaded Goto–Kakizaki rats. Journal of Functional Foods 2009;1(4) 366-374.

[98] López-Fandiño R, Recio I, Ramos M. Egg-Protein-Derived Peptides with Antihypertensive Activity. In: Huopalahti R, López-Fandiño R, Anton M, Schade R. (eds.) Bioactive egg compounds. New York; Springer-Verlag Heidelberg; 2007. p199-211.

[99] Yamada Y, Matoba M, Usui H, Onishi K, Yoshikawa M. Design of a highly potent anti-hypertensive peptide based on ovokinin(2-7). Bioscience, Biotechnology and Biochemistry 2002;66(6) 1213-1217.

[100] da Costa EL, da Rocha Gontijo JA, Netto FM. Effect of heat and enzymatic treatment on the antihypertensive activity of whey protein hydrolysates. International Dairy Journal 2007;17(6) 632-640.

[101] Boelsma E, Kloek J. Lactotripeptides and antihypertensive effects:a critical review. British Journal of Nutrition 2008;101(6) 776-786.

[102] Jauhiainen T, Vapaatalo H, Poussa T, Kyrönpalo S, Rasmussen S, Korpela R. *Lactobacillus helveticus* fermented milk lowers blood pressure in hypertensives subjects in 24-h ambulatory blood pressure measurement. American Journal of Hypertension 2005;18(12 Pt 1) 1600-1605.

[103] Yamasue K, Morikawa N, Mizushima S, Tochikubo O. The blood pressure lowering effect of lactotripeptides and salt intake in 24-h ambulatory blood pressure measurements. Clinical and Experimental Hypertension 2010;32(4) 214-220.

[104] Germino FW, Neutel J, Nonaka M, Hendler SS. The impact of lactotripeptides on blood pressure response in stage 1 and stage 2 hypertensives. Journal of Clinical Hypertension 2010;12 153-159.

[105] Kurosawa MT, Nakamura Y, Yamamoto N, Yamada K, Iketani T. Effects of Val-Pro-Pro and Ile-Pro-Pro on nondipper patients: a preliminary study. Journal of Medicinal Food, 2011;14(5) 538-542.

[106] Cicero AFG, Rosticci M, Gerocarni B, Bacchelli S, Veronesi M, Strocchi E, Borghi C. Lactotripeptides effect on office and 24-h ambulatory blood pressure, blood pressure stress response, pulse wave velocity and cardiac output in patients with high-normal blood pressure or first-degree hypertension: a randomised double-blind clinical trial. Hypertension Research 2011;34 1035-1040.

[107] Jauhiainen T, Rönnback M, Vapaatalo H, Wuolle K, Kautiainen H, Groop PH, Korpela R. Long-term intervention with *Lactobacillus helveticus* fermented milk reduces augmentation index in hypertensive subjects. European Journal of Clinical Nutrition 2010;64 424–431.

[108] Nakamura T, Mizutani J, Sasaki K, Yamamoto N, Takazawa K. Beneficial potential of casein hydrolysate containing Val-Pro-Pro and Ile-Pro-Pro on central blood pressure and hemodynamic index: A preliminary study. Journal of Medicinal Food, 2009;12 1-6.

[109] Hirota T, Ohki K, Kawagishi R, Kajimoto Y, Mizuno S, Nakamura Y, Kitakaze M. Casein hydrolysate containing the antihypertensive tripeptides Val-Pro-Pro and Ile-Pro-Pro improves vascular endothelial function independent of blood pressure-lowering effects: contribution of the inhibitory action of angiotensin-converting enzyme. Hypertension Research 2007;30 489-496.

[110] Yoshizawa M, Maeda S, Miyaki A, Misono M, Choi Y, Shimojo N, Ajisaka R, Tanaka H. Additive beneficial effects of lactotripeptides and aerobic exercise on arterial

compliance in postmenopausal women. American Journal of Physiology. Heart and Circulatory Physiology 2009;297 H1899-903.

[111] Cadée JA, Chang CY, Chen CW, Huang CN, Chen SL, Wang CK. Bovine casein hydrolysate (C12 Peptide) reduces blood pressure in prehypertensive subjects. American Journal of Hypertension 2007;20(1) 1-5.

[112] Pins JJ, Keenan JM. The antihypertensive effects of a hydrolysed whey protein isolate supplement (BioZate® 1): a pilot study. FASEB Journal 2003;17 A1110.

[113] Fujita H, Yamagami T, Ohshima K. Effects of an ACE-inhibitory agent, katsuobushi oligopeptide, in the spontaneously hypertensive rat and in borderline and mildly hypertensive subjects. Nutrition Research 2001;21(8) 1149-1158.

[114] Inoue K, Shirai T, Ochiai H, Kasao M, Hayakawa K, Kimura M, Sansawa H. Blood-pressure-lowering effect of a novel fermented milk containing gamma-aminobutyric acid (GABA) in mild hypertensives. European Journal of Clinical Nutrition, 2003;57(3) 490-5.

[115] Kawasaki T, Seki E, Osajima K, Yoshida M, Asada K, Matsui T, Osajima Y. Antihypertensive effect of valyl-tyrosine, a short chain peptide derived from sardine muscle hydrolyzate, on mild hypertensive subjects. Journal of Human Hypertension 2000;14(8) 519-523.

[116] Suetsuna K, Nakano T. Identification of an antihypertensive peptide from a peptic digest of wakame (*Undaria pinnatifida*). *Journal of Nutritional* Biochemistry 2000;11(9) 450-454.

[117] Pripp AH. Effect of peptides derived from food proteins on blood pressure: a meta-analysis of randomized controlled trials. Food & Nutrition Research 2008;52 doi: 10.3402/fnr.v52i0.1641.

[118] Xu, J-Y., Qin, L.Q., M.S.M., Wang, P.Y., Li, W. & Chang, C. Effect of milk tripeptides on blood pressure: a meta-analysis of randomized controlled trials. Nutrition 2008; 24(10) 933-940.

[119] Cicero AFG, Gerocarni B, Laghi L, Borghi C. Blood pressure lowering effect of lactotripeptides assumed as functional foods: a meta-analysis of current available clinical trials. Journal of Human Hypertension 2011;1 1-12.

[120] Contreras MM, Sevilla MA, Monroy-Ruix J, Amigo L, Gómez-Sala B, Molina E, Ramos M, Recio I. Food-grade production of an antihypertensive casein hydrolysate and resistance of active peptides to drying and storage. International Dairy Journal 2011;21 470-476.

[121] Smithers GW, Ballard FJ, Copeland AD, DeSilva KJ, Dionysius DS, Francis GL, Goddard C, Grieve PA, McIntosh GH, Mitchell IR, Pearce RJ, Regester GO. New opportunities from the isolation and utilization of whey proteins. Journal of Dairy Science 1996;79(8) 1454-1459.

[122] O' Loughlin IB, Murray BA, Kelly PM, FitzGerald RJ, Brodkorb A. Enzymatic hydrolysis of heat-induced aggregates of whey protein isolate. Agricultural & Food Chemistry 2012;60(19) 4895-4904.

[123] Agyei D, Danquah MK. Industrial-scale manufacturing of pharmaceutical-grade bioactive peptides. Biotechnology Advances 2011; 29(3) 272–277.

[124] Pedroche J, Yust MM, Lqari H, Megias C, Girón-Calle J, Alaiz M, Vioque J, Millán F. Obtaining of *Brassica carinata* protein hydrolysates enriched in bioactive peptides using immobilized digestive proteases. Food Research International 2007;40(7) 931–938.

[125] Kim SK, Senevirathne M. Membrane bioreactor technology for the development of functional materials from sea-food processing wastes and their potential health benefits. Membranes 2011;1(4) 327-344.

[126] FitzGerald RJ, Cuinn GO. Enzymatic debittering of food protein hydrolysates. Biotechnology Advances 2006;24 234– 237.

[127] Ministry of Health, Labour and Welfare: Food for Specified Health Uses (FOSHU). http://www.mhlw.go.jp/english/topics/foodsafety/fhc/02html
(accessed 16 May 2012).

[128] European Food Safety Authority: Scientific Opinion on the Substantiation of Health Claims Related to a C12 Peptide (Phe-Phe-Val-Ala-Pro-Phe-Pro-Glu-Val-Phe-Gly-Lys) and Maintenance of Normal Blood Pressure (ID 1483, 3130) Pursuant to Article 13(1) of Regulation (EC) No 1924/2006.
http://www.efsa.europa.eu/en/efsajournal/doc/1478.pdf (assessed 15 May 2012).

[129] European Food Safety Authority: Scientific Opinion on the Substantiation of Health Claims Related to Bonito Protein Peptide and Maintenance of Normal Blood Pressure (ID 1716) Pursuant to Article 13(1) of Regulation (EC) No 1924/2006. http://www.efsa.europa.eu/en/efsajournal/doc/1730.pdf [accessed 15 May 2012]. U.S. Food and Drug Administration. Nutritional Labeling and Education Act (NLEA) Requirements (8/94 - 2/95). [Online], available:
http://www.fda.gov/ICECI/Inspections/InspectionGuides/ucm074948.htm?utm_campai gn=Google2&utm_source=fdaSearch&utm_medium=website&utm_term=nutrition labelling and education act&utm_content=1 (accessed 15 May 2012).

[130] U.S. Food and Drug Administration: Dietary Supplement Health and Education Act of 1994.
http://www.fda.gov/RegulatoryInformation/Legislation/FederalFoodDrugand CosmeticActFDCAct/SignificantAmendmentstotheFDCAct/ucm148003.htm? utm_campaign=Google2&utm_source=fdaSearch&utm_medium=website&utm_term=D ietarySupplement Health and Education Act&utm_content=1
(accessed 15 May 2012).

[131] Phelan, M. & Kerins, D. The potential of milk-derived peptides in cardiovascular disease. Food & Function 2011; 2,153-167.

[132] Usinger L, Reimer C, Ibsen H. Fermented milk for hypertension. Cochrane Database of Systematic Reviews 2012;4 Art. No.: CD008118. DOI: 10.1002/14651858.CD008118.pub2.

[133] Shimizu M, Tsunogai M, Arai S. Transepithelial transport of oligopeptides in the human intestinal cell, caco-2. Peptides 1997;18(5) 681-687.

Section 2

Foods as Source of Bioactive Peptides

Chapter 4

Whey Proteins as Source of Bioactive Peptides Against Hypertension

Tânia G. Tavares and F. Xavier Malcata

Additional information is available at the end of the chapter

http://dx.doi.org/10.5772/52680

1. Introduction

A food can be considered as functional if, beyond its nutritional outcomes, it provides benefits upon one or more physiological functions, thus improving health while reducing the risk of illness [1]. This definition – originally proposed by the European Commission Concerted Action on Functional Food Science in Europe (FuFoSE), should be refined in that: (i) the functional effect is different from the nutritional one; and (ii) the benefit provided requires scientific consubstantiation in terms of improvement of physiological functions, or reduction of occurrence of pathological conditions. The concept of functional food emerged in Japan during the 80's, chiefly because of the need to improve the quality of life of a growing elderly population – who typically incurs in much higher health costs [2]. Nowadays, a growing consumer awareness of the relationship between nutrition and health has made the market of functional foods to boom.

Bioactive peptides can be commercially sold as nutraceuticals; a nutraceutical is an edible substance possessing health benefits that may accordingly be used to prevent or treat a disease. However, a distinction should be made between nutraceuticals taken to prevent diseases – and which are present as natural ingredients of functional foods consumed as part of the daily diet, and nutraceuticals used as adjuvants for treatment of diseases – which require pharmacologically active compounds.

Milk and dairy products have been concluded to be functional foods; a number of studies have indeed shown that many peptides from milk proteins play a role in several metabolic processes, so a considerable interest has arisen from the part of the dairy industry towards large-scale production of dairy proteins in general, and bioactive peptides in particular. Manufacture of bioactive peptides is usually carried out through hydrolysis using digestive, microbial, plant or animal enzymes, or by fermentation with lactic starter cultures. In some cases, a combination of these processes has proven crucial to obtain functional peptides of small size [3,4]. Proteins recovered from whey released by cheese manufacture already found a role as current ingredients on industrial scale. Use of these proteins (concentrated or isolated), and mainly of biologically active peptides derived therefrom as dietary supplements, pharmaceutical preparations or functional ingredients is of the utmost interest for the pharmaceutical and food industries – while helping circumventing the pollution problems associated with plain whey disposal.

2. Cheese whey

Despite having been labeled over the years as polluting waste owing to its high lactose and protein contents [5], whey is a popular protein supplement in various functional foods and the like [6]. In fact, whey compounds exhibit a number of functional, physiological and nutritional features that make them potentially useful for a wide range of applications (Table 1).

Advantageous features	Disadvantageous features
High nutritional value of protein fraction in terms of amino acid residues (e.g. Lys, Thr, Leu, Ser) Possibility of lactose production in parallel Reduction in pollution owing to biochemical oxygen demand of proteins	High dilution requiring costly dehydration High salt content (ca. 10 % of dry matter) High sugar content requiring delactosation Highly perishable raw material Widely dispersed cheese production facilities Technical innovation needed in separation (e.g. ultrafiltration and diafiltration)

Table 1. Major features associated with use of whey (adapted from Alais [62])

Whey can be converted into lactose-free whey powder, condensed whey, whey protein concentrates and whey protein isolates [7] – all of which are commercially available at present. In the case of bovine milk, ca. 9 L of whey is produced from 10 L of milk during cheesemaking; estimates of worldwide production of cheese in 2011 point at ca. 15 million tonnes (United States Department of Agriculture – Foreign Agricultural Service). For environmental reasons, whey cannot be dumped as such into rivers due to its high chemical and biological oxygen demands. On the other hand, whey can be hardly used as animal feed or fertilizer due to economic unfeasibility.

2.1. Physicochemical composition

There are two types of whey, depending on how it is obtained; when removal of casein is via acid precipitation at its isoelectric point (pH 4.6 at room temperature) [8], it is called acid whey; however, the most common procedure is coagulation via enzymatic action, so the product obtained is called sweet whey [9-10].

Despite containing ca. 93 % water, whey is a reservoir of milk components of a high value: it indeed contains ca. half of the nutrients found in whole milk. Said composition depends obviously on how the cheese is produced and the milk source; the compound found to higher level is lactose (4.5-5 %, w/v), followed by soluble proteins (0.6-0.8 %, w/v), lipids (0.4-0.5 %, w/v) and minerals (8-10 %, $w/w_{dry\ extract}$) – particularly calcium, and vitamins such as thiamine, riboflavin and pyridoxin [11-13]. In fact, whey is now considered as a co-product rather than a by-product of cheese production, in view of its wide range of potential applications [13-15].

2.2. Protein composition

Milk has been recognized as one of the main sources of protein [16] in feed for young animals and food for humans of all ages [17]. Bovine milk contains ca. 3 % protein [9] – of which 80 % is caseins and 20 % is whey proteins [18]. Whey comprises a heterogeneous group of proteins that remain in the supernatant after precipitation of caseins; they are characterized by genetic polymorphisms that usually translate into replacement of one or more amino acid residues in their original peptide sequence.

Two major types of proteinaceous material can be found in whey: β-lactoglobulin (β-Lg) and α-lactalbumin (α-La); and proteose-peptone (derived from hydrolysis of β-casein, β-CN), small amounts of blood-borne proteins (including bovine serum albumin, BSA, and immunoglobulins, Igs), and low molecular weight (MW) peptides derived from casein hydrolysis (soluble at pH 4.6 and 20 °C) [16, 19]. Whey proteins have a compact globular structure that accounts for their solubility (unlike caseins that exist as a micellar suspension, with a relatively uniform distribution of non-polar, polar and charged groups). These proteins have amino acid profiles quite different from caseins: they have a smaller fraction of Glu and Pro, but a greater fraction of sulfur-containing amino acid residues (i.e. Cys and Met). These proteins are dephosphorylated, easily denatured by heat, insensitive to Ca^{2+}, and susceptible to intramolecular bond formation via disulfide bridges between Cys sulfhydryl groups. Selected physicochemical parameters typical of whey proteins are tabulated in Table 2.

Proteins	Concentration (gL^{-1})	MW (kDa)	Isoelectric point (pI)
β-Lg	3 – 4	18.4	5.2
α-La	1.5	14.2	4.7 – 5.1
BSA	0.3 – 0.6	69	4.7 – 4.9
IgG, IgA, IgM	0.6 – 0.9	150 – 1000	5.5 – 8.3
Lactoperoxidase	0.006	89	9.6
Lactoferrin	0.05	78	8.0
Protease-peptone	0.5	4 – 20	
Caseinomacropeptide		7	

Table 2. Characteristics of major whey proteins (adapted from Zydney [186])

2.2.1. *β-Lactoglobulin (β-Lg)*

The major protein in ruminant whey is β-Lg, which represents ca. 50 % of the total whey protein inventory in cow's milk and 12 % of the total milk proteins [9, 20-21]. Although it can be found in the milk of many other mammals, it is essentially absent in human milk [22]. This is a globular protein, with 162 amino acid residues in its primary structure and a MW of 18.4 kDa. There are at least twelve genetic variants of β-Lg (A, B, C, D, DR, DYAK/E, F, G, H, I, W and X) – of which A is the most common.

The monomer of β-Lg has a free thiol group and two disulfide bridges – which makes β-Lg exhibit a rigid spacial structure [8]; however, its conformation is pH-dependent [23] – and at temperatures above 65 °C (at pH 6.5), β-Lg denatures, thus giving rise to aggregate formation [24]. Between pH values 5.2 and 7.2, that protein appears as a dimmer – with a MW of 36.0 kDa [8]. At low pH, association of monomers leads to octamer formation; and, at high temperatures, the dimer dissociates into its monomers. The solubility of β-Lg depends on pH and ion strength – but it does not precipitate during milk acidification [25].

A number of useful nutritional and functional features have made β-Lg become an ingredient of choice for food and beverage formulation: in fact, it holds excellent heat-gelling [26] and foaming features – which can be used as structuring and stabilizer agents in such dairy products as yogurts and cheese spreads. This protein is resistant to gastric digestion, as is stable in the presence of acids and proteolytic enzymes [22, 27-30]; hence, it tends to remain intact during passage through the stomach. It is also a rich source of Cys, an amino acid bearing a key role in stimulating synthesis of glutathione (GSH) – composed by three amino acids, Glu, Cys and Gly [31].

Many techniques have been developed for purification of β-Lg – which normally rely on its precipitation [32-35]; when large scale purification is intended, precipitation is usually complemented by ion exchange [35-36].

2.2.2. *α-Lactalbumin (α-La)*

α-La appears quantitatively second in whey; it comprises ca. 20 % of all proteins in bovine whey, and 3.5 % of the total protein content of whole milk [9]. It is a calcium metalloprotein composed of 123 amino acids, with a MW of 14.4 kDa [37]; and appears as three genetic variants (A, B and C), with B being the most common [38]. Chromatographic and electrophoretic analysis within stability studies carried out at various times and temperatures (pH 6.7) indicated that α-La is more heat resistant than β-Lg – in part due to its denaturation being reversible below 75 °C [39]. Owing to such a relatively high thermal stability, it holds a poor capacity to gel; however, it can be easily incorporated in fluid or viscous products to increase their nutritional value. This protein is commercially used in supplements for infant formulae, because of its similarity in structure and composition to human milk proteins – coupled with its higher content of Cys, Trp, Ile, Leu and Val residues, which make it also the ingredient of choice in sport supplements [13, 40-41]. Regarding tertiary structure, α-La is a compact globular protein consisting of 26 % α-helix,

14 % β-sheet and 60 % other motifs; it is also very similar to lysozyme [9]. This protein is one of the most studied proteins with regard to understanding the mechanism of protein stability and folding/unfolding [42]; at low pH [43], high temperature [44] or moderate concentrations of denaturants – e.g. guanidine hydrochloride [45], α-La adopts a conformational structure called molten globule. A partially unfolded state, the apo-state, is formed at neutral pH upon removal of Ca^{2+} by ethylenediamine tetracetic acid (EDTA) [46-47]; this state preserves its secondary, but not its tertiary structure [48].

The molten globule state of α-La retains a high fraction of its native secondary structure, as well as a flexible tertiary structure [45, 48-49]; it accordingly appears as an intermediate in the balance between native and unfolded states [50-51]. This structure of α-La is highly heterogeneous, with proeminence of α-helix driven mainly by weak hydrophobic interactions – while the β-sheet domain is significantly unfolded.

2.2.3. *Caseinomacropeptide (CMP)*

CMP is a heterogeneous polypeptide fraction derived from cleavage of Phe_{105}-Met_{106} in κ-casein (κ-CN). When milk is hydrolyzed with chymosin during cheesemaking, κ-CN is hydrolyzed into two portions: one remains in the cheese (*para*-κ-CN) and the other (CMP) is lost in whey; the latter is relatively small, with 63 residues and a MW of ca. 8 kDa [52]. Further to its polymorphisms, CMP may exist in various forms depending on the extent of post-transcriptional changes: it glycosylates through an *O*-glycoside bridge, and phosphorylates via a Ser residue. Note that post-transcriptional modifications of κ-CN occur exclusively in the CMP portion of the molecule.

The amino acid sequence of CMP is well-known; it lacks aromatic amino acid residues (Phe, Trp and Tyr) and Arg, but several acidic and hydroxyl amino acids are present [53]. CMP from cow is soluble at pH in the range 1-10, with a minimum solubility (88 %) between pH 1 and 5 [54-55]. CMP appears to remain essentially soluble following heat treatment at 80-95 °C for 15 min at pH 4 and 7 [55]. Its emulsifying activity exhibits a minimum near the isoelectric point [54]. Dziuba and Minkiewicz [56] showed that a decrease in pH leads to a decrease in CMP volume, owing to reduction of internal electrostatic forces and steric repulsion; this apparently has a significant influence upon its emulsifying capacity.

2.2.4. *Bovine serum albumin (BSA)*

BSA is derived directly from the blood, and represents 0.7-1.3 % of all whey proteins [8]. Its molecule has 582 amino acid residues and a MW of 69 kDa – and contains 17 disulfide bonds and one free sulphydryl group [9]. Because of its size and higher levels of structure, BSA can bind free fatty acids and other lipids, as well as flavor compounds [57] – but this feature is severely hampered upon denaturation. Its heat-induced gelation at pH 6.5 is initiated by intermolecular thiol-disulphide interchange – similar to what happens with β-Lg [58].

2.2.5. *Immunoglobulins (IGs)*

IGs represent 1.9-3.3 % of the total milk proteins, and are derived from blood serum [8]; they constitute a complex group, the elements of which are produced by β-lymphocytes. Igs encompass three distinct classes: IgM, IgA and IgG (IgG_1 and IgG_2) – with IgG_1 being the major Ig present in bovine milk and colostrum [8], whereas IgA is predominant in human milk. The physiological function of Igs is to provide various types of immunity to the body; they consist of two heavy (53 kDa) and two light (23 kDa) polypeptide chains, linked by disulfide bridges [9]. The complete Ig, or antibody molecule has a MW of about 180 kDa [59]. Igs are partially resistant to proteolytic enzymes, and are in particular not inactivated by gastric acids [59].

2.2.6. *Lactoferrin (LSs)*

LFs are single-chain polypeptides of ca. 80 kDa, containing 1-4 glycans depending on the species. Bovine and human LFs consist of 989 and 691 amino acids, respectively [60]: the former is present to a concentration of ca. 0.1 mg mL^{-1} [25, 61], and is an iron-binding glycoprotein - so it is thought to play a role in iron transport and absorption in the gut of young people.

2.2.7. *Proteose-peptones (PPs)*

The total PP fraction (TPP) of bovine milk represents ca. 10 % of the whole whey protein content; it is accounted for by the whey protein fraction soluble after heating at 95 °C for 30 min, followed by acidification to pH 4.6 [62]. The TPP fraction is often divided in two main groups: the first one includes PPs originated from casein hydrolysis; its principal components have been labeled as 5 (PP5), 8 fast (PP8 fast) and 8 slow (PP8 slow), according to their electrophoretic mobility [62, 63]. PP3 constitutes the second group, and it is not derived from casein (it is actually found only in whey); it is known for its extreme hydrophobicity.

2.3. Functional ingredients from whey proteins

Whey proteins have unique characteristics [64] beyond their great importance in nutrition; they exhibit chemical, physical, physiological, functional and technological features also useful for food processing [14]. Based on these properties, more and more individual proteins and protein concentrates of whey have been incorporated in food at industrial scale. Therefore, whey proteins address two major issues in practice: nutritionally, they supply energy and essential amino acids, besides being important for growth and cellular repair; in terms of functionality, they play important roles upon texture, structure and overall appearance of food – e.g. gel formation, foam stability and water retention.

A few physiological properties useful in therapies have been found [65]: a number of reviews have accordingly examined to some length the bioactive properties of whey proteins in general [66-67], or of β-Lg and α-La in particular [26]; other authors have covered mainly such biological activities as anticarcinogenic [68] and immunomodulatory [69]. It was observed that whey proteins trigger immune responses that are significantly higher than those by diets containing casein or soy protein. Antimicrobial and antiviral actions, immune system stimulation and anticarcinogenic activity (among other metabolic features) have indeed been associated with ingestion of β-Lg and α-La, as well as LF, LP, BSA and CMP; the main biological activities of whey proteins are listed in Table 3.

With regard to bioactive peptides, research has undergone a notable intensification during the past decade [4, 70]. Advances in nutritional biochemistry and biomedical research have in fact helped unravel the complex relationships between nutrition and disease, thus suggesting that food proteins and peptides originated during digestion (or from *in vitro* proteolysis) may play important roles in preventing or treating diseases associated with malnutrition, pathogens and injuries [71-72].

Protein/Peptide	Treatment	Biological function	Reference
Whole whey protein		Prevention of cancer	[74]
		Breast and intestinal cancer;	[14, 75]
		Chemically-induced cancer	[76-77]
		Increment of gluthatione levels	[64]
		Increase of tumour cell vulnerability	[78-79]
		Antimicrobial activities	[80]
		Increment of satiety response	
		Increment in plasma amino acids, cholecystokinin and glucagon-like peptide	[81]
	Enzyme hydrolysis	**ACE[a]-inhibitor**	[82]
		Antiulcerative	
		Prostaglandin production	[83-85]
	Enzyme hydrolysis	Antiulcerative	[83, 86]
β-Lactoglobulin		Transporter	
		Retinol	[9, 41, 87, 88]
		Palmitate	[89]
		Fatty acids	[90]
		Cellular defence against oxidative stress and detoxification	[31, 65, 91-93]
		Enhancement of pregastic esterase activity	[94]
		Transfer of passive immunity	[95]
		Regulation of mammary gland phosphorus metabolism	[96]
	Enzyme hydrolysis; Fermentation	**ACE[a]-inhibitor**	[97-107]
	Enzyme hydrolysis	Antimicrobial against several gram-positive bacteria	[108-111]
	Enzyme hydrolysis	Antimicrobial (bactericidal)	[112-113]
	Enzyme hydrolysis	Hypocholesterolemic	[113-114]

Protein/Peptide	Treatment	Biological function	Reference
	Enzyme hydrolysis	Opioid agonist	[73, 97, 115]
	Enzyme hydrolysis	**Antihypertensive**	[99, 116-117]
	Enzyme hydrolysis	Ileum contracting	[97, 99]
	Enzyme hydrolysis	Antinociceptive	[118]
		Prevention of cancer	
	Enzyme hydrolysis	Intestinal cancer	[14]
α-Lactalbumin		Prevention of cancer	[119]
		Apoptosis of tumoral cells	[120-122]
		Lactose synthesis	[25, 123]
		Treatment of chronic stress-induced disease	[124]
		Antimicrobial (bactericidal)	
		Streptococcus pneumonia	[125]
		Stress reduction	[123, 126]
		Immunomodulation	[127]
	Enzyme hydrolysis	Antimicrobial against several gram-positive bacteria	[108-110]
	Enzyme hydrolysis	Opioid agonist	[97, 115, 128]
	Enzyme hydrolysis	**ACE[a]-inhibitor**	[26, 97-98, 101, 107]
	Enzyme hydrolysis	**Antihypertensive**	[117, 129]
	Enzyme hydrolysis	Ileum contracting	[97]
		Antiulcerative	
		Prostaglandin production	[130-132]
Bovine serum albumin		Fatty acid binding	[13]
		Antioxidant	[133-134]
		Prevention of cancer	[135]
	Enzyme hydrolysis	**ACE[a]-inhibitor**	[136-137]
	Enzyme hydrolysis	Ileum contracting	[138]
	Enzyme hydrolysis	Opioid agonist	[97, 128, 139]
Immunoglobulins		Immunomodulation	[140]
		Disease protection through passive immunity	[141-142]
		Antibacterial	[143-145]
		Antifungal	[146]
		Opioid agonist	[147]
Caseinomacropeptide		Antithrombotic	[148-153]
		ACE[a]-inhibitor	[154-156]
		Antimicrobial	[56, 111, 157-160]
	Enzyme hydrolysis	Prebiotic	[161]
		Increment in plasma amino acids and cholecystokinin peptide	[162-165]

ACE[a]- angiotensin-converting enzyme

Table 3. Biological functions of whey proteins/peptides (adapted from Madureira *et al.* [87])

Although inactive within the primary structure of their source proteins, hydrolysis (e.g. mediated by a protease) may release peptides with specific amino acid sequences possessing biological activity. A number of chemical and biological methods of screening have accordingly been developed to aid in search for specific health effects; however, only some of those found *in vitro* have eventually been confirmed in studies encompassing human volunteers [73].

Scientific evidence has shown that whey proteins contain a wide range of peptides that can play crucial physiological functions and modulate some regulatory processes (see Table 3). Due to its high biological value, coupled with excellent functional properties and clean flavor, whey has earned the status of a recommended source of functional ingredients [71] – designed to reduce or control chronic diseases and promote health, thus eventually reducing the costs of health care [3, 166].

Favorable health effects have indeed been claimed for some peptides derived from food proteins – being able to affect the cardiovascular, nervous, digestive or immune systems; these encompass antimicrobial properties, blood pressure-lowering (or angiotensin-converting enzyme (ACE)-inhibitory) effects, cholesterol-lowering ability, antithrombotic and antioxidant activities, enhancement of mineral absorption and/or bioavailability thereof, cyto- or immunomodulatory effects, and opioid features. With regard to the mechanisms underlying the physiological roles of bioactive peptides, a few involve action only upon certain receptors, whereas others are enzyme inhibitors; they may also regulate intestinal absorption, and exhibit antimicrobial or antioxidant activities. Recall that oxidative metabolism is essential for survival of cells, but it generates free radicals (and other reactive oxygen species) as side effect – which may cause oxidative damage. Antioxidant activity has been found specifically in whey proteins, probably via scavenging of such radicals via Tyr and Cys amino acid residues – which is predominantly based on proton-coupled single electron or hydrogen atom transfer mechanisms; or else chelation of transition metals [167-168].

On the other hand, bioactive peptides derived from food proteins differ in general from endogenous bioactive peptides in that they can entail multifunctional features [98]. Furthermore, bioactive peptides that cannot be absorbed though the gastrointestinal tract may exert a direct role upon the intestinal lumen, or through interaction with receptors in the intestinal wall itself; some of these receptors have been implicated in such diseases as cancer, diabetes, osteoporosis, stress, obesity and cardiovascular complications.

3. Production of bioactive peptides in whey

Bioactive peptides derived from whey proteins constitute a new concept, and have open up a wide range of possibilities within the market for functional foods [4, 169]; of special interest are those released via enzymatic action – as happens during clotting in cheesemaking.

The enzymes used to bring about milk coagulation are selected protein preparations that provide in general a high clotting activity – i.e. a considerable, but selective proteolytic activity. Animal rennet obtained from the calf stomach, composed by 88-94 % and 6-12 % chymosin and pepsin, respectively, has been the coagulant of choice for cheesemaking. However, due to increased world production of cheese, the supply of animal rennet has lied below its demand; the increased prices have driven a search for alternative coagulants (including plant and microbial sources). With regard to animal rennet substitutes, pig pepsin has enjoyed a remarkable commercial success; with regard to rennet from microbial origin, the proteases from *Mucor miehei, Mucor pusillus* and *Endothia parasitica* are the most successful [170]. Recombinant bovine chymosin is, nowadays, one of the proteinases with greater commercial expression – even though its use is still prohibited in certain countries [171].

Chymosin and the other rennet substitutes are aspartic proteases, with optimal activity at acidic pH, and possessing high degree of homology in primary and 3-dimensional structures, 3-dimensional structure and catalytic mechanism. The specificity towards the substrate is, however, rather variable; although they have a greater tendency to break peptide bonds between hydrophobic amino acids having bulky side residues, they hydrolyze a large number of bond types [172]. Of particular interest is vegetable rennet, which – with few exceptions, enjoys a still limited use worldwide. Many plant enzyme preparations proved indeed to be excessively proteolytic for manufacture of cheese, causing defects in terms of flavor and texture of the final product. These difficulties arise from the presence of non-specific enzymes that belong to complex enzyme systems (which, as such, are difficult to control). An exception to the poor suitability of vegetable coagulants is the proteinases in aqueous extracts of plants of the *Cynara* genus – which have been employed for traditional cheesemaking in Portugal and Spain since the Roman period.

Bioactive peptides derived from whey proteins can be released at industrial scale via enzyme-mediated hydrolysis with digestive enzymes – and pepsin, trypsin and chymotrypsin have been the most frequent vectors therefor [4, 169, 173]. However, whey proteins are not easily broken down by proteases in general – a realization that also explains their tendency to cause allergies upon ingestion [174]. Hence, less conventional sources of proteolytic enzymes have been sought that can cleave the whey protein backbone at specific and usual sites. This is the case of aspartic proteinases present in the flowers of *Cynara cardunculus* – a plant related to the (common) globe artichoke. They can cleave the whey protein backbone next to hydrophobic amino acid residues, especially Phe, Leu, Thr and Tyr [82, 175], and act mainly on α-La, either in whole whey or following concentration to whey protein concentrate (WPC) [176-178]; conversely β-Lg appears not to be hydrolyzed thereby to a significant extent [82, 175].

4. Recovery of proteins/peptides from whey

The relatively low concentration of proteins in whey requires concentration processes to assure high hydrolysis productivity. Development of membrane separation techniques has been essential toward this endeavor – and food industry has taken advantage of its relatively easy scale-up, as well as its being inexpensive when compared with preparative chromatographic techniques [41]. Furthermore, the absence of heat treatment allows the bioactive components to remain intact (or become only slightly affected) during processing. Recall that membrane separation allows differential concentration of a liquid, provided that the solute of interest is larger in molecular diameter than the membrane pores – so the liquid that percolates the membrane (filtrate) contains only components smaller than that size threshold [179].

The dairy industry has pioneered development of equipment and techniques for membrane filtration, which recovers whey proteins in a non-denatured state. Typical procedures include: (i) basic membrane separation, e.g. reverse osmosis, ultrafiltration and diafiltration [180-186], that permits fractionation of proteins, as well as concentration and purification thereof; (ii) nanofiltration (or ultraosmosis) that allows removal of salts or low MW contaminants; and (iii) microfiltration to remove suspended solid particles or microorganisms [179, 187]. Note that isolation of individual whey proteins on laboratory scale has resorted chiefly to salting out, ion exchange chromatography and/or crystallization [188]; such a fractionation allows fundamental studies of their immunological properties to be carried out, which are necessary to establish and support industrial interest [189-190].

5. Activity of peptides from whey upon hypertension

Hypertension is a major public health issue worldwide that affects nearly one fourth of the population; and it is usually associated with such other disorders as obesity, pre-diabetes, renal disease, atherosclerosis and heart stroke [191-194]. Its specific treatment will likely reduce the risk of incidence of cardiovascular diseases, which currently account for 30 % of all causes of death [195].

Blood pressure can be regulated through diet changes and physical exercise, as well as administration of calcium T channel antagonists, angiotensin II receptor antagonists, diuretics and ACE inhibitors [104]. A few mechanisms have been described that rationalize how peptides lower blood pressure. Traditionally, control of hypertension has focused on the renin-angiotensin system, via inhibition of ACE [173]. Captopril, enalapril and lisinopril have accordingly been used as antihypertensive drugs that act essentially as ACE inhibitors; they found a widespread application in treatment of patients with hypertension, heart failure or diabetic nephropathies [193, 196-197]. However, they bring about undesirable side effects, so safer (and, hopefully, less expensive) alternatives are urged [198-199].

In fact, increasing evidence has been provided that mechanisms other than ACE inhibition may be involved in blood pressure decrease arising from consumption of many food-derived peptides [200]; although there are few studies to date with antihypertensive peptides obtained from whey. One of them corresponds to interaction with opioid receptors that are present in the central nervous system and in peripheral tissues, while another is based on release of nitric oxide (NO) that causes vasodilatation and thus affects blood pressure. Those peptides hold the advantage of no side effects, unlike happens with such other opiates as morphine [102]. One example is α-lactorfin, a tetrapeptide derived from α-La [129, 201], for which studies showed that antihypertensive effects are mediated through the vasodilatory action of binding to opioid receptors. Furthermore, endothelium-dependent relaxation of mesenteric arteries in spontaneously hypertensive rats (SHR, which is the animal model normally accepted to study human hypertension) that was inhibited by an endothelial nitric oxide synthase (eNOS) inhibitor was also observed [202]. That peptide may even chelate minerals, and thus facilitates calcium absorption [200].

Alternative mechanisms are other routes of vasoregulator substance synthesis – e.g. kallikrein-kinin, neutral endopeptidase and endothelin-converting enzyme systems. The release of vasodilator substances like prostaglandin I_2 or carbon monoxide could be implied in dependent and independent mechanisms of ACE inhibition responsible for antihypertensive effects [203-205]; an example is the peptide ALPMHIR, which inhibits release of an endothelial factor (ET-1) that causes contractions in smooth muscle cells [206].

In the last decade, production of antioxidant peptides from whey has been reported [207]. Experimental evidence – including SHR and human studies, claimed that oxidative stress is one of the causes of hypertension and several vascular diseases, via increase production of reactive oxygen species and reduction of NO synthesis and bioavailability of antioxidants [208].

Nevertheless, the most intensively studied peptides – i.e. VPP and IPP derived from caseins, showed possible mechanisms of action that could be found also in other peptides. In studies performed with rats, VPP and IPP increased plasma renin levels and activity [202]; and decreased ACE activity in the serum; they also showed endothelial function protection in mesenteric arteries [208]. The influence of VPP and IPP on gene expression of SHR abdominal aorta unfolded a significant increase of genes related with blood pressure regulation – the eNOS and connexin 40 genes [208]. Other studies have highlighted the peptide effects on the vasculature itself, showing that the antihypertensive activity of the peptide rapakinin is induced mainly by CCK_1 and IP-receptor-dependent vasorelaxation; this peptide relaxes the mesenteric artery of SHR via prostaglandin I_2-IP receptor, followed by CCK-CCK_1 receptor pathway; other peptides improve aorta and mesenteric acetylcholine relaxation, and decrease left ventricular hypertrophy, accompanied by significant decrease in interstitial fibrosis [209]. In order to prevent hypertension, two alternative enzyme inhibitors were suggested: renin (a protease recognized as the initial compound of the renin–angiotensin system) and platelet-activating factor acetylhydrolase (PAF-AH) (a circulating enzyme secreted by inflammatory cells and involved in atherosclerosis) [208].

5.1. Inhibition of angiotensin-converting enzyme (ACE)

Since diet has a direct relationship to hypertension, the food industry (in association with research and public health institutions) has promoted development of novel functional ingredients that can contribute to keep a normal blood pressure – thus avoiding the need to take antihypertensive drugs [73, 173, 209-212]. Various investigators have accordingly hypothesized that certain peptides formed through hydrolysis of food proteins have the ability to inhibit ACE; López-Fandiño [173], FitzGerald [104, 137], Gobetti [213], Meisel

[214], Korhonen and Pihlanto [4], Silva and Malcata [215], Vermeirssen [216], and Martínez-Maqueda [208] have comprehensively reviewed this subject. In general, it has been claimed that a diet rich in foods containing antihypertensive peptides is effective toward prevention and treatment of hypertension [173, 201].

ACE-inhibitory peptides may be obtained from precursor food proteins via enzymatic hydrolysis, using viable or lysed microorganisms or specific proteases [3, 73, 137, 169]. Although *in vitro* studies are useful at screening stages, the efficacy and safety of such peptides requires *in vivo* testing – first in animals, and then in human volunteers [217]. This issue is particularly relevant because *in vitro* ACE inhibition does not necessarily correlate with *in vivo* antihypertensive features, as peptides often undergo breakdown during gastrointestinal digestion that hampers manifestation of their potential physiological function. Conversely, antihypertensive activity may be promoted after long-chain peptide precursors release bioactive fragments by gastrointestinal enzymes [73].

In the latest two decades, various active peptides have been identified from animal proteins, including some with antihypertensive effects in animals (e.g. SHR) and even in humans [3, 73, 137, 169, 173, 201, 208, 212, 217, 299]: bovine plasma proteins [218], egg proteins [203, 219] and tuna proteins [220]; but also plant proteins, e.g. from soy [221], wine [222] and maize [223]. Nevertheless, milk proteins still appear to be the best source of ACE-inhibitory peptides.

Recall that caseins are the most abundant proteins in milk, and have an open and flexible structure that makes them susceptible to attack by proteases; hence, many ACE-inhibitors have been obtained via enzyme-mediated approaches [224-225] – e.g. casokinins. Studies on peptides with ACE-inhibitory activity obtained from whey proteins (called lactokinins) are more limited – which may be due to the rigid structure of β-Lg (the major whey protein) that makes it particularly resistant to digestive enzymes. However, bioactive protein fragments with ACE-inhibitory activity have been found in whey protein hydrolyzates [107, 217, 226-228]; and Manso and López-Fandiño [155] also identified this activity in CMP hydrolyzates. Characterization of hydrolyzates of the main whey proteins – including the amino acid sequences of peptides therein that exhibit *in vitro* ACE-inhibiting activity, is provided in Table 4.

The ACE-inhibitory activity depends on the protein substrate and the proteolytic enzymes used to break it down. ACE (i.e. a dipeptidyl carboxypeptidase) is an enzyme ubiquitous in tissues and biological fluids – where it plays an important physiological role upon regulation of the cardiovascular function, including a basic role in regulation of peripheral blood pressure via the renin-angiotensin system [229-230]. ACE inhibitors and angiotensin II receptor blockers [231-232] have been therapeutically important, since they act as efficient drugs and bring about very few collateral effects.

ACE-inhibitor peptide can reduce blood pressure in a process regulated (in part) by the renin-angiotensin system: renin — a protease secreted in response to various physiological stimuli, cleaves the protein angiotensinogen to produce the inactive decapeptide

Source protein	Enzyme	Peptide fragment	Amino acid sequence	IC_{50} (μM)[a]	Reduction in SBP[b] (mm Hg) (Dose (mg kg^{-1}_{bw}))	References
Whole whey protein	Fermentation + trypsin + chymotrypsin	β-Lg f9-14	GLDIQK	580		[104, 233]
	Yogurt starter + trypsin + pepsin	β-Lg f15-20	VAGTWY	1682		[100]
	Fermentation with lactic acid bacteria + prozyme 6	β-Lg f17-19	GTW	464.4		[105]
	Cardosins	β-Lg f33-42	DAQSAPLR VY[c]	12.2	10 (5)	[107, 117]
	Proteinase K	β-Lg f78-80	IPA[c]	141	31 (8)	[136]
	Cardosins	α-La f16-26	KGYGGVSL PEW[c]	0.7	20 (5)	[107, 117]
	Cardosins	α-La f97-103	DKVGINY[c]	99.9		[107]
	Cardosins	α-La f97-104	DKVGINY W[c]	25.4	15 (5)	[107, 117]
	Fermentation + trypsin + chymotrypsin	α-La f105-110	LAHKAL	621		[100]
	Fermentation by cheese microflora	α-La f104-108	WLAHK	77		[233]
	Neutrase	α-La f105-110	INYWL	11		[234]
β-Lactoglobulin	Trypsin	f7-9	MKG	71.8		[103]
	Trypsin	f10-14	LDIQK	27.6		[103]
	Pepsin + trypsin + chymotrypsin	f15-19	VAGTW	1054		[233]
	Trypsin	f22-25	LAMA	556		[233]
	Trypsin	f32-40	LDAQSAPL R	635		[233]
	Protease N Amano	f36-42	SAPLRVY	8		[235]
	Thermolysin	f58-61	LQKW[c]	34.7	18.1 (10)	[103, 236]
	Trypsin	f81-83	VKF	1029		[233]
	Pepsin + trypsin + chymotrypsin	f94-100	VLDTDYK	946		[106, 233]
	Pepsin + trypsin + chymotrypsin	f102-103	YL[c]	122		[214]
	Pepsin + trypsin + chymotrypsin	f102-105	YLLF[c]	172		[113]
	Thermolysin	f103-105	LLF[c]	79.8	29 (10)	[113, 236]
	Pepsin + trypsin + chymotrypsin	f106-111	CMENSA	788		[233]
	Pepsin + trypsin + chymotrypsin	f142-145	ALPM[c]	928	21.4 (8)	[116]
	Pepsin + trypsin + chymotrypsin	f142-146	ALPMH[c]	521		[233]
	Trypsin	f142-148	ALPMHIR[c]	43		[226]
α-Lactalbumin	Thermolysin	f15-26	LKGYGGVS LPEW	83		[237]
	Thermolysin	f18-26	YGGVSLPE W	16		[237]
	Thermolysin	f20-26	GVSLPEW	30		[237]
	Thermolysin	f21-26	VSLPEW	57		[237]
	Synthetic	f50-51 or f18-	YG[c]	1522		[98]

Source protein	Enzyme	Peptide fragment	Amino acid sequence	IC_{50} (μM)[a]	Reduction in SBP[b] (mm Hg) (Dose (mg $kg^{-1}{}_{bw}$))	References
		19				
	Pepsin + trypsin + chymotrypsin	f50-52	YGL	409		[233]
	Pepsin	f50-53	YGLF[c]	733	23.4 (0.1)	[129, 233]
	Synthetic	f52-53	LF[c]	349		[26]
	Trypsin	f99-108	VGINYWLA HK	327		[233]
	Trypsin	f104-108	WLAHK	77		[233]
Bovine serum albumin	Proteinase K	f208-216	ALKAWSV AR[c]	3		[238]
	Proteinase K	f221-222	FP	315	27 (8)	[136]
Caseinomacropeptide	Trypsin	f106-112	MAIPPKK		28 (10)	[239]
Lactoferrin	Pepsin	f20-25	RRWQWR		16.7 (10)	[240]
	Pepsin	f22-23	WQ		11.4 (10)	[240]

[a] Concentration of peptide needed to inhibit 50 % of original ACE activity.
[b] Systolic blood pressure.
[c] Synhetic peptides used.

Table 4. Primary structural characteristics of whey peptides with ACE-inhibitory activity and antihypertensive activity in spontaneously hypertensive rats, and vectors of generation thereof.

angiotensin I. Cleavage of angiotensin I – via removal of two amino acid residues from the C-terminal end by ACE, produces the active octapeptide angiotensin II that is a potent vasoconstrictor; however, there are alternative routes to generate angiotensin II [198, 241-242]. Angiotensin II activates angiotensin II type 1 (AT_1) receptor — a member of the G-protein-coupled-receptor superfamily, which plays various roles, e.g. vasoconstriction, as well as stimulation of aldosterone synthesis and release (which leads to sodium retention, and thus increases blood pressure) [198, 217, 242]. In addition, ACE acts on the kallikrein-kinin system, catalyzing degradation of the nonapeptide bradykinin – which is a vasodilator [241]. ACE-inhibitor peptides exert a hypotensive effect by preventing angiotensin II formation and degradation of bradykinin, thus reducing blood pressure in hypertensive patients [217].

Several tests on SHRs – probably the best experimental model for antihypertensive studies because they exhibit vascular reactivity and renal function similar to those in human beings [243], have been described that prove control of arterial blood pressure following a single oral administration of known ACE-inhibitory hydrolyzates or/and peptides derived from whey proteins. The antihypertensive effect associated with some of those peptides is comparable to that exhibited by VPP – an antihypertensive peptide included in functional foods that is already available in the market [117, 129, 137, 154, 201, 210-212, 242, 244-247]. To measure ACE-inhibitory activity, distinct biological, radiochromatographic, colorimetric and radioimmunologic methods have been employed – using angiotensin I as substrate. Chemical methods are sensitive, and resort to a tripeptide with a substituted amino-terminus, Z-Phe-His-Leu, as ACE-substrate – from which the dipeptide His-Leu is released and quantified by specific fluorometric procedures. A similar tripeptide used as substrate of ACE is Bz-Gly-His-Leu, or Hippuryl-His-Leu (HHL); upon incubation with the enzyme, hippuric acid is formed and the dipeptide His-Leu is released, which is subsequently measured by one of several colorimetric [248] or fluorometric methods [249], or even by capillary electrophoresis [250].

One of the most performing methods to measure ACE-inhibitory activity was developed by Cushman and Cheung [251], and is based on spectrophotometric measurement at 228 nm of hippuric acid formed by incubating the substrate HHL with ACE – in the presence of selected inhibitory substances. More recently, a modified tripeptide, furanacriloil Gly-Phe-Gly, has been chosen as substrate for a spectrophotometric method [252]. The ACE-inhibitory activity is usually measured in terms of IC_{50} (i.e. the concentration of inhibitory substance required to inhibit 50 % of ACE activity); a low IC_{50} value means that a small concentration of inhibitory substance is required to produce enzyme inhibition, so that substance displays a potent inhibitory activity.

As shown in Table 4, ACE-inhibitor peptides are produced mainly by enzymatic hydrolysis, but active sequences have also been obtained via chemical synthesis [253]. Starter and non-starter bacteria are commonly used in cheese manufacture – taking advantage namely of their proteolytic system, which contains at least 16 different peptidases that have already been characterized. Some of these bacteria were found to have ACE-inhibitory activity, or release peptides with this activity. For instance, *Lactobacillus helveticus* is able to release ACE-inhibitory peptides; the best-known ACE-inhibitory peptides – *viz.* VPP and IPP, have indeed been identified in milk fermented with *L. helveticus* strains [154, 244, 254]. More recently, an ACE-inhibitory peptide derived from β-CN – FFVAPFPEVFGK, was successfully expressed by genetic engineering in *Escherichia coli* [255].

5.1.1. Structure/activity relationships

ACE-inhibitor peptides contain usually between 2 and 12 amino acid residues – even though larger peptides may also exhibit such an activity [173]. Ondetti [229] rationalized the interaction of competitive inhibitors for the ACE active site based on enzyme homology with carboxypeptidase A; the first ACE-inhibitor (i.e. captopril), which is one of the oral drugs widely used to treat hypertension, was designed based on this model. Recently, this model was reviewed and used to design even more potent ACE inhibitors [229, 256-257]. The base model proposes that residues of the carboxy-terminal (C-terminal) tripeptide interact with the S_1, S'_1 and S'_2 subunits of the enzyme active site. One of the subunits has a positively charged group that forms an ionic bond with the C-terminal peptide group. The following subunit contains a group capable of interacting with the peptidic bond of the C-terminal amino acid – probably through hydrogen bonding. The third subunit has a Zn^{2+} atom able to carry the carbonyl group of the peptidic bond between the one before the last and the last amino acid residue of the substrate – thus making it more susceptible to hydrolysis [256].

Although the relationships between structure and activity have not been fully elucidated, ACE-inhibitory peptides possess a number of analogies with each other. The tripeptide at the C-terminus is crucial – because this is where the peptide binds to the active site of the enzyme [256]. ACE prefers substrates (or competitive inhibitors) with hydrophobic residues (e.g. Trp, Tyr, Phe and Pro) at the C-terminus, and shows poorer affinity to substrates containing dicarboxylic amino acids in the final position, or those that have a Pro residue in the one before the last position. However, presence of Pro as the last residue [258], or in the third position from the terminus [259] favors binding of peptide to enzyme, in much the same way as when Leu appears in the last position [260,261].

Bioinformatics has been used more recently to find the structural requirements of ACE-inhibitor peptides; these are termed quantitative structure/activity relationship (QSAR) models. Through a QSAR model, Pripp [262] concluded – for milk-derived peptides up to six amino acids in length, that there is a relationship between ACE-inhibitory activity and presence of a hydrophobic (or positively charged) amino acid residue in the last position of the sequence; however, no special relation was found with the structure of the N-terminus. Based on the QSAR model for peptides containing between 4 and 10 amino acid residues, Wu [263] claimed that the residue of the C-terminal tetrapeptide may determine the potency of ACE inhibition – with preference for Tyr and Cys in the first C-terminal position; His, Trp and Met in the second; Ile, Leu, Val and Met in the third; and Trp in the fourth position. Results from other QSAR-based studies aimed at finding ACE-inhibitory activity of di- and tripeptides derived from food proteins have shown that dipeptides with hydrophobic chains, as well as tripeptides with an aromatic amino acid residue at the C-terminus, a positively charged residue at the intermediate position and a hydrophobic amino acid residue at the N-terminus are likely to exhibit ACE-inhibitory power [263].

On the other hand, a biopeptide may adopt a different configuration depending on the prevailing environmental conditions; but the final structural conformation may be crucial for its ACE-inhibitory activity. The fact that the catalytic center of ACE has different structural requirements may unfold the need to develop complex mixtures of peptides, with different structural conformations, so as to produce more complete inhibition than a single peptide [264]. Meisel [265] postulated that the mechanism of ACE inhibition may involve interaction of inhibitor with the subunits that are not normally occupied by substrate, or with the anionic bond site that is different from the enzyme catalytic center. Moreover, somatic ACE has two homologous domains – each of which has an active site with distinct biochemical characteristics. *In vitro* ACE-inhibition studies showed that it is necessary to block the two active centers for complete inhibition of its action upon angiotensin I and bradykinin. Nevertheless, *in vivo* studies in rats showed that the selective inhibition of the N- or C-terminal domains of ACE prevents conversion of angiotensin I to II, but not of bradykinin [266].

Despite the importance arising from the three amino acids in their C-terminus, it was shown that peptides with identical sequences at the C-terminus may exhibit quite different ACE-inhibitory activities from each other. One example is VRYL and VPSERYL, both identified in

Manchego cheese; despite having the same C-terminal tripeptide sequence, they exhibit IC_{50} values of 24.1 μM and 249.5 μM, respectively – i.e. the latter is 10-fold less active than the former. If Val were replaced by a dicarboxylic amino acid at the fourth position of the C-terminus, e.g. via synthesis of ERVL, the IC_{50} measured would be 200.3 μM, which corresponds to an ACE-inhibitory activity 8-fold lower than VRYL – hence demonstrating the crucial role of Val in that position for the intended bioactivity [261].

5.1.2. Bioavailability

Among the several bioactive peptides studied to date, ACE-inhibitory peptides have received particular attention because of their beneficial effects upon hypertension [226, 233, 267]. Note that such effects depend on their ability to reach the target organs without having undergone decay or transformation. Tests encompassing hypertensive animals and human clinical trials have shown that certain sequences can lower blood pressure; however, it is difficult to establish a direct link between the ability to inhibit ACE *in vitro* and the actual antihypertensive activity *in vivo*. Knowledge of the mechanism of action of such bioactive peptides is obviously crucial before manufacture of functional foods with physiological properties is in order [268].

Some peptides with ACE-inhibitory and antihypertensive activities can be transported through the intestinal mucosa via the $PepT_1$ transporter [269]; likewise, there is evidence that other peptides may exert a direct role upon the intestinal lumen [151, 270-271]. Digestive enzymes, absorption through the intestinal tract and blood proteases can bring about hydrolysis of ACE-inhibitor peptides, thus producing fragments with lower or greater activity than their precursor sequences [216]. Hence, for ACE-inhibitor peptides exert an *in vivo* effect, they should not act as substrates of the enzyme. Peptides may accordingly be classified into three groups based on their behavior regarding ACE: (1) true inhibitors, for which IC_{50} is not modified when incubated with the enzyme; (2) ACE-substrates, which are hydrolyzed during incubation, thus giving rise to fragments with a lower ACE-inhibition activity; and (3) peptides that are converted to real inhibitors by ACE and gastrointestinal protease action. Note that only sequences belonging to groups 1 and 3 may show an antihypertensive effect [245].

Effective inclusion of ACE-inhibitory peptides in the diet consequently requires them to somehow resist the strong stomach hydrolysis that may cause loss of bioactivity [104], and afterwards be able to pass into the blood stream – where they should be resistant to peptidases therein, so as to eventually reach the target sites where they are supposed to exert their physiological effects *in vivo*. The structure and bioactivity of short-chain peptides are more easily preserved through gastrointestinal passage than those of their long-chain counterparts [272] – whereas sequences containing Pro residue(s) are generally more resistant to degradation by digestive enzymes [273]. Furthermore, peptides absorbed following digestion may accumulate in specific organs, and then exert their action in a systematic and gradual manner [274, 275]. However, antihypertensive peptides that cannot be absorbed from the digestive tract may still exert their function directly in the intestinal lumen – e.g. via interaction with receptors on the intestinal wall [97, 265, 276].

Besides carrying out protein degradation to varying extents, gastrointestinal digestion plays a key role in formation of ACE-inhibitory peptides [216, 277]; hence, it is relevant to assess the gastrointestinal bioavailability of any potentially interesting peptides. Several studies have accordingly provided evidence for this realization – as happened with Manchego cheese, as well as with other fermented solutions and infant formulae [100, 261, 278-281]; for instance, a potent antihypertensive peptide was released via gastrointestinal digestion from a precursor with poor ACE-inhibitory activity *in vitro* [282] – and some peptides possess a remarkable intrinsic stability, whereas others are susceptible to unwanted degradation [136, 261, 281]; however, whether of any of those options will apply cannot be known in advance.

Animal and human trials are therefore nuclear when assessing bioactivity of peptides; peptides that do not show *in vitro* activity may exhibit *in vivo* antihypertensive activity, and vice versa. For instance, YKVPQL identified in a casein hydrolyzate and released by a proteinase from *L. helveticus* CP790, had a high *in vivo* ACE-inhibitory activity (IC_{50} 22 μM) but did not show any antihypertensive one [282] – probably as a consequence of degradation during the digestion process [137]. When the hydrolyzate was purified, another peptide sequence (KVLPVPQ) was found. Unlike the previous case – with a low *in vitro* ACE-inhibitory activity ($IC_{50} > 1000$ μM), the latter showed a potent *in vivo* antihypertensive activity. It was claimed that this was due to pancreatic digestion that releases Gln, thus forming KVLPVP; furthermore, this fragment showed ACE-inhibitory activity *in vitro*, characterized by an IC_{50} of only 5 μM. Finally, there are reports on peptides with a low ACE-inhibitory activity *in vitro* that possess antihypertensive activity *in vivo* – owing to a hypotensive mechanism of action distinct from that of ACE inhibition. One example is YP, the IC_{50} of which is 720 μM; however, it significantly decreases blood pressure between 2 and 8 h after oral administration to SHR [283]. It should be emphasized that *in vivo* tests of (putatively) promising bioactive peptides should not come into play before careful *in vitro* models have been checked – as they can provide useful preliminary information on the stability of such peptides upon exposure to the various peptidases and proteinases that they will likely find in the gastrointestinal tract, prior to eventual transport across the intestinal barrier [278-279].

Simulated (physiological) digestion is a useful tool to assess the stability of peptides with ACE-inhibitory activity against digestive enzymes. However, the degree of hydrolysis of a given peptide depends not only on its size and nature, but also on the presence of other peptides in its vicinity [272] – which would make it difficult to test the required number of possibilities in a rather limited experimental program. Several *in vitro* studies were carried out that show the importance of digestion upon formation and degradation of ACE-inhibitor peptides [107, 272, 278-280, 284]. In these studies, peptides were subjected to two stages of hydrolysis that mimic digestion in the body. First, hydrolysis with pepsin, at acidic pH, intended to simulate the digestion process prevailing in the stomach; and second, digestion with a pancreatic extract, at basic pH as prevailing during intestinal digestion.

Results encompassing prior or subsequent hydrolysis of peptides showed that *in vitro* digestion controls bioavailability of ACE-inhibitor peptides [162, 278]

Some authors used whey proteins, fermented (or not at all) with *L. helveticus* and *Saccharomyces cerevisiae,* and then subjected them to gastrointestinal digestion; they reached a maximum ACE-inhibitory activity, and unfermented samples were the most active. However, some peptides with *in vivo* antihypertensive activity – as is the sequence KVLPVPQ, and which did not show *in vitro* ACE-inhibitory activity, could be transformed to active forms via gastrointestinal digestion [282]. Simulation of digestion is also useful in studies of the mechanism of action of antihypertensive peptides with demonstrated *in vivo* activity. For example, Miguel [277] found that YAEERYPIL derived from ovalbumin – which is a powerful ACE-inhibitor (IC_{50} = 4.7 μM) and exhibits antihypertensive activity, was susceptible to degradation by digestive enzymes; that peptide was indeed fully hydrolyzed during simulated gastrointestinal digestion, thus giving rise to fragments YAEER and YPI. Tests on mice showed that YAEER could not significantly lower blood pressure, but the peptide YPI exhibited a significant antihypertensive effect. This fragment may possibly be the active form hidden in the sequence YAEERYPIL, and may exert its action via a different mechanism of ACE-inhibition [285].

In vitro models provide useful information to assess the stability of bioactive peptides to different peptidases and proteinases of the body, yet transport across the intestinal barrier raises an extra resistance – so they have limitations. *In vitro* simulated digestion is in fact not entirely reliable; the degree of hydrolysis depends on the size, nature and neighborhood of the peptide [272], so *in vivo* studies (with laboratory animals and human volunteers) are eventually necessary to ascertain in full the behavior of the peptide. Another example is the release of *potent* ACE-inhibitory peptides from WPC brought about by aqueous extracts from the plant *C. cardunculus.* A peptide mixture – in which 3 peptides were pinpointed: α-La f(16-26), with the sequence KGYGGVSLPEW; α-La f(97-104), with the sequence DKVGINYW; and β-Lg f(33-42), with the sequence DAQSAPLRVY, produced ACE-inhibition (see Table 4). Such peptides were then exposed to simulated gastrointestinal digestion: no peptide was able to keep its integrity, but even total hydrolysis to smaller peptides did not significantly compromise the overall ACE-inhibitory activity observed. In view of their ACE-inhibitory activities, both in the absence or following gastrointestinal digestion, peptides KGYGGVSLPEW and DAQSAPLRVY are expected to eventually exhibit notable antihypertensive activities *in vivo* [107].

6. Concluding remarks

Processing of whey proteins yields several bioactive peptides able to trigger physiological effects in the human body. Such peptides, in concentrated form, can be commercially appealing because their claimed health-promoting features are nowadays an important driver for consumers' food choices. Hence, they may constitute an excellent alternative for whey upgrade. Use of selective membranes to isolate, and eventually purify whey proteins and peptides has substantially increased the number and depth of studies encompassing those molecules and their hydrolysates. The technology developed is not excessively expensive, and can easily be implemented in dairy plants – of either small or large dimension. Most whey peptides bearing biological activity are released by enzymatic hydrolysis, so new alternatives to enzymes of animal origin have been under scrutiny.

This chapter focused on studies of whey peptides with antihypertensive activity – including their mechanisms of action (especially ACE inhibition), as well as the bioavailability of these peptides, and highlighting the main *in vitro* and *in vivo* results, as well as clinical trials in humans.

Although a good deal of data have been generated encompassing food bioactive peptides, much is still left to do with whey peptides. Hence, several opportunities for further research exist, on incorporation of said ingredients in food products for human consumption. However, several scientific, technological and regulatory issues should be addressed before such peptide concentrates (and pure peptides) will have a chance to be marketed at large, aiming at both human nutrition and health.

More detailed studies are indeed welcome for a better understanding of antihypertensive mechanisms. In particular, the antihypertensive activity should be checked with extra detail – including deep studies on the blood pressure-reducing mechanisms, such as the effects of peptides on neutral endopeptidases and their putative beneficial activity upon cardiovascular diseases. The pharmacological effect of said peptides should be determined both on post- and prejunctional receptors. More extensive clinical trials should also be performed – after thorough bioavailability studies *in vitro*, such as stability to gastrointestinal digestion and passage through the blood barrier, have taken place.

Author details

Tânia G. Tavares
Instituto de Tecnologia Química e Biológica, Universidade Nova de Lisboa, Oeiras, Portugal

F. Xavier Malcata*
Instituto de Tecnologia Química e Biológica, Universidade Nova de Lisboa, Oeiras, Portugal
Department of Chemical Engineering, University of Porto, Porto, Portugal

Acknowledgement

This work received partial finantial support from project NEW PROTECTION – NativE, Wild PRObiotic sTrain EffecCT In Olives in briNe (ref. PTDC/AGR-ALI/117658/2010), from FCT, Portugal, coordinated by author F. X. M.

* Coresponding Author

7. References

[1] Diplock AT, Aggett PJ, Ashwell M, Bornet F, Fern EB, Roberfroid MB. Scientific concepts of functional foods in Europe consensus document. British Journal of Nutrition 1999; 81, 1-27.

[2] Arai S. Studies on functional foods in Japan–state of the art. Bioscience, Biotechnology and Biochemistry 1996; 60, 9-15.

[3] Korhonen H, Pihlanto A. Food-derived bioactive peptides – opportunities for designing future foods. Current Pharmaceutical Design 2003; 9, 1297-1308.

[4] Korhonen H, Pihlanto A. Bioactive peptides: production and functionality. International Dairy Journal 2006; 16, 945-960.

[5] Pintado ME, Pintado AE, Malcata FX. Controlled whey protein hydrolysis using two alternative proteases. Journal of Food Engineering 1999; 42, 1-13.

[6] Recio I, López-Fandiño R. Ingredientes y productos lácteos funcionales, bases científicas de sus efectos en la salud. In: Alimentos funcionales (ed.) Fundación Espanõla para la Ciencia y la Tecnología; 2005.

[7] Mulvihill DM. Production, functional properties and utilization of milk protein products. In: Fox PF. (ed.) Advanced dairy chemistry – proteins, vol 1. London, UK: Elsevier; 1992. p369-404.

[8] Nakai S, Modler HW. Food proteins, properties and characterization. New York, USA: Wiley – VCH Publishers; 1996.

[9] Fox PF, McSweeney PLH. Milk Proteins. In: Dairy chemistry and biochemistry. London, UK: Blackie Academic and Professional; 1998.

[10] Pintado ME, Macedo AC, Malcata FX. Review: technology, chemistry and microbiology of whey cheeses. Food Science and Technology International 2001; 7, 105-116.

[11] Blenford DE. Whey: from waste to gold. International Food Ingredients 1996; 1, 27-29.

[12] Barth CA, Behnke U. Nutritional significance of whey and whey components. Nahrung-Food 1997; 41, 2-12.

[13] Walzem RL, Dilliard CJ, German JB. Whey components: millenia of evolution create functionalities for mammalian nutrition: what we know and what we may be overlooking. Critical Reviews in Food Science and Nutrition 2002; 42, 353-375.

[14] McIntosh GH, Royle PJ, le Leu RK, Regester GO, Johnson MA, Grinsted RL, Kenward RS, Smithers GW. Whey proteins as functional food ingredients? International Dairy Journal 1998; 8, 425-434.

[15] Balagtas JV, Hutchinson FM, Krochta JM, Sumner DA. Anticipating market effects of new uses for whey and evaluating returns to research and development. Journal of Dairy Science 2003; 86, 1662-1672.

[16] Miller GD, Jarvis JK, Mcbeon LD. Handbook of Dairy Foods and Nutrition. Boca Raton, USA: CRC, 2000.

[17] Sgarbieri VC. Proteinas em alimentos protéicos: propriedades, degradações, modificações. Varela, São Paulo 1996; 139-157.

[18] Pihlanto-Leppälä A, Korhonen H. Bioactive peptides and proteins. Advances in Food and Nutrition Research 2003; 47, 175-276.

[19] Whitney RM. Milk Proteins In: Fundamentals of dairy chemistry. van Nostrand Reinhold. New York, USA; 1988.

[20] Law AJR, Leaver J, Banks JM, Horne DS. Quantitative fractionation of whey proteins by gel permeation FPL. Milchwissenschaft 1993; 48, 663-666.

[21] Creamer L. K., Sawyer L. β-Lactoglobulin. In: Roginski H, Fuquay JW, Fox PF (ed) Encyclopedia of Dairy Sciences. New York: Academic Press; 2003.

[22] Sawyer L, Kontopidis G. The core lipocalin, bovine β-lactoglobulin. Biochimica et Biophysica Acta 2000; 1482, 136-148.

[23] Imafidon IG, Farkye YN, Spanier MA. Isolation, purification, and alteration of some functional groups of major milk proteins: a review. Food Science and Nutrition 1997; 37, 663-689.

[24] Gough P, Jenness R. Heat denaturation of β-lactoglobulins A and B. Journal of Dairy Science 1961; 44, 1163-1168.

[25] Walstra P, Geurts TJ, Noomen A, Tellema A, van Bockel. MAJS. Principles of milk properties and processes – Dairy Technology. New York, USA: Marcel Dekker; 1999.

[26] Chatterton DEW, Smithers G, Roupas P, Brodkorb A. Bioactivity of β-lactoglobulin and α-lactalbumin – technological implications for processing. International Dairy Journal 2006; 16, 1229-1240.

[27] Papiz MZ, Sawyer L, Eliopoulos EE, North AC, Findlay JB, Sivaprasadarao R, Jones TA, Newcomer ME, Kraulis PJ. The structure of β-lactoglobulin and its similarity to plasma retinol-binding protein. Nature 1986; 324, 383-385.

[28] Mansouri A, Haertle T, Gerard A, Gerard H, Gueant JL. Retinol free and retinol complexed β-lactoglobulin binding sites in bovine germ cells. Biochimica et Biophysica Acta 1997; 1357, 107-114.

[29] Mansouri A, Gueant JL, Capiaumont J, Pelosi P, Nabet P, Haertle T. Plasma membrane receptor for β-lactoglobulin and retinol-binding protein in murine hybridomas. Biofactors 1998; 7, 287-298.

[30] Barros RM, Ferreira CA, Silva SV, Malcata FX. Quantitative studies on the enzymatic hydrolysis of milk proteins brought about by cardosins precipitated by ammonium sulfate. Enzyme and Microbial Technology 2001; 29, 541-547.

[31] Anderson ME. Glutathione: an overiew of biosynthesys and modulation. Chemico-Biological Interaction – Limerick 1998; 111, 1-14.

[32] Armstrong JM, McKenzie HA, Sawyer WH. On the fractionation of β-lactoglobulin and α-lactalbumin. Biochimica et Biophysica Acta 1967; 147, 60-72.

[33] Monaco HL, Zanotti G, Spadon P, Bolognesi M, Sawyer L, Eliopoulos EE. Crystal structure of the trigonal form of bovine β-lactoglobulin and of its complex with retinol at 2.5 Å resolution. Journal of Molecular Biology 1987; 197, 695-706.

[34] Felipe X, Law AJ. Preparative-scale fractionation of bovine, caprine and ovine whey proteins by gel permeation chromatography. Journal of Dairy Research 1997; 64, 459-464.

[35] de Jongh HH, Groneveld T, de Groot J. Mild isolation procedure discloses new protein structural properties of β-lactoglobulin. Journal of Dairy Science 2001; 84, 562-571.

[36] Outinen M, Tossavainen O, Tupasela T, Koskela P, Koskinen H, Rantamaki P, Syvaoja EL, Antila P, Kankare V. Fractionation of proteins from whey with different pilot scale processes. Food Science and Technology 1996; 29, 411-417.
[37] Hiraoka Y, Segawa T, Kuwajima K, Sugai S, Murai N. α-Lactalbumin: a calcium metalloprotein. Biochemical and Biophysical Research Communication 1980; 95, 1098-1104.
[38] Eigel WN, Butler JE, Ernstrom CA, Farrell H, Harwalkar VR, Jenness R, Whitney RM. Nomenclature of proteins of cows milk – 5th revision. Journal of Dairy Science, 1984; 67, 1599-1631.
[39] Law AJR, Horne DS, Banks JM, Leaver J. Heat-induced changes in the whey proteins and caseins. Milchwissenschaft 1994; 49, 125-129.
[40] Heine WE, Klein PD, Reeds PJ. The importance of α-lactalbumin in infantil nutrition. Journal of Nutrition 1991; 121, 277-283.
[41] Tolkach A, Kulozik, U. Fractionation of whey proteins and caseinomacropeptide by means of enzymatic crosslinking and membrane separation techniques. Journal of Food Engineering 2005; 67, 13-20.
[42] Chang J, Bulychev A, Li L. A stabilized molten globule protein. FEBS Letters 2000; 487, 298-300.
[43] Dolgikh DA, Abaturov IA, Bolotina, Brazhnikov EV, Bychkova VE, Gilmanshin RI, Lebedev YO, Semisotnov GV, Tiktopulo EI, Ptitsyn OB. Compact state of a protein molecule with pronounced small-scale mobility. European Biophysics Journal 1985; 13, 109-121.
[44] Vanderheeren G, Hanssens I. Thermal unfolding of bovine α-lactalbumin. Comparison of circular dichroism with hydrophobicity measurements. Journal of Biological Chemistry 1994; 269, 7090-7094.
[45] Kuwajima K. The molten globule state as a clue for understanding the folding and cooperativity of globular-protein struture. Protein: Structure, Function and Genetics 1989; 6, 87-103.
[46] Kuwajima K, Hiraoka Y, Ikeguchi M, Sugai S. Comparison of the transient folding intermediates in lysozyme and α-lactalbumin. Biochemistry 1985; 24, 874-881.
[47] Kuwajima K. The molten globule state of α-lactalbumin. The FASEB Journal 1996; 1, 102-109.
[48] Dolgikh DA, Gilmanshin RI, Brazhnikov EV, Bychkova VE, Semisotnov GV, Venyaminov S, Ptitsyn OB. Alpha-lactalbumin: compact state with fluctuating terciary structure? FEBS Letters 1981; 136, 311-315.
[49] Ptitsyn OB. Molten globule and protein folding. Advances in Protein Chemistry 1995; 47, 83-229.
[50] Arai M, Kuwajima K. Role of the molten globule state in protein folding. Advances in Protein Chemistry 2000; 53, 209-282.
[51] Leandro P, Gomes CM. Protein misfolding in conformational disorders: rescue of folding defects and chemical chaperoning. Mini-Reviews in Medicinal Chemistry 2008; 8, 901-911.

[52] Delfour A, Jollès J, Alais C, Jollès P. Caseino-glycopeptides: characterization of a methionine residue and of the N-terminal sequence. Biochemical and Biophysical Research Communications 1965; 19, 452-455.

[53] Manso MA, López-Fandiño R. κ-Casein macropeptides from cheese whey, physicochemical, biological, nutritional, and technological features for possible uses. Food Reviews International 2004; 20, 329-355.

[54] Chobert JM, Touati A, Bertrand-Harb C, Dalgalarrondo M, Nicolas MG. Solubility and emulsifying properties of κ-casein and its caseinomacropeptide. Journal of Food Biochemistry 1989; 13, 457-473.

[55] Moreno FJ, López-Fandiño R, Olano A. Characterization and functional properties of lactosyl caseinomacropeptide conjugates. Journal of Agricultural and Food Chemistry, 2002; 50, 5179-5184.

[56] Dziuba J, Minkiewicz P. Influence of glycosylation on micelle-stabilizing ability and biological properties of C-terminal fragments of cow's κ-casein. International Dairy Journal 1996; 6, 1017-1044.

[57] Kinsella E, Whitehead DM. Proteins in whey: chemical, physical and functional properties. Advances in Food and Nutrition Research 1989: 33, 343-438.

[58] de Wit JN. The use of whey protein products. In: Fox PF. (ed.) Developments in dairy chemistry 4. Functional milk proteins. Barking, UK: Elsevier Science Publishers; 1989. p323-345.

[59] Korhonen H, Marnila P, Gill H. Milk immunoglobulins and complement factors, a review. British Journal of Nutrition 2000; 84, 75-80.

[60] Steijns JM. Milk ingredients as nutraceuticals. International Journal of Dairy Technology 2001; 54, 81-88.

[61] Fonseca ML, Fonseca CSP, Brandão SCC. Propriedades anticarcinogênicas de componentes do leite. Indústria de Laticínios 1999; 4, 5-55.

[62] Alais C (ed.). Science du lait: principes des techniques laitières 4. Paris: SEPAIC; 1984.

[63] Innocente N, Corradini C, Blecker C, Paquot M. Emulsifying properties of the total fraction and the hydrophobic fraction of bovine milk proteose-peptones. International Dairy Journal 1998; 8, 981-985.

[64] Parodi PW. A role for milk protein in cancer prevention. Australian Journal of Dairy Technology 1998; 53, 37-47.

[65] Regester GO, Mcintosh GH, Lee VWK, Smithers GW. Whey proteins as nutritional and functional food ingredients. Food Australia 1996; 48, 123-127.

[66] Ha E, Zemel MB. Functional properties of whey, whey components, and essential amino acids: mechanisms underlying health benefits for active people (review). Journal of Nutritional Biochemistry 2003; 14, 251-258.

[67] Smithers GW. Whey and whey proteins – 'from gutter-to-gold'. International Dairy Journal 2008; 18, 695-704.

[68] Bouchard D, Morisset D, Bourbonnais Y, Tremblay GM. Proteins with whey acidic-protein motifs and cancer. Lancet Oncology 2006; 7, 167-174.

[69] Gauthier SF, Pouliot Y, Saint-Sauveur D. Immunomodulatory peptides obtained by the enzymatic hydrolysis of whey proteins. International Dairy Journal 2006; 16, 1315-1323.

[70] Xu RJ. Bioactive peptides in milk and their biological and health implications. Food Reviews International 1998; 14, 1-16.

[71] Prates JAM, Mateus CMRP. Functional foods from animal sources and their physiologically active components. Revue de Médicine Vétérinaire 2002; 153, 155-160.

[72] Amaya-Farfán J. Avanços no conhecimento sobre a função das proteínas nas dietas para desempenho físico. 5º Simpósio Latino Americano de Ciência de Alimentos – Desenvolvimento Científico e Tecnológico e Inovação na Indústria de Alimentos. Campinas-Unicamp, 2003.

[73] Hartmann R, Meisel H. Food-derived peptides with biological activity: from research to food applications. Current Opinion in Biotechnology 2007; 18, 163-169.

[74] Gill HS, Rutherford KJ, Cross ML. Bovine milk: a unique source of immunomodulatory ingredients for functional foods. In: Buttriss J, Saltmarsh M. (eds.) Functional foods II – claims and evidence. Cambridge, UK: Royal Society of Chemistry Press; 2000. p82-90.

[75] Badger TM, Ronis MJJ, Hakkak R. Developmental effects and health aspects of soy protein isolate, casein and whey in male and female rats. International Journal of Toxicology 2001; 20, 165-174.

[76] Hakkak R, Korourian S, Shelnutt SR, Lensing S, Ronis MJJ, Badger TM. Diets containing whey proteins or soy protein isolate protect against 2,12-dimethylbenzanthracene-induced mammary tumours in female rats. Cancer Epidemiology, Biomarkers and Prevention 2000; 9, 113-117.

[77] Rowlands JC, He L, Hakkak R, Ronis MJJ, Badger TM. Soy and whey proteins down regulate DMBA-induced liver and mammary gland CYP1 expression in female rats. Journal of Nutrition 2001; 131, 3281-3287.

[78] Micke P, Beeh KM, Schlaak JF, Buhl R. Oral supplementation with whey proteins increases plasma GSH levels of HIV-infected patients. European Journal of Clinical Investigation, 2001; 31, 171-178.

[79] Micke P, Beeh KM, Buhl R. Effects of long-term supplementation with whey proteins on plasma GSH levels of HIV-infected patients. European Journal of Nutrition 2002; 41, 12-18.

[80] Clare DA, Catignani GL, Swaisgood HE. Biodefense properties of milk, the role of antimicrobial proteins and peptides. Current Pharmaceutical Design 2003; 9, 1239-1255.

[81] Hall WL, Millward DJ, Long SJ, Morgan LM. Casein and whey exert different effects on plasma amino acid profiles, gastrointestinal hormone secretion and appetite. British Journal of Nutrition 2003; 89, 339-348.

[82] Tavares TG, Contreras MM, Amorim M, Martín-Álvarez PJ, Pintado ME, Recio I, Malcata FX. Optimization, by response surface methodology, of degree of hydrolysis, antioxidant and ACE-inhibitory activities of whey protein hydrolyzates obtained with cardoon extract. International Dairy Journal 2011; 21, 926-933.

[83] Rosaneli CF, Bighetti AE, Antônio MA, Carvalho JE, Sgarbieri VC. Efficacy of a whey protein concentrate on the inhibition of stomach ulcerative lesions caused by ethanol ingestion. Journal of Medicinal Food 2002; 5, 221-228.

[84] Pacheco MTB, Bighetti EA, Antônio M, Carvalho JE, Possenti A, Sgarbieri VC. Antiulcerogenic activity of fraction and hydrolyzate obtained from whey protein concentrate. Brazilian Journal of Food Technology, III JIPCA 2006; 15-22.

[85] Mezzaroba LFH, Carvalho JE, Ponezi AN, Antônio MA, Monteiro KM, Possenti A, Sgarbieri, VC. Antiulcerative properties of bovine α-lactalbumin. International Dairy Journal 2006; 16, 1005-1012.

[86] Tavares TG, Monteiro KM, Possenti A, Pintado ME, Carvalho JE, Malcata FX. Antiulcerogenic activity of peptide concentrates obtained from hydrolysis of whey protein brought about by proteases from *Cynara cardunculus*. International Dairy Journal 2011; 21, 934-939.

[87] Madureira AR, Pereira CI, Gomes AMP, Pintado ME, Malcata FX. Bovine whey proteins – overview on their main biological properties. Food Research International 2007; 40, 1197-1211.

[88] Puyol P, Dolores-Perez M, Sanchez L, Ena JM, Calvo M. Uptake and passage of β-lactoglobulin, palmitic acid and retinol across the Caco-2 monolayer. Biochimica et Biophysica Acta – Biomembranes 1995; 1236, 149-154.

[89] Wu SY, Pérez MD, Puyol P, Sawyer L. β-Lactoglobulin binds palmitate within its central cavity. Journal of Biological Chemistry 1999; 274, 170-177.

[90] Puyol P, Pérez MD, Ena JM, Calvo M. Interaction of β-lactoglobulin and other bovine and human whey proteins with retinol and fatty acids. Agricultural and Biological Chemistry 1991; 10, 2515-2520.

[91] Bounous G, Batist G, Gold P. Whey proteins in cancer prevention. Cancer Letters 1991; 57, 91-94.

[92] Baruchel S, Wang T, Farah R, Jamali M, Batist G. *In vivo* modulation of tissue glutathione in rat mammary carcinoma model. Biochemical Pharmacology 1995; 50, 1505-1508.

[93] Bounous G. Whey protein concentrate (WPC) and glutathione modulation in cancer treatment. Anticancer Research 2000; 20, 4785-4792.

[94] Perez MD, Sanchez L, Aranda P, Ena JM, Oria R, Calvo M. Effect of β-lactoglobulin on the activity of pregastric lipase. A possible role for this protein in ruminant milk. Biochimica et Biophysica Acta 1992; 1123, 151-155.

[95] Warme PKF, Momany A, Rumball SV, Tuttle RW, Scherag HA. Computation of structures of homologous proteins. α-Lactalbumin from lysozyme. Biochemistry 1974; 13, 768-782.

[96] Farrel HM, Bede MJ, Enyeart JA. Binding of *p*-nitrophenyl phosphate and other aromatic compounds by β-Lg. Journal of Dairy Science 1987; 70, 252-258.

[97] Meisel H, Schlimme E. Bioactive peptides derived from milk proteins: ingredients for functional foods? Kieler Milchwirtschaftliche und Forschungsberichte 1996; 48, 343-357.

[98] Mullally MM, Meisel H, Fitzgerald RJ. Synthetic peptides corresponding to α-lactalbumin and β-lactoglobulin sequences with angiotensin-I-converting enzyme inhibitory activity. Biological Chemistry Hoppe-Seyler 1996; 377, 259-260.

[99] Pihlanto-Leppälä A, Paakkari I, Rinta-Koski M, Antila P. Bioactive peptide derived from *in vitro* proteolysis of bovine β-lactoglobulin and its effect on smooth muscle. Journal of Dairy Research 1997; 64, 149-155.

[100] Pihlanto-Leppälä A, Rokka T, Korhonen H. Angiotensin I converting enzyme inhibitory peptides derived from bovine milk proteins. International Dairy Journal 1998; 8, 325-331.

[101] Pihlanto L. Bioactive peptides derived from bovine whey proteins: opioid and ACE-inhibitory peptides. Trends in Food Science and Technology 2000; 11, 347-356.

[102] Ijäs H, Collin M, Finckenberg P, Pihlanto-Leppälä A, Korhonen H, Korpela R, Vapaatalo H, Nurminen ML. Antihypertensive opioid-like milk peptide α-lactorphin: lacks effect on behavioural tests in mice. International Dairy Journal 2004; 14, 201-205.

[103] Hernández-Ledesma B, López-Expósito I, Ramos M, Recio I. Bioactive peptides from milk proteins. In: Pizzano R (ed.) Immunochemistry in dairy research. Kerala, India: Trivandrum; 2006. p37-60.

[104] FitzGerald RJ, Murray BA. Bioactive peptides and lactic fermentations. International Journal of Dairy Technology 2006; 59, 118-125.

[105] Chen G-W, Tsai J-S, Pan BS. Purification of angiotensin I-converting enzyme inhibitory peptides and antihypertensive effect of milk produced by protease-facilitated lactic fermentation. International Dairy Journal 2007; 17, 641-647.

[106] Roufik S, Gauthier S, Turgeon S. Physicochemical characterization and *in vitro* digestibility of β-lactoglobulin/β-Lg f142-148 complexes. International Dairy Journal 2007; 17, 471-480.

[107] Tavares TG, Contreras MM, Amorim M, Pintado ME, Recio I, Malcata FX. Novel whey-derived peptides with inhibitory activity against angiotensin-converting enzyme: *in vitro* activity and stability to gastrointestinal enzymes. Peptides 2011; 32, 1013-1019.

[108] Expósito IL, Récio I. Antibacterial activity of peptides and holding variants from milk proteins. International Dairy Journal 2006; 16, 1294-1305.

[109] Pihlanto-Leppälä A, Marnila P, Hubert L, Rokka T, Korhonen HJT, Karp M. The effect of α-lactalbumin and β-lactoglobulin hydrolysates on the metabolic activity of *Escherichia coli* JM103. Journal of Applied Microbiology 1999; 87, 540-545.

[110] Pellegrini A, Thomas U, Bramaz N, Hunziker P, Fellenberg R. Isolation and identification of three bactericidal domains in the bovine α-lactalbumin molecule. Biochimica et Biophysica Acta 1999; 1426, 439-448.

[111] Bruck WM, Graverholt G, Gibson GR. A two-stage continuous culture system to study the effect of supplemental α-lactalbumin and glycomacropeptide on mixed populations of human gut bacteria challenged with enteropathogenic *Escherichia coli* and *Salmonella* serotype *Typhimurium*. Journal of Applied Microbiology 2003; 95, 44-53.

[112] Pellegrini A, Dettling C, Thomas U, Hunziker P. Isolation and characterization of four bactericidal domains in the bovine β-lactoglobulin. Biochimica et Biophysica Acta 2001; 1526, 131-140.

[113] Groleau PE, Morin P, Gauthier SF, Pouliot Y. Effect of physicochemical conditions on peptide-peptide interactions in a tryptic hydrolysate of β-lactoglobulin and

identification of aggregating peptides. Journal of Agricultural and Food Chemistry 2003; 51, 4370-4375.

[114] Nagaoka S, Futamura Y, Miwa K, Awano T, Yamauchi K, Kanamaru Y, Kojima T, Kuwata T. Identification of novel hypocholesterolemic peptides derived from bovine milk β-lactoglobulin. Biochemical and Biophysical Research Communications 2001; 281, 11-17.

[115] Antila P, Paakkari I, Järvinen A, Mattila MJ, Laukkanen M, Pihlanto-Leppälä A, Mäntsälä P, Hellman J. Opioid peptides derived from *in vitro* proteolysis of bovine whey proteins. International Dairy Journal 1991; 1, 215-229.

[116] Murakami M, Tonouchi H, Takahashi R, Kitazawa H, Kawai Y, Negishi H, Saito T. Structural analysis of a new anti-hypertensive peptide (β-lactosin B) isolated from a commercial whey product. Journal of Dairy Science 2004; 87, 1967-1974.

[117] Tavares TG, Sevilla MA, Montero MJ, Carrón R, Malcata FX. Acute effects of whey peptides upon blood pressure of hypertensive rats, and relationship with their angiotensin-converting enzyme inhibitory activity. Molecular Nutrition and Food Research 2011; doi: 10.1002/mnfr.201100381.

[118] Yamauchi R, Sonoda S, Jinsmaa Y, Yoshikawa M. Antinociception induced by β-lactotensin, a neurotensin agonist peptide derived from β-lactoglobulin, is mediated by NT2 and D1 receptors. Life Sciences 2003; 73, 1917-1923.

[119] de Wit JN. Nutritional and functional characteristics of whey proteins in food products. Journal of Dairy Science 1998; 81, 597-608.

[120] A, Zhivotovsky B, Orrenius S, Sabharwal H, Svanborg C. Apoptosis induced by a human milk protein. Proceedings of the National Academy of Sciences of USA 1995; 92, 8064-8068.

[121] Svensson M, Sabharwal H, Hakansson A, Mossberg AK, Lipniunas P, Leffler H, Svanborg, C, Linse S. Molecular characterization of α-lactalbumin folding variants that induce apoptosis in tumor cells. Journal of Biological Chemistry 1999; 274, 6388-6396.

[122] Svensson M, Hakansson A, Mossberg AK, Linse S, Svanborg C. Conversion of α-lactalbumin to a protein inducing apoptosis. Proceedings of the National Academy of Sciences of USA 2002; 97, 4221-4226.

[123] Markus CR, Olivier B, Haan E. Whey protein rich in α-lactalbumin increases the ratio of plasma tryptophan to the sum of the other large neutral amino acids and improves cognitive performance in stress-vulnerable subjects. American Journal of Clinical Nutrition 2002; 75, 1051-1056.

[124] Ganjam LS, Thornton WH, Marshall RT, MacDonald RS. Antiproliferative effects of yoghurt fractions obtained by membrane dialysis on cultured mammalian intestinal cells. Journal of Dairy Science 1997; 80, 2325-2329.

[125] Hakansson A, Svensson M, Mossberg AK, Sabharwal H, Linse S, Lazou I, Lönnerdal B, Svanborg C. A folding variant of α-lactalbumin with bactericidal activity against *Streptococcus pneumoniae*. Molecular Microbiology 2000; 35, 589-600.

[126] Markus CR, Olivier B, Panhuysen GEM, Gugten J, van Der, Alles MS, Tuiten A, Westenberg HGM, Fekkes D, Kopperschaar HF, Haan E. The bovine α-lactalbumin increases the plasma ratio of tryptophan to the other large neutral amino acids, and in

vulnerable subjects raises brain serotonin activity, reduces cortisol concentration, and improves mood under stress. American Journal of Clinical Nutrition 2000; 71, 1536-1544.

[127] Montagne PM, Cuiliere ML, Mole CM, Bene MC, Faure GC. Dynamics of the main immunologically and nutritionally available proteins of human milk during lactation. Journal of Food Composition and Analysis 2000; 13, 127-137.

[128] Meisel H, FitzGerald RJ. Opioid peptides encrypted in milk proteins. Journal of Nutrition 2000; 84, 27-31.

[129] Nurminen ML, Sipola M, Kaarto H, Pihlanto-Leppälä, Piilola K, Korpela R, Tossavainen O, Coronen H, Vapaatalo H. α-Lactorphin lowers blood pressure measured by radiotelemetry in normotensive and spontaneously hypertensive rats. Life Science 2000; 66, 1535-1543.

[130] Matsumoto H, Shimokawa Y, Ushida Y, Toida T, Hayasawa H. New biological function of bovine α-lactalbumin: protective effect against ethanol- and stress-induced gastric mucosal injury in rats. Bioscience, Biotechnology and Biochemistry 2001; 65, 1104-1111.

[131] Uchida K, Tateda T, Takagi S. Hypothetical mechanism of prostaglandin E1-induced bronchoconstriction. Medical Hypotheses 2003; 61, 378-384.

[132] Rosaneli CF, Bighetti AE, Antônio MA, Carvalho JE, Sgarbieri VC. Protective effect of bovine milk whey protein concentrate on the ulcerative lesions caused by subcutaneous administration of indomethacin. Journal of Medicinal Food 2004; 7, 309-314.

[133] Tong LM, Sasaki S, McClements DJ, Decker EA. Mechanisms of the antioxidant activity of a high molecular weight fraction of whey. Journal of Agricultural and Food Chemistry 2000; 48, 1473-1478.

[134] Smith C, Halliwell B, Aruoma O I. Protection of albumin against the pro-oxidant actions of phenolic dietary components. Food Chemistry and Toxicology 1992; 30, 483-489.

[135] Laursen I, Briand P, Lykkesfeldt AE. Serum albumin as a modulator of the human breast cancer cell line MCF-7. Anticancer Research 1990; 10, 343-351.

[136] Abubakar A, Saito T, Kitazawa H, Kawai Y, Itoh T. Structural analysis of new antihypertensive peptides derived from cheese whey protein by proteinase K digestion. Journal of Dairy Science 1998; 81, 3131-3138.

[137] FitzGerald RJ, Murray BA, Walsh DJ. Hypotensive peptides from milk proteins. Journal of Nutrition 2004; 134, 980-988.

[138] Yamauchi K. Biologically functional proteins of milk and peptides derived from milk proteins. IDF Bulletin 1992; 272, 51-57.

[139] Tani F, Shiota, Chiba H, Yosliikawa M. Serophin an opioid peptide derived from bovine serum albumin. In: Brantl V. (ed.) β-Casomorphins and relate peptides: recent developments. Weinheim, Germany: VCH-Verlag; 1993. p49-53.

[140] Ormrod DJ, Miller TE. The anti-inflammatory activity of a low molecular weight component derived from the milk of hyperimmunized cows. Agents and Actions 1991; 32, 160-166.

[141] Mitra AK, Mahalanabis D, Unicomb L, Eechels R, Tzipori S. Hyperimmune cow colostra reduces diarrhoea due to rotavirus: a double-blind, controlled clinical trial. Acta Paediatrica 1995; 84, 996-1001.

[142] Tomita M, Todhunter DA, Hogan JS, Smith KL. Immunisation of dairy cows with an *Escherichia coli* J5 lipopolysaccharide vaccine. Journal of Dairy Science 1995; 78, 2178-2185.

[143] Oona M, Rägö T, Maaroos HI, Mikelsaar M, Loivukene K, Salminen S, Korhonen H. *Helicobacter pylori* in children with abdominal complaints: has immune bovine colostrum some influence on gastritis? Alpe Adria Microbiology Journal 1997; 6, 49-57.

[144] Freedman DJ, Tacket CO, Delehanty A, Maneval DR, Nataro J, Crabb JH. Milk immunoglobulin with specific activity against purified colonization factor antigens can protect against oral challenge with enterotoxigenic *Escherichia coli*. Journal of Infectious Diseases 1998; 177, 662-667.

[145] Loimaranta V, Laine M, Söderling E, Vasara E, Rokka S, Marnila P, Korhonen H, Tossavainen O, Tenovuo J. Effects of bovine immune- and non-immune whey preparations on the composition and pH response of human dental plaque. European Journal of Oral Science 1999; 107, 244-250.

[146] Okhuysen PC, Chappell CL, Crabb J, Valdez LM, Douglass E, DuPont HL. Prophylactic effect of bovine anti-*Cryptosporidium hyperimmune* colostra immunoglobulin in healthy volunteers challenged with *Cryptosporidium parvum*. Clinical Infectious Diseases 1998; 26, 1324-1329.

[147] Sharpe SJ, Gamble GD, Sharpe DN. Cholesterol-lowering and blood pressure effects of immune milk. American Journal of Clinical Nutrition 1994; 59, 929-934.

[148] Jollès P, Fiat AM. The carbohydrate portions of milk glycoproteins. Journal of Dairy Research 1979; 46, 187-191.

[149] Jollès P, Levy-Toledano S, Fiat AM, Soria C, Cillessen D, Thomaidis A, Dunn FW, Caen JP. Analogy between fibrinogen and casein. Effect of an undecapeptide isolated from κ-casein on platelet function. European Journal of Biochemistry 1986; 158, 379-382.

[150] Shebuski RJ, Berry DE, Bennett DB, Romoff T, Storer BL, Ali F, Samanen J. Demonstration of Ac-Arg-Gly-Asp-Ser-NH_2 as an antiaggregatory agent in the dog by intracoronary administration. Journal of Thrombosis and Haemostasis 1989; 61, 183-188.

[151] Chabance B, Marteau P, Rambaud JC, Migliore-Samour D, Boynard M, Perrotin P, Guillet R, Jollès P, Fiat AM. Casein peptide release and passage to the blood in humans during digestion of milk or yogurt. Biochimie 1998; 80, 155-165.

[152] Rutherford KJ, Gill H. Peptides affecting coagulation. British Journal of Nutrition 2000; 84, 99-102.

[153] Manso MA, Escudero C, Alijo M, López-Fandiño R. Platelet aggregation inhibitory activity of bovine, ovine and caprine κ-casein macropeptides and their tryptic hydrolyzates. Journal of Food Protection 2002; 65, 1992-1996.

[154] Nakamura Y, Yamamoto N, Sakai K, Okubo A, Yamazaki S, Takano T. Purification and characterization of angiotensin I-converting enzyme inhibitors from sour milk. Journal of Dairy Science 1995; 78, 777-783.

[155] Manso MA, López-Fandiño R. Angiotensin I converting enzyme-inhibitory activity of bovine, ovine, and caprine κ-casein macropeptides and their tryptic hydrolysates. Journal of Food Protection 2003; 66, 1686-1692.

[156] Mizuno S, Matsuura K, Gotou T, Nishimura S, Kajimoto O, Yabune M, Kajimoto Y, Yamamoto N. Antihypertensive effect of casein hydrolysate in a placebo-controlled study in subjects with high-normal blood pressure and mild hypertension. British Journal of Nutrition 2005; 94, 84-91.

[157] Neeser JR, Chambaz A, del Vedovo S, Prigent MJ, Guggenheim B. Specific and nonspecific inhibition of adhesion of oral actinomyces and streptococci to erythrocytes and polystyrene by caseinoglycopeptide derivatives. Infection and Immunity 1988; 56, 3201-3208.

[158] Kawasaki Y, Isoda H, Tanimoto M, Dosako S, Idota T, Ahiko K. Inhibition by lactoferrin and κ-casein glycomacropeptide of binding of cholera toxin to its receptor. Bioscience, Biotechnology and Biochemistry 1992; 56, 195-198.

[159] Schupbach P, Neeser JR, Golliard M, Rouvet M, Guggenheim B. Incorporation of caseinoglycomacropeptide and caseinophosphopeptide into the salivary pellicle inhibits adherence of mutant streptococci. Journal of Dental Research 1996; 75, 1779-1788.

[160] Oh S, Worobo RW, Kim B, Rheem S, Kim S. Detection of cholera toxin-binding activity of κ-casein macropeptide and optimization of its production by the response surface methodology. Bioscience, Biotechnology and Biochemistry 2000; 64, 516-522.

[161] Bouhallab S, Favrot C, Maubois JL. Growth-promoting activity of tryptic digest of caseinomacropeptide of *Lactococcus lactis* subsp. *lactis*. Le Lait 1993; 73, 73-77.

[162] Beucher S, Levenez F, Yvon M, Corring T. Effects of gastric digestive products from casein on CCK release by intestinal-cells in rat. Journal of Nutritional Biochemistry 1994; 5, 578-584.

[163] Yvon M, Beucher S, Guilloteau P, le Huerou-Luron I, Corring T. Effects of caseinomacropeptide (CMP) on digestion regulation. Reproduction, Nutrition and Development 1994; 34, 527-537.

[164] Pederson NLR, Nagain-Domaine C, Mahe S, Chariot J, Roze C, Tome D. Caseinomacropeptide specifically stimulates exocrine pancreatic secretation in the anesthetized rat. Peptides 2000; 21, 1527-1535.

[165] Dartey C, Leveille G, Sox TE. (2003). Compositions for appetite control and related methods. US Patent 0,059,495 A1.

[166] Hasler C. A new look at an ancient concept. Chemistry and Industry 1998; 2, 84-89.

[167] Pihlanto-Leppälä A. Antioxidative peptides derived from milk proteins. International Dairy Journal 2006; 16, 1306-1314.

[168] Udenigwe CC, Aluko RE. Food protein-derived bioactive peptides: production, processing, and potential health benefits. Journal of Food Science 2012; 77, R11–R24.

[169] Korhonen H. Milk-derived bioactive peptides: from science to applications. Journal of Functional Foods 2009; 1, 177-187.

[170] Ustunol Z, Zeckzer T. Relative proteolytic action of milk-cloting enzyme preparations on bovine β and κ casein. Journal of Food Science 1996; 61, 1136-1138.

[171] Fox PF. Exogeneous enzymes in dairy technology. A review. Journal of Food Biochemistry 1993; 17, 173-199.

[172] Simões IIG. Caracterização molecular da acção das cardosinas A e B sobre caseínas β- e κ-bovinas. Tese de Mestrado, Universidade de Coimbra; 1998.

[173] López-Fandiño R, Otte J, van Camp J. Physiological, chemical and technological aspects of milk-protein-derived peptides with antihypertensive and ACE-inhibitory activity. International Dairy Journal 2006; 16, 1277-1293.

[174] Schmidt DG, Meijer RJ, Slangen CJ, van Beresteijn EC. Raising the pH of the pepsin-catalysed hydrolysis of bovine whey proteins increases the antigenicity of the hydrolyzates. Clinical and Experimental Allergy 1995; 25, 1007-1017.

[175] Barros RM, Malcata FX. Molecular characterization of peptides released from β-lactoglobulin and α-lactalbumin via cardosins A and B. Journal of Dairy Science 2006; 89, 483-494.

[176] Lamas EM, Barros RM, Balcão VM, Malcata FX. Hydrolysis of whey proteins by proteases extracted from *Cynara cardunculus* and immobilized onto highly activated supports. Enzyme and Microbial Technology 2000; 28, 642-652.

[177] Barros RM, Malcata FX. Modelling the kinetics of whey protein hydrolysis brought about by enzymes from *Cynara cardunculus*. Journal of Agricultural and Food Chemistry 2002; 50, 4347-4356.

[178] Barros RM, Malcata FX. A kinetic model for hydrolysis of whey proteins by cardosin A extracted from *Cynara cardunculus*. Food Chemistry 2004; 88, 351-359.

[179] Saboya LV, Maubois JL. Current developments of microfiltration technology in the dairy industry. Le Lait 2000; 80, 541-553.

[180] Muller A, Daufin G, Chaufer B. Ultrafiltration modes of operation for the separation of α-lactalbumin from acid casein whey. Journal of Membrane Science 1999; 153, 9-21.

[181] Cheang B, Zydney AL. Separation of α-lactalbumin and β-lactoglobulin using membrane ultrafiltration. Biotechnology and Bioengineering 2003; 83, 201-209.

[182] Cheang B, Zydney AL. A two-stage ultrafiltration process for fractionation of whey protein isolate. Journal of Membrane Science 2004; 231, 159-167.

[183] Brans G, Schroën CGPH, van der Sman RGM, Boom RM. Membrane fractionation of milk: state of the art and challenges. Journal of Membrane Science 2004; 243, 263-272.

[184] Díaz O, Pereira CD, Cobos A. Functional properties of ovine whey protein concentrates produced by membrane technology after clarification of cheese manufacture by-products. Food Hydrocolloids 2004; 18, 601-610.

[185] Atra R, Vatai G, Bekassy-Molnar E, Balint A. Investigation of ultra- and nanofiltration for utilization of whey protein and lactose. Journal of Food Engineering 2005; 67, 325-332.

[186] Zydney AL. Protein separations using membrane filtration: new opportunities for whey fractionation. International Dairy Journal 1998; 8, 243-250.

[187] Rektor A, Vatai G. Membrane filtration of Mozzarella whey. Desalination 2004;162, 279-286.

[188] Smithers GW, Ballard FJ, Copeland AD, Silva KJ, Dionysius DA, Francis GL, Goddard C, Grieve PA, Mcintosh GH, Mitchell IR, Pearce RJ, Regester GO. New opportunities

from the isolation and utilization of whey proteins. Journal of Dairy Science 1996; 79,1454-1459.
[189] Wong CW, Seow HF, Husband AJ, Regester GO, Watson DL. Effects of purified bovine whey factors on cellular immune functions in ruminants. Veterinary Immunology and Immunopathology 1997; 56, 85-96.
[190] Tavares TG, Amorim M, Gomes D, Pintado ME, Pereira CD, Malcata FX. Bioactive peptide-rich concentrates from whey: pilot process characterization. Journal of Food Engineering 2012; 110, 547-552.
[191] Eastern Stroke and Coronary Heart Disease Collaborative Research Group. Blood pressure, cholesterol and stroke in Eastern Asia. Lancet 1998; 352, 1801-807.
[192] Chalmers J. Blood pressure burden: vascular changes and cerebrovascular complications. Journal of Hypertension 2000; 18, S1-S2.
[193] Chobanian AV, Bakris GL, Black HR, Cushman WC, Green LA, Izzo JL, Jones DW, Materson BJ, Oparil S, Wright JT, Roccella EJ. The seventh report of the Joint National Committee on Prevention, Detection, Evaluation, and Treatment of High Blood Pressure. JAMA 2003; 289, 2560-2571.
[194] Tsai JS, Chen TJ, Pan BS, Gong SD, Chung MY. Antihypertensive effect of bioactive peptides produced by protease-facilitated lactic acid fermentation of milk. Food Chemistry 2008; 106, 552-558.
[195] Murray CJL, Lopez AD (eds.). The global burden of disease: a comprehensive assessment of mortality and disability from disease, injuries and risk factors in 1990 and projected to 2020. Global Burden of Disease and Injury Series, vol 1. Cambridge: Harvard School of Public Health; 1996.
[196] Geisterfer AA, Peach MJ, Owens GK. Angiotensin II induces hypertrophy, not hyperplasia, of cultured rat aortic smooth muscle cells. Circulation Research 1988; 62, 749-756.
[197] Daemen MJ, Lombardi DM, Bosman FT, Schwartz SM. Angiotensin II induces smooth muscle cell proliferation in the normal and injured rat arterial wall. Circulation Research 1991; 68, 450-456.
[198] Geerlings A, Villar IC, Zarco FH, Sánchez M, Vera R, Gomez AZ, Boza J, Duarte J. Identification and characterization of novel angiotensin-converting enzyme inhibitors obtained from goat milk. Journal of Dairy Science 2006; 89, 3326-3335.
[199] Hong F, Ming L, Yi S, Zhanxia L, Yongquan W, Chi L. The antihypertensive effect of peptides: a novel alternative to drugs? Peptides 2008; 29, 1062-1071.
[200] Scholz-Ahrens K, Schrezenmeir J. Effects of bioactive substances in milk on mineral and trace element metabolism with special reference to casein phosphopeptides. British Journal of Nutrition 2000; 84, S147-S153.
[201] Jauhiainen T, Korpela R. Milk peptides and blood pressure. Journal of Nutrition 2007; 137, 825S-829S.
[202] Sipola M, Finckenberg P, Korpela R, Vapaatalo H, Nurminen ML. Effect of long-term intake of milk products on blood pressure in hypertensive rats. Journal of Dairy Research 2002; 69, 103-111.

[203] Fujita H, Usui H, Kurahashi K, Yoshikawa M. Isolation and characterization of ovokinin, a bradykinin b-1 agonist peptide derived from ovalbumin. Peptides 1995; 16, 785-790.

[204] Matoba N, Usui H, Fujita H, Yoshikawa M. A novel anti-hypertensive peptide derived from ovalbumin induces nitric oxide-mediated vasorelaxation in an isolated SHR mesenteric artery. FEBS Letters 1999; 452, 181-184.

[205] Erdmann K, Grosser N, Schipporeit K, Schröder H. The ACE inhibitory dipeptide Met-Tyr diminishes free radical formation in human endothelial cells via induction of heme oxygenase-1 and ferritin. Journal of Nutrition 2006; 136, 2148-2152.

[206] Maes W, Van Camp J, Vermeirssen V, Hemeryck M, Ketelslegers JM, Schrezen-meir J, Van Oostveldt P, Huyghebaert. Influence of the lactokinin Ala-Leu-Pro-Met-His-Ile-Arg (ALPMHIR) on the release of endothelin-1 by endothelial cells. Regulatory Peptides 2004; 118, 105-109.

[207] Hernández-Ledesma B, Miralles B, Amigo L, Ramos M, Recio I. Identification of antioxidant and ACE-inhibitory peptides in fermented milk. Journal of the Science of Food and Agriculture 2005; 85, 1041–1048.

[208] Martínez-Maqueda D, Miralles B, Recio I, Hernández-Ledesma B. Antihypertensive peptides from food proteins: a review. Food and Function 2012; 3, 350-361.

[209] Sánchez D, Kassan M, Contreras MM, Carrón R, Recio I, Montero MJ, Sevilla MA. Long-term intake of a milk casein hydrolysate attenuates the development of hypertension and involves cardiovascular benefits. Pharmacological Research 2011; 63, 398-404.

[210] Quirós A, Ramos M, Muguerza B, Delgado MA, Miguel M, Aleixandre A, Recio I. Identification of novel antihypertensive peptides in milk fermented with *Enterococcus faecalis*. International Dairy Journal 2007; 17, 33-41.

[211] Miguel M, Contreras MM, Recio I, Aleixandre A. ACE-inhibitory and antihypertensive properties of a bovine casein hydrolysate. Food Chemistry 2009; 112, 211-214.

[212] Hernández-Ledesma B, Contreras MM, Recio I. Antihypertensive peptides: production, bioavailability and incorporation into foods. Advances in Colloid and Interface Science 2011; 165, 23-35.

[213] Gobetti M, Minervini F, Grizzello C. Angiotensin-I-converting-enzyme-inhibitory and antimicrobial bioactive peptides. International Journal of Dairy Technology 2004; 57, 173-188.

[214] Meisel H. Biochemical properties of peptides encrypted in bovine milk proteins. Current Medicinal Chemistry 2005; 12, 1905-1919.

[215] Silva SV, Malcata FX. Caseins as source of bioactive peptides. International Dairy Journal 2005; 15, 1-15.

[216] Vermeirssen V, Verstraete W, van Camp J. Bioavailability of angiotensin I converting enzyme inhibitory peptides. British Journal of Nutrition 2004; 92, 357-366.

[217] Pihlanto-Leppälä A. Bioactive peptides derived from bovine whey proteins: opioid and ACE-inhibitory peptides. Trends in Food Science and Technology 2001; 11, 347-356.

[218] Lee SH, Song KB. Isolation of an angiotensin-converting enzyme inhibitory peptide from irradiated bovine blood plasma protein hydrolysates. Food Chemistry and Toxicology, 2003; 68, 2469-2472.

[219] Mine Y, Kovacs-Nolan J. New insights in biologically active proteins and peptides derived from hen egg. World Poultry Science Journal 2006; 62, 87-96.

[220] Matsumura N, Fujii M, Takeda Y, Shimizu T. Isolation and characterization of angiotensin I-converting enzyme-inhibitory peptides derived from bonito bowels. Bioscience, Biotechnology and Biochemistry 1993; 57, 1743-1744.

[221] Wung J, Ding X. Hypotensive and physiological effects of angiotensin converting enzyme inhibitory peptides derived from soy protein on spontaneously hypertensive rats. Journal of Agricultural and Food Chemistry 2001; 49, 501- 506.

[222] Takayanagi T, Yokotsuka K. Angiotensin I converting enzyme-inhibitory peptides from wine. American Journal of Enology and Viticulture 1999; 50, 65-68.

[223] Miyoshi S, Ishikawa H, Kaneko, Fukui F, Tanaka H, Maruyama S. Structures and activity of angiotensin-converting enzyme inhibitors in α-zein hydrolysate. Agricultural and Biological Chemistry 1991; 55, 1313-1318.

[224] Maruyama S, Suzuki H. A peptide inhibitor of angiotensin I converting enzyme in the tryptic hydrolysate of casein. Agricultural and Biological Chemistry 1982; 46, 1393-1394.

[225] Maruyama S, Mitachi H, Tanaka H, Tomizuka N, Suzuki H. Studies on the active site and antihypertensive activity of angiotensin I-converting enzyme inhibitors derived from casein. Agricultural and Biological Chemistry 1987; 51, 1581-1586.

[226] Mullally MM, FitzGerald RJ, Meisel H. Identification of a novel angiotensin-I-converting enzyme inhibitory peptide corresponding to a tryptic fragment of bovine β-lactoglobulin. FEBS Letters 1997; 402, 99-101.

[227] FitzGerald RJ, Meisel H. Lactokinins, whey protein derived ACE inhibitory peptides. Nährung-Food 1999; 43, 165-167.

[228] Hernández-Ledesma B, Recio I, Ramos M, Amigo L. Preparation of ovine and caprine β-lactoglobulin hydrolysates with ACE-inhibitory activity. Identification of active peptides from caprine β-lactoglobulin hydrolysed with thermolysin. International Dairy Journal 2002; 12, 805-812.

[229] Ondetti MA, Rubin B, Cushman DW. Design of specific inhibitors of angiotensin-converting enzyme: new class of orally active antihypertensive agents. Science 1977; 196, 441-444.

[230] Jeunemaitre X, Soubrier F, Kotelevtsev YV, Lifton RP, Williams CS, Charru A, Hunt SC, Hopkins PN, Williams RR, Lalouel JM, Corvol P. Molecular basis of human hypertension: role of angiotensinogen. Cell 1992; 71, 169-180.

[231] Timmermans PBMWM, Wong PC, Chiu AT, Herblin WF, Benfield P, Carini DJ, Lee RJ, Wexler RR, Saye JAM, Smith RD. Angiotensin II receptors and angiotensin II receptor antagonists. Pharmacological Reviews 1993; 45, 205-251.

[232] Nelson KM, Yeager BF. What is the role of angiotensin-converting enzyme inhibitors in congestive heart failure and after myocardial infarction? Annals of Pharmacotherapy 1996; 30, 986-993.

[233] Pihlanto-Leppälä A, Koskinen P, Piilola K, Tupasela T, Korhonen H. Angiotensin-I converting enzyme inhibitory properties of whey proteins digests: concentration and characterization of active peptides. Journal of Dairy Research 2000; 67, 53-64.

[234] Schlothauer RC, Schollum LM, Reid JR, Harvey SA, Carr A, Fanshawe RL. Improved bioactive whey protein hydrolyzate. Patent PCT/NZ01/00188 (WO 02/19837 A1), New Zealand 2002.

[235] Ortiz-Chao P, Gómez-Ruiz JA, Rastall RA, Mills D, Cramer R, Pihlanto A, Korhonen H, Jauregi P. Production of novel ACE inhibitory peptides from β-lactoglobulin using Protease N Amano. International Dairy Journal 2009; 19, 69-76.

[236] Hernández-Ledesma B, Miguel M, Amigo L, Aleixandre MA, Recio I. Effect of simulated gastrointestinal digestion on the antihypertensive properties of synthetic β-lactoglobulin peptide sequences. Journal of Dairy Research 2007; 74, 336-339.

[237] Otte J, Shalaby S, Zakora M, Nielsen MS. Fractionation and identification of ACE-inhibitory peptides from α-lactalbumin and β-casein produced by thermolysin-catalysed hydrolysis. International Dairy Journal 2007; 17, 1460-1472.

[238] Chiba H, Yoshikawa M. Bioactive peptides derived from food proteins. Kagaku to Seibutsu (in Japanese) 1991; 29, 454-458.

[239] Miguel M, Manso MA, López-Fandiño R, Alonso MJ, Salaices M. Vascular effects and antihypertensive properties of κ-casein macropeptide. International Dairy Journal 2007; 17, 1473-1477.

[240] Ruiz-Giménez P, Ibáñez A, Salom JB, Marcos JF, López-Díez JJ, Vallés S, Torregrosa G, Alborch E, Manzanares P. Antihypertensive properties of lactoferricin B-derived peptides. Journal of Agricultural and Food Chemistry 2010; 58, 6721-6727.

[241] Kang DG, Kim YC, Sohn EJ, Lee YM, Lee AS, Yin MH, Lee HS. Hypotensive effect of butein via the inhibition of angiotensin-converting enzyme. Biological and Pharmaceutical Bulletin 2003; 26, 1345-1347.

[242] Miguel M, López F, Ramos M, Aleixandre A. Short-term effect of egg-white hydrolysate products on the arterial blood pressure of hypertensive rats. British Journal of Nutrition 2005; 94, 731-737.

[243] Okamoto K, Aoki K. Development of a strain of spontaneously hypertensive rats. Japanese Circulation Journal 1963; 27, 282-293.

[244] Nakamura Y, Yamamoto N, Sakai K, Takano T. Antihypertensive effect of sour milk and peptides isolated from it that are inhibitors to angiotensin I-converting enzyme. Journal of Dairy Science 1995; 78, 1253-1257.

[245] Fujita H, Yokoyama K, Yoshikawa M. Classification and antihypertensive activity of angiotensin I-converting enzyme inhibitory peptides derived from food proteins. Journal of Food Science 2000; 65, 564-569.

[246] Muguerza B, Ramos M, Sánchez E, Manso MA, Miguel M, Aleixandre A, Delgado MA, Recio I. Antihypertensive activity of milk fermented by *Enterococcus faecalis* strains isolated from raw milk. International Dairy Journal 2006; 16, 61-69.

[247] Contreras MM, Carrón R, Montero MJ, Ramos M, Recio I. Novel casein-derived peptides with antihypertensive activity. International Dairy Journal 2009; 19, 566-573.

[248] Matsui T, Matsufugi HY, Osajima Y. Colorimetric measurement of angiotensin I-converting enzyme inhibitory activity with trinitrobenzene sulfonate. Bioscience, Biotechnology and Biochemistry 1992; 56, 517-518.

[249] Friedland J, Silverstein E. A sensitive fluorometric assay for serum angiotensin converting enzyme. American Journal of Clinical Pathology 1976; 66, 416-424.

[250] Shihabi ZK. Analysis of angiotensin-converting enzyme by capillary electrophoresis. Journal of Chromatography A 1999; 853, 185-188.

[251] Cushman DW, Cheung HS. Spectrophotometric assay and properties of the angiotensin-converting enzyme of rabbit lung. Biochemical Pharmacology 1971; 20, 1637-1648.

[252] Vermeirssen V, van Camp J, Verstraete W. Optimisation and validation of an angiotensin-converting enzyme inhibition assay for the screening of bioactive peptides. Journal of Biochemical and Biophysical Methods 2002; 51, 75-87.

[253] Kohmura M, Nio N, Kubo K, Minoshima Y, Munekata E, Ariyoshi Y. Inhibition of angiotensin-converting enzyme by synthetic peptides of human β-casein. Agricultural and Biological Chemistry 1989; 53, 2107-2114.

[254] Seppo L, Jauhiainen T, Poussa T, Korpela R. A fermented milk high in bioactive peptides has a blood pressure-lowering effect in hypertensive subjects. American Journal of Clinical Nutrition 2003; 77, 326-330.

[255] Lv GS, Huo GC, Fu XY. Expression of milk-derived antihypertensive peptide in *Escherichia coli*. Journal of Dairy Science 2003; 86, 1927-1931.

[256] Ondetti MA, Cushman DW. Enzymes of the renin-angiotensin system and their inhibitors. Annual Reviews of Biochemistry 1982; 51, 283-308.

[257] Cushman DW, Cheung HS, Sabo EF, Ondetti MA. Development and design of specific inhibitors of angiotensin-converting enzyme. American Journal of Cardiology 1982; 49, 1390-1394.

[258] Cheung HS, Wang FL, Ondetti MA, Sabo EH, Cushman DW. Binding of peptide substrates and inhibitors of angiotensin-converting enzyme. Journal of Biological Chemistry 1980; 25, 401-407.

[259] Stevens RL, Micalizzi ER, Fessler DC, Pals DT. Angiotensin I converting enzyme of calf lung. Method of assay and partial purification. Biochemistry 1972; 11, 2999-3007.

[260] Kim SK, Byun HG, Park PY, Fereidoon S. Angiotensin I converting enzyme inhibitory peptides purified from bovine skin gelatin hydrolysate. Journal of Agricultural and Food Chemistry 2001; 49, 2992-2997.

[261] Gómez-Ruiz JA, Ramos M, Recio I. Angiotensin-converting enzyme-inhibitory activity of peptides isolated from Manchego cheese. Stability under simulated gastrointestinal digestion. International Dairy Journal 2004; 14, 1075-1080.

[262] Pripp AH, Isaksson T, Stepaniak L, Sørhaug T. Quantitative structure-activity relationship modelling of ACE-inhibitory peptides derived from milk proteins. European Food Research and Technology 2004; 219, 579-583.

[263] Wu JP, Aluko RE, Nakai S. Structural requirements of antiotensin I-converting enzyme inhibitory peptides. Quantitative structure-activity relationship study of di- and tripeptides. Journal of Agricultural and Food Chemistry 2006; 54, 732-738.

[264] Gobbetti M, Stepaniak L, de Angelis M, Corsetti A, Cagno RD. Latent bioactive peptides in milk proteins, proteolytic activation and significance in dairy processing. Critical Reviews in Food Science and Nutrition 2002; 42, 223-239.

[265] Meisel H. Biochemical properties of regulatory peptides derived from milk proteins. Biopolymers 1997; 43, 119-128.

[266] Georgiadis D, Beau F, Czarny B, Cotton J, Yiotakis A, Dive V. Roles of the two active sites of somatic angiotensin-converting enzyme in the cleavage of angiotensin-I and bradykinin, insights from selective inhibitors. Circulation Research 2003; 93, 148-154.

[267] Saito T. Antihypertensive peptides derived from bovine casein and whey proteins. In: Bösze, Z (ed.). Advances in experimental medicine and biology: bioactive components of milk, vol. 606. New York, USA: Springer 2008; 295-317.

[268] Shimizu M. Food-derived peptides and intestinal functions. Biofactors 2004; 21, 43-47.

[269] Yang CY, Dantzig AH, Pidgeon C. Intestinal peptide transport systems and oral drug availability. Pharmaceutical Research 1999; 16, 1331-1343.

[270] Gardner MLG. Intestinal assimilation of intact peptides and proteins from the diet – a neglected field. Biological Reviews of the Cambridge Philosophical Society 1984; 59, 289-331.

[271] Roberts PR, Burney JD, Black KW, Zaloga GP. Effects of chain length on absorption of biologically active peptides from the gastrointestinal tract. Digestion 1999; 60, 332-337.

[272] Roufik S, Gauthier SF, Turgeon SL. *In vitro* digestibility of bioactive peptides derived from bovine β-lactoglobulin. International Dairy Journal 2006; 16, 294-302.

[273] Kim YS, Bertwhistle W, Kim YW. Peptide hydrolyses in the brush borde rand soluble fractions of small intestinal mucosa of rat and man. Journal of Clinical Investigation 1972; 51, 1419-1430.

[274] Masuda O, Nakamura Y, Takano T. Antihypertensive peptides are present in aorta after oral administration of sour milk containing these peptides in spontaneously hypertensive rats. Journal of Nutrition 1996; 126, 3063-3068.

[275] Matsui T, Yukiyoshi A, Doi S, Sugimoto H, Yamada H, Matsumoto K. Gastrointestinal enzyme production of bioactive peptides from royal jelly protein and their antihypertensive ability in SHR. Journal of Nutritional Biochemistry 2002; 13, 80-86.

[276] Langley-Danysz P. Des hydrolysats protéiques pour développer des aliments santé. RIA Technology Veille 1998; 581, 38-40.

[277] Miguel M, Recio I, Ramos I, Delgado MA, Aleixandre MA. Antihypertensive effect of peptides obtained from *Enterococcus faecalis*-fermented milk in rats. Journal of Dairy Science 2006; 89, 3352-3359.

[278] Vermeirssen V, van Camp J, Decroos K, van Wijmelbeke L, Verstraete W. The impact of fermentation and *in vitro* digestion on the formation of angiotensin-I-converting enzyme inhibitory activity from pea and whey protein. Journal of Dairy Science 2003; 86, 429-438.

[279] Hernández-Ledesma B, Amigo L, Ramos M, Recio I. Angiotensin converting enzyme inhibitory activity in commercial fermented products. Formation of peptides under simulated gastrointestinal digestion. Journal of Agricultural and Food Chemistry, 2004; 52, 1504-1510.

[280] Hernández-Ledesma B, Amigo L, Ramos M, Recio I. Release of angiotensin converting enzyme-inhibitory peptides by simulated gastrointestinal digestion of infant formulas. International Dairy Journal 2004; 14, 889-898.

[281] Quirós A, Contreras MM, Ramos M, Amigo L, Recio I. Stability to gastrointestinal enzymes and structure-activity relationship of β-casein-peptides with antihypertensive properties. Peptides 2009; 30, 1848-1853.

[282] Maeno M, Yamamoto N, Takano T. Identification of an antihypertensive peptide from casein hydrolyzate produced by a proteinase from *Lactobacillus helveticus* CP790. Journal of Dairy Science 1996; 79, 1316-1321.

[283] Yamamoto N, Maeno M, Takano T. Purification and characterization of an antihypertensive peptide from a yogurt-like product fermented by *Lactobacillus helveticus* CPN4. Journal of Dairy Science 1999; 82, 1388-1393.

[284] Gómez-Ruiz JA, Ramos M, Recio I. Identification and formation of angiotensin-converting enzyme-inhibitory peptides in Manchego cheese by high-performance liquid chromatography-tandem mass spectrometry. Journal of Chromatography A, 2004; 1054, 269-277.

[285] Miguel M, Aleixandre MA, Ramos I, López-Fandiño R. Effect of simulated gastrointestinal digestion on the antihypertensive properties of ACE-inhibitory peptides derived from ovalbumin. Journal of Agricultural and Food Chemistry 2006; 54, 726-731.

Chapter 5

Functional Proteins and Peptides of Hen's Egg Origin

Adham M. Abdou, Mujo Kim and Kenji Sato

Additional information is available at the end of the chapter

http://dx.doi.org/10.5772/54003

1. Introduction

Hen's egg has long history as a food. It contains a great variety of nutrients to sustain both life and growth. Egg provides an excellent, inexpensive and low calorie source of high-quality proteins. Moreover, Eggs are a good source of several important nutrients including protein, total fat, monounsaturated fatty acids, polyunsaturated fatty acids, cholesterol, choline, folate, iron, calcium, phosphorus, selenium, zinc and vitamins A, B_2, B_6, B_{12}, D, E and K [1]. Eggs are also a good source of the antioxidant carotenoids, lutein and zeaxanthin [2]. The high nutritional properties of eggs make them ideal for many people with special dietary requirements.

Egg proteins are nutritionally complete with a good balance of essential amino acids which are needed for building and repairing the cells in muscles and other body tissues [3]. Egg proteins are distributed in all parts of the egg, but most of them are present in the egg white and egg yolk amounting to 50% and 40%, respectively. The remaining amount of protein is in the egg shell and egg shell membranes.

In addition to excellent nutritional value, egg proteins have unique biological activities. Hyperimmunized hens could provide a convenient and economic source of specific immunoglobulin in their yolks (IgY) that have been found to be effective in preventing many bacteria and viruses infections [4]. Proteins in the egg white as lysozyme, ovotransferrin, and avidin have proven to exert numerous biological activities. Moreover, a specific protein in eggshell matrix shows unique activity; enhancement of calcium transportation in the human intestinal epithelial cells.

It is well-known that egg proteins are a source of biologically active peptides. Many researches are aiming to unlock the hidden biological functions of peptides hidden in egg proteins. These peptides are inactive within the sequence of parent proteins and can be released during gastrointestinal digestion or food processing and exerting biological activities. Once bioactive peptides are liberated, they may act as regulatory compounds, and exhibit various activities such as anti-hypertensive, bone growth promoting, anticancer or exaggerated antimicrobial activities.

For development of bioactive peptides from parent proteins, following techniques have been conventionally used; the establishment of an assay system of biological activities, hydrolysis of proteins by digestive enzymes, isolation of peptides, determination of structures and synthesis of peptides. Recently, bioengineer technique; synthesis of peptides within egg proteins based on sequence similarities of peptides having known biological activity, has been used. The functional characteristics of either natural or modified egg proteins and the use of eggs components as "functional ingredients" are relatively new applications. A truly impressive volume of researches is now available for the egg industry to apply these new applications. Herein, some aspects concerning biologically functional egg proteins or peptides, biochemical and physiological properties as well as possible applications of egg proteins or peptides are discussed.

2. Egg yolk bio proteins

The major portion of egg yolk exists as lipoproteins, which can be separated by centrifugation into a plasma fraction (which remains soluble) and a granular fraction (which precipitates). Lipovitellenin, lipovitellin, phosvitin, livetin, yolk immunoglobulins (IgY), and some minor components have been isolated and identified in egg yolk.

2.1. Lipoproteins

Low density lipoprotein (LDL), which contains between 80 and 90% lipids, characterized by its emulsifying capacity. LDL is the major protein in egg yolk, accounting for 70% of yolk proteins. When LDL is treated with ether, residual fraction is referred to lipovitellenin containing 40% lipid. [5]. Shinohara et al. (1993) studied the effect of some constituents of egg yolk lipoprotein on the growth and IgM production of human-human hybridoma cells and other human-derived cells [6]. LDL-rich fractions were found to enhance the growth and IgM secretion of HB4C5 cells. The promoting activity was found in the commercial LDL.

High density lipoprotein (HDL) or lipovitellin comprise about one sixth of egg yolk solids in the granular yolk proteins. It has a molecular weight of $4X10^5$ and composed of 80% protein and 20% lipid. HDL exists as a complex with a phosphoprotein referred to phosvitin [5]. The addition of one or two eggs a day to a healthy person's diet does not adversely affect lipoprotein levels, and can actually increase plasma HDL levels [1].

2.2. Phosvitin

The name phosvitin comes from both its high phosphorus content (10%) and its source in the egg yolk. The emulsification properties of phosvitin, particularly emulsion-stabilizing activity, were found to be higher than those of other food proteins. Phosvitin is water insoluble but under low ionic strength and acidic conditions, it becomes soluble and can become complex with various metal ions (*e.g.* Ca^{++}, Mg^{++}, Mn^{++}, Co^{++}, Fe^{++} and Fe^{+++}) [5]. It could be used as a potent natural antioxidant on the basis of its potential to inhibit metal-catalyzed lipid oxidation [7]. The conjugation of egg yolk phosvitin with galactomannan produces a novel macromolecular antioxidant with significantly improved emulsifying activity and emulsion-stabilizing activity [8]. The antibacterial activity of phosvitin has been investigated against *Escherichia coli* and suggested that a significant part of the bactericidal activity of phosvitin could be attributed to the synergistic effects of the high metal-chelating ability and the high surface activity under the influence of thermal stress [9]. Phosvitin and the phosvitin-galactomannan conjugate may represent safe anti-bacterial agents for foods.

2.3. Vitamin-binding protein

A riboflavin-binding protein exists in the egg yolk. It is a hydrophilic phosphoglycoprotein with a molecular weight of $3.6x10^6$ Da, and it can conjugate one mole of riboflavin per mole of apoprotein. Also, biotin and cobalamin-binding proteins could be found in egg yolk.

2.4. Livetin

Livetin is a water-soluble, non-lipid, globular glycoprotein, which is immunologically analogous to the plasma proteins of mammals. α-Livetin is analogous to serum albumin, β-livetin to 2-glycoprotein and γ-livetin to γ-globulin [10]. Most of the research effort has been focused on the immune proteins found in the egg yolk (IgY). Recent advances in IgY technology will be discussed.

2.5. Egg yolk immunoglobulin (IgY)

Immunoglobulin from yolk (IgY) is the major antibody found in hen eggs. In 1893, Klemperer first described the acquisition of passive immunity in birds, by demonstrating the transfer of immunity against tetanus toxin from the hen to the chick [11]. Three immunoglobulin classes analogues to the mammalian immunoglobulin classes; IgA, IgM, and IgG, have been shown to exist in chicken. In the egg, IgA and IgM are present in the egg white, while IgG is present in the egg yolk [12]. IgG in egg yolk has been referred to as IgY to distinguish it from its mammalian counterpart [13].

The concentration of IgY in the yolk is essentially constant (10-20 mg/mL) through the oocyte maturation. Approximately 100-400 mg IgY is packed in an egg. The concentration of IgY in the yolk is 1.23 times to the serum concentration [14]. A delay of 3 to 4 days is observed for the appearance of specific IgY in yolk after first appearance of specific IgG in the serum of a hen.

2.5.1. Structure and Characteristics of Avian IgY Versus Mammalian IgG

Composition differences. General structure of IgY molecule is the same as mammalian IgG with 2 heavy (Hv) chains with a molecular mass of 67–70 kDa each and two light (L) chains with the molecular mass of 25 kDa each (Figure 1). The major difference is the number of constant regions (C) in H chains: IgG has 3 C regions (Cy1–Cy3), while IgY has 4 C regions (Cv1–Cv4). Due to occurrence of one additional C region with two corresponding carbohydrate chains, molecular mass of IgY (180 kDa) is larger than *mammalian* IgG (150 kDa). IgY is less flexible than mammalian IgG due to the absence of the hinge between Cy1 and Cy2. There are some regions in IgY (near the boundaries of Cv1–Cv2 and Cv2–Cv3) containing proline and glycine residues enabling only limited flexibility. IgY has isoelectric point 5.7–7.6 and is more hydrophobic than IgG [13, 15].

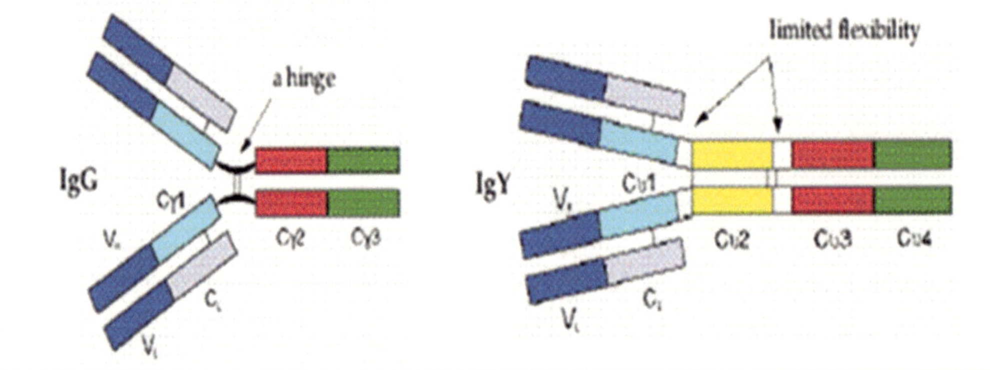

Figure 1. Structure of IgG and IgY.

Advantage of IgY. Most biological effectors' functions of immunoglobulin are activated by the Fc region, where the major structural difference between IgG and IgY is located. Therefore, Fc-dependent functions of IgY are essentially different from those of mammalian IgG. First, IgY does not activate the complement system[16], second, IgY does not bind to protein-A and G [17], third, IgY is not recognized by mammalian antibodies [18] *i.e.* rheumatoid factors (RF, an autoantibody reacting with the Fc portion of IgG) or HAMA (human anti-murine antibodies), and fourth, it does not bind to cell surface Fc receptor [19]. These differences in molecular interactions bring great advantages to the application of IgY antibodies. Then that were IgY has been successfully applied into a variety of methods in different areas of research, diagnostics, and medical areas. For these applications, IgY can successfully compete with antibodies (IgG) isolated from the blood of mammals [20]. The advantage of usage of the immunized hen is that it produces a large number of eggs. Approximately 40g of IgY could be collected from egg of one hen each year, compared with about 1.3g from blood of one rabbit [21]. An industrial scale production of IgY is possible because of the availability of large number of chicken farms and automation of egg breaking and processing.

2.5.2. Applications and uses of IgY

Oral administration of antibodies specific to host pathogens is an attractive approach to establish protective immunity, especially against gastrointestinal pathogens both in human and animals. Eggs are normal dietary components and there is practically no risk of toxic side effects of IgY given orally. As mentioned above, IgY does not activate mammalian complement system nor interact with mammalian Fc-receptors that could mediate inflammatory response in the gastrointestinal tract.

On the basis of these facts, IgY has been used for suppression of growth of food-bone pathogens [11]. Whole egg yolks and water soluble fractions were prepared from the egg of the hens that had been immunized with pathogens such as *E. coli* O157:H7, *Salmonella enteritidis, Salmonella typhimurium, Campylobacter jejuni, Staphylococcus aureus,* and *Listeria monocytogenes*. It has been demonstrated that pathogen-specific IgY is bound to the surface of bacteria, resulting in structural alterations of cell wall and consequently kills bacteria. Sarker et al. (2001) performed a study for children with proven rotavirus diarrhea. The patients were treated with IgY from the eggs of chickens immunized with human rotavirus strains [22]. The treatment moderated diarrhea, which was characterized by an earlier clearance of rotavirus from the stools. Recently, IgY has been applied to cancer therapy. Hens were immunized with an antigen purified from human stomach cancer cells. The purified IgY recognized gastrointestinal cancer cells. Conjugation of antibodies and drugs may be an important agent for cancer treatment [23].

These IgY treatments have been shown to provide a safer, more efficient and less expensive method than those using conventional mammalian antibiotics for managing disease-causing pathogens. Recently, successful progresses in industrialization of IgY has been achieved in Japan, where IgY as a bioactive ingredient in food, nutraceuticals, cosmetics and other sectors is applied . The followings are the most recent applications.

2.5.2.1. Anti-Helicobacter pylori IgY

Helicobacter pylori, a spiral gram-negative microaerophilic pathogen, has been shown to be a common inhabitant of the gastric and duodenal mucosa. The microorganism is recognized as one of the most prevalent human pathogens. It infects over 50% of the population worldwide [24], and is recognized as the etiologic agent of gastritis, peptic ulcer, and has been linked to the development of gastric adenocarcinoma and mucosa associated lymphoid tissue lymphoma [25, 26]. The eradication of *H. pylori* by administration of oral antimicrobials is not always successful and may be associated with adverse effects [27]. For colonization in gastric mucosa, *H. pylori* abundantly produces urease enzyme, which degrades urea into ammonia. *Helicobacter pylori* organism uses the ammonia to neutralize microenvironment in gastric mucosa. Accordingly, a novel approach in prevention and reduction of *H. pylori* infection using urease-specific IgY has been developed. It has been reported that oral administration of anti-*H. pylori* urease IgY (IgY-urease) could suppress the bacterial colonization.

Preparation of IgY-Urease Yogurt. Today, consumers prefer foods that promote good health and could reduce risk of diseases. Dairy products are excellent media to generate an array of products that fit into the current consumer demand for functional foods [28]. Scientific and clinical evidence is mounting to corroborate the consumer perception of health from yogurt. Designing a yogurt fortified with IgY-urease could supply passive immunization with a natural and highly specific attempt to decrease the *H. pylori* infection. In order to suppress *H. pylori* infection, a yogurt fortified with IgY-urease has been designed and developed. Three clinical studies were done to examine the efficacy of a specially designed functional yogurt containing IgY-urease on the suppression of *H. pylori* in humans. IgY-urease containing yogurt (plain and drinking) have been prepared and in markets in Japan, Korea, and Taiwan (Figure 2). IgY-urease was pasteurized and then added to yogurt mix at specific dose after all heat-treatment steps. Yogurts were cooled and stored at 4° C for up to 3 weeks. IgY-urease activity remained in the product throughout the 3 weeks of storage.

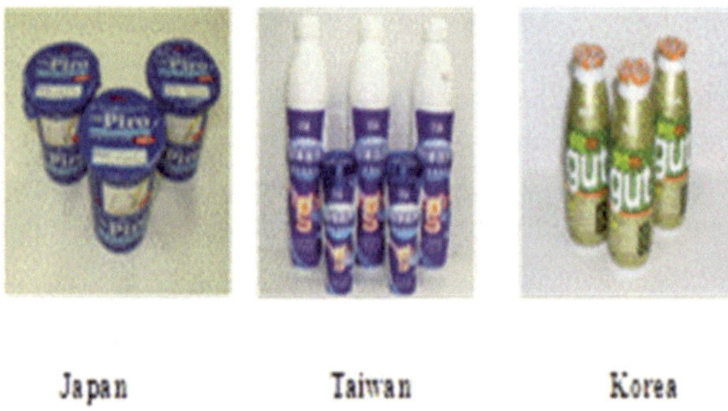

Figure 2. Different IgY-urease yogurt products in Japan, Korea, and Taiwan.

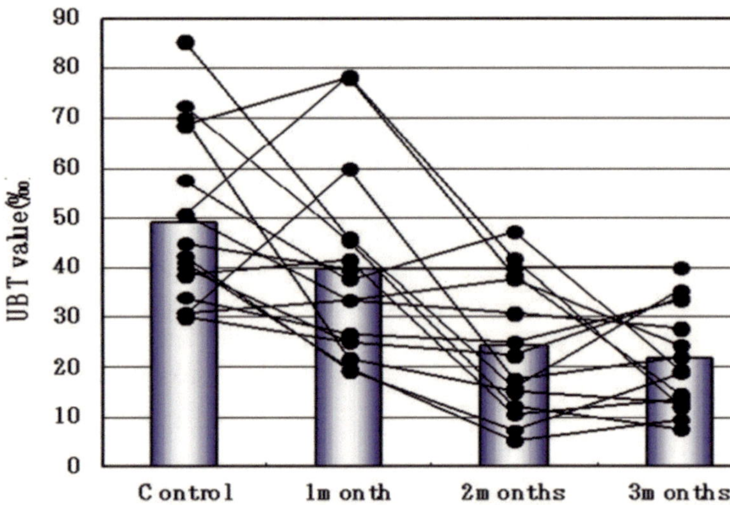

Figure 3. UBT Value Change of volunteers (clinical study in Japan).

Clinical studies. Plain yogurt containing 2g IgY-urease egg yolk was produced commercially in Japan. A clinical study was conducted to determine the effect of IgY-urease yogurt to decrease *H. pylori* in humans [29]. To assess presence of *H. pylori* in stomach, UBT method has been extensively used. This method based on the invasive detection of exhaled ^{13}C-labeled carbon dioxide resulting from *H. pylori* urease activity [30]. One hundred seventy-four volunteers were screened using a ^{13}C-urea breath test (UBT). Heavily infected volunteers (with UBT values over 30‰) were selected (16 subjects) and recruited. Each volunteer consumed 1 cups of yogurt twice daily (4 g/d egg yolk containing 40 mg IgY-urease) for 12 wk. Volunteers were tested after 4, 8 and 12 weeks. The UBT values obtained at week 8 and 12 were significantly different from those obtained at week 0 ($P < 0.001$), showing a 55.1% and 57.2% reduction in UBT values after 8 and 12 weeks, respectively (Figure 4). Other clinical studies using IgY-urease containing drinking yogurt were carried out in Taiwan [31] and Korea [32] showed nearly similar results.

The three different studies demonstrated that administration of a specially designed yogurt with highly specific antibodies from egg yolk could effectively decreases number of *H. pylori* in humans. During the study period of the three clinical studies, the ingestion regimen was well-tolerated and no adverse effects or any complications were observed.

The use of probiotics for the suppression of *H. pylori* in humans has been studied by some investigators [33, 34]. However, none of these studies were able to show a significant suppression of *H. pylori* in humans, and others showed a slight but no significant trend toward a suppressive effect of drinking yogurt containing specific lactic acid bacteria. Anti-*H. pylori* effect of yogurt containing specific lactic acid bacteria has been examined. However, no significant reduction of *H. pylori* in human stomach has been observed, although trend for decrease was observed [35, 36].

The use of IgY against a pathogenic factor of *H. pylori* would be a prudent way to suppress the infection. It was demonstrated that IgY-urease was highly specific and had a significant effectiveness against *H. pylori* because of its ability to inhibit *H. pylori* from adhering to the gastric mucosa [4, 37, 38]. Because IgY-urease binds urease only, the functional efficacy observed was presumably via capture of bacterium-associated urease within the gastric mucus layer, which resulting in bacterial aggregation and clearance via the constant washing action of the gut. By such a mechanism, consumption of IgY-urease yogurt may play a dual role in suppression and prophylaxis against *H. pylori* in humans (Figure 4).

These findings opened new gate of applications of IgY-urease against *H. pylori* in the food industry to prevent *H. pylori*. Recently, a specially designed egg containing IgY-urease was produced in Japan. This egg has been on the Japanese market under a trade name of "stomach friendly egg". Moreover, IgY-urease was incorporated in neutraceutical formulations that launched recently in the Japanese market aiming to prevent and reduce *H. pylori* infection.

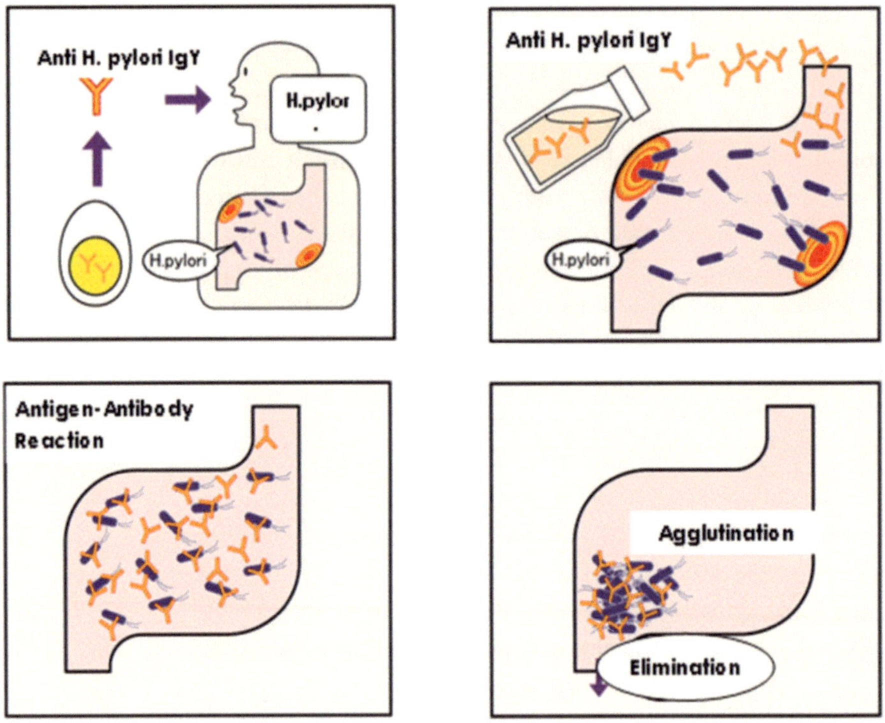

Figure 4. Suppressive mechanism of anti-*H. pylori* urease IgY.

2.5.2.2. Anti-Streptococcus mutans IgY

Dental caries is still one of the most widespread diseases of mankind. Human are frequently infected with cariogenic microorganisms in early life. The cariogenic microorganisms survive in dental biofilm and can emerge under favorable environmental condition and consequently cause dental disease [39]. *Streptococcus mutans* is the main etiologic agent of dental caries and that infection is transmissible [40]. Abilities of mutants' streptococci to adhere tooth surface in the presence of sucrose and release acids by fermention play a significant role in development of dental caries [41]. Initial attachment of *S. mutans* to the saliva-coated enamel surface occurs through the surface protein of *S. mutans*. For the colonization of *S. mutans*, synthesis of water-insoluble and adherent glucan from sucrose by the glucosyl transferases (GTases) is essential. *Streptococcus mutans* produces both cell-associated (CA) and cell-free (CF) forms of GTase; the former primarily synthesizes water-insoluble glucan, while the latter produces water soluble glucan. The combined action of these two GTases on the cell surface of *S. mutans* during its growth in the presence of sucrose is critically important in allowing firm adherence. The GTase system of *S. mutans* has therefore been considered an important virulence factor promoting caries development. Administration of IgY against *S. mutans* CA-GTase specifically inhibited insoluble glucan-synthesizing CA-GTase, resulting in a significant reduction in the development of dental caries. Otake et al. (1991) reported that anti-*S. mutans* CA-GTase IgY suppressed development of dental carries in rat model [42]. Hatta et al. (1997) reported that the effectiveness of IgY with specificity to *S. mutans* prevented the colonization of mutans streptococci in the oral cavity of humans [43]. Recently, food products such as candies, chocolates and gums containing fourth- or anti-*S. mutans* IgY have been launched in the Japanese market for oral care [37].

2.5.2.3. Anti-Influenza virus IgY (Anti-influenza biofilter)

Influenza caused by a virus is called the influenza virus. Influenza or "flu" is an infection of the respiratory tract that can affect millions of people every year. It is highly contagious and occurs mainly in the late fall, winter, or early spring. Influenza is spread from person-to-person through mists or sprays of infectious respiratory secretions caused by coughing and sneezing. Influenza affects all age groups and causes severe illness, loss of school and work, and complications such as pneumonia, hospitalization, and death.

Recently, specific anti-influenza IgY was successfully produced from hens immunized with inactivated influenza virus strain. This specific IgY significantly reacts virus in vitro [44]. Subsequently, an anti-influenza IgY biofilter which trap the influenza virus has been developed by Daikin Environment Laboratories, Japan, a research arm of Daikin Industries, in cooperation with five Japanese research institutions. It was found that 99.99% of the influenza virus sprayed over the biofilter was captured within 10 minutes. An air cleaner with anti-influenza filter is recently launched in Japan (Figure 5 and 6). Moreover, a facemask with anti-influenza IgY was developed and will be available in the Japanese market.

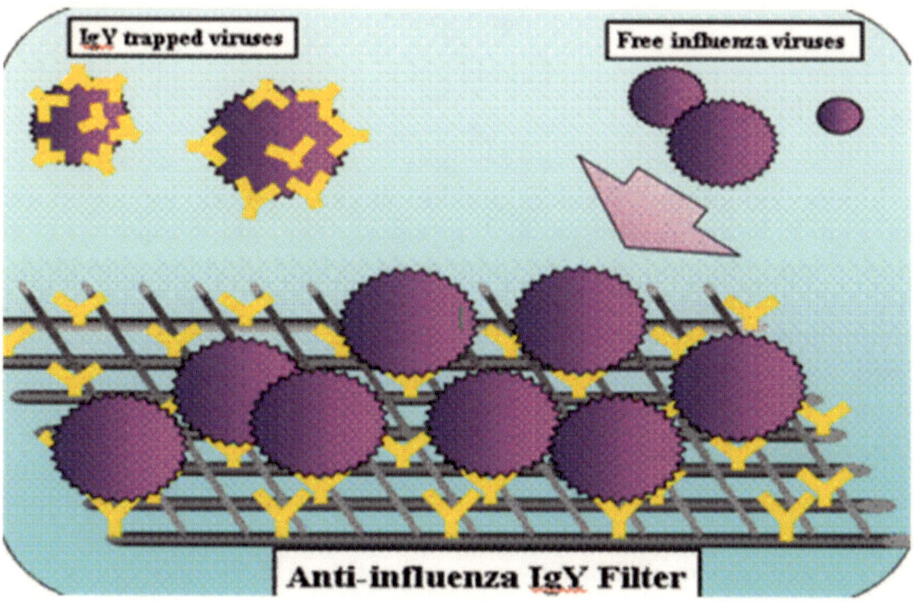

Figure 5. Diagram for the Anti-influenza IgY biofilter.

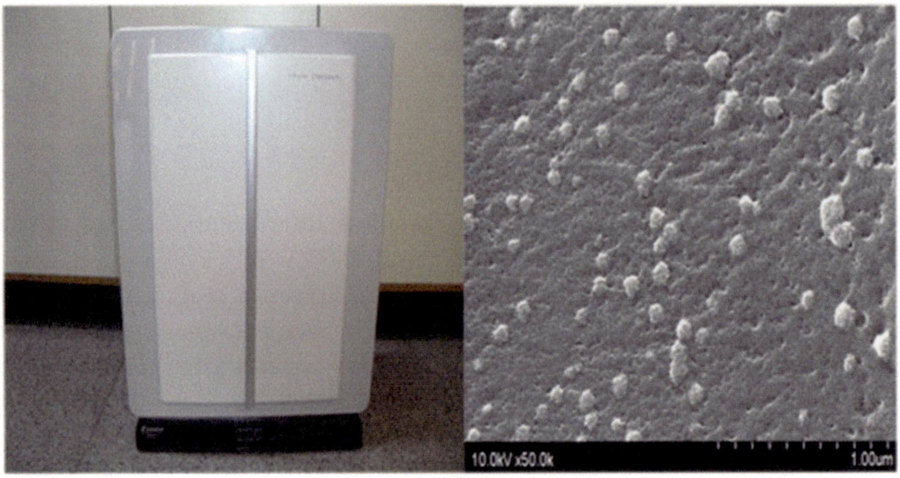

Figure 6. An air cleaner with anti-Influenza virus biofilter available in Japanese market.

2.5.2.4. *Future prospects of IgY applications*

Many research activities and proposals are going on in order to provide new applications for IgY technology. An anti-*Bacteriodes gengivalis* is under development for improving the oral health. For cosmetics sector, IgY against *Propionibacterium acnes* and its lipases has been developed to prevent and treat acne that is the most common skin disease [45]. For the medical sector, many researchers are working on using transgenic chicken to produce human antibodies in the transgenic hen's eggs in the form of IgY to help for treating human diseases.

3. Egg yolk bio peptides

Recently, it has become clear that proteins are a source for biologically active peptides. These peptides are inactive within the sequence of parent protein and can be released during gastrointestinal digestion or food processing. Egg yolk proteins could be an important source of bioactive peptides. The resultant peptides could show biologically new function with improved stability and/or solubility. In this section, some biologically functional egg yolk derived peptides are introduced and their underlying mechanisms are discussed.

3.1. Lipovitellenin peptide

Vitellenin is the apoprotein of lipovitellenin. Digestion of vitellenin with pronase gives two glycopeptides. Glycopeptide A has high content of sialic acid and glycopeptide B contains most of the carbohydrates of vitellenin but devoid of sialic acid (N-Acetylneuraminic acid). Sialic acid is naturally occurring carbohydrate with numerous biological functions, including blood protein half-life regulation, variety of toxin neutralization, regulation of cellular adhesion and glycoprotein lytic protection.

3.2. Phosvitin peptide

Phosvitin phosphopeptides are new functional bioactive peptides derived from egg yolk with molecular masses of 1-3 kDa prepared from tryptic hydrolysate by partial dephosphorylation [46]. The phosvitin phosphopeptides were shown to be effective for enhancing the calcium binding capacity and inhibiting the formation of insoluble calcium phosphate. The results suggest a potential application of phosvitin peptides as novel functional peptides for the prevention of osteoporosis.

3.3. Egg yolk bone peptides (Bonepep®)

Hen egg turns into a full skeleton chick within 3 weeks; based on this fact, some investigations were carried out to explore the biologically active substances in hen egg that would initiate and enhance bone growth. It has been found that a specific yolk water-soluble protein (YSP) has a bone growth promotion activity *in vitro* [47] and *in vivo* [48]. These findings encouraged to search the functional yolk peptides that would promote bone growth. Different enzymes were used to hydrolyze YSP and the effect of the peptide preparations on osteoblast MC3T3-E1 cell proliferation was investigated. A novel peptides preparation (Bonepep®) with bone growth promotion activity was obtained. An *in vitro* study showed that Bonepep enhances the osteoblast MC3T3-E1 cell proliferation (Figure 7). Furthermore, an *in vivo* study showed that it promotes the elongation of rat tibia bone [49]. Compared with control and YSP fed rats; rats fed on Bonepep® showed marked and significant increase of chondrocytes proliferation and the bone formation in the tibia and consequently significantly increased elongation of their bone.

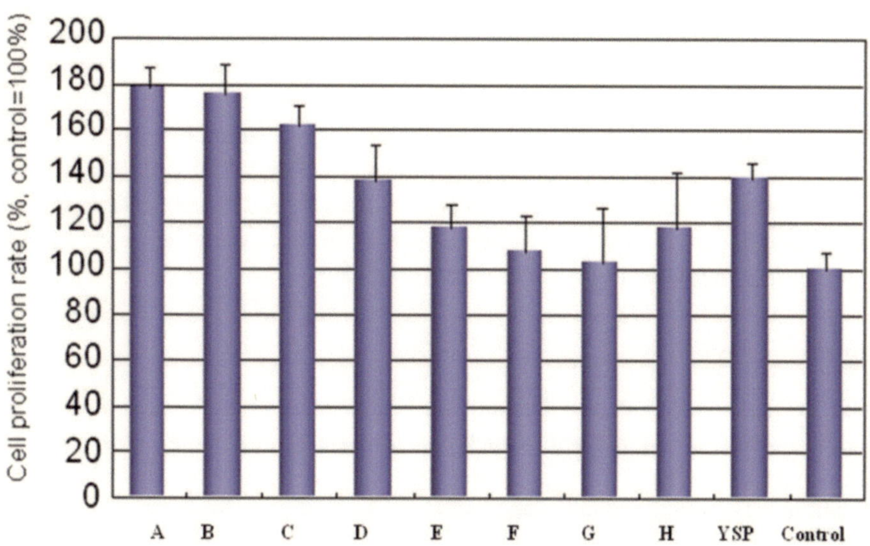

(A: Bonepep, B~H: peptides by other enzymes, YSP: Yolk Soluble Protein).

Figure 7. Promotion of osteoblast MC3T3-E1 cell proliferation by Bonepep®.

4. Egg white bio proteins

Well-known biological functions of egg white proteins are the prevention of microorganisms' penetration into the yolk and supply of nutrients to the embryo during the late stages of development. Most of the egg white proteins appear to possess antimicrobial properties or certain physiological functions to interfere with the growth and spread of invading microorganisms.

Most of egg white proteins are soluble and can easily be isolated. Egg white contains approximately 40 different proteins. Egg white proteins possess unique functional properties, such as antimicrobial, enzymatic and anti-enzymatic, cell growth stimulatory, metal binding, vitamin binding, and immunological activities [50].

4.1. Ovalbumin

Ovalbumin is a predominant protein contributing to the functional properties of egg white [51]. Ovalbumin is a monomeric phosphoglycoprotein with a molecular weight of 44.5 kDa and an isoelectric point of 4.5. Ovalbumin is a key reference protein in biochemistry. As a carrier, stabilizer, blocking agent or standard, highly purified ovalbumin has served the fundamentalists as well as the food industry. It has long been the subject of physical and chemical studies as a convenient protein model.

It is believed that ovalbumin, especially its unphosphorylated form, serves as a source of amino acids for the developing embryo. Despite the intensive investigations undertaken on ovalbumin, its function remains largely unknown. Ovalbumin is the only egg white protein which contains free sulfhydryl groups. The complete amino acid sequence of hen ovalbumin (which comprises 385 residues) and its crystal structure have been reported [52]. The unexpected finding that this protein belongs to the serpin superfamily has stimulated new interest in the structure and function of ovalbumin. The serpins are a family of more than 300 homologous proteins with diverse functions found in animals, plants, insects and viruses, but not in prokaryotes [53]. They include the major serine protease inhibitors of human plasma that control enzymes of the coagulation, fibrinolytic, complement and kinin cascades, as well as proteins without any known inhibitory properties such as hormone binding globulins, angiotensinogen and ovalbumin [54].

4.2. Ovotransferrin

Ovotransferrin (also known as conalbumin) has been identified as the iron-binding protein from avian egg white. Ovotransferrin, which constitutes 12% of the egg white protein, has a molecular weight of 77.7 kDa and a pI of about 6.1. It contains 686 amino acid residues and has 15 disulfide bridges [55]. It is glycosylated and contains a single glycan chain (composed of mannose and N-acetylglucosamine residues) in the C-terminal domain. Ovotransferrin is a neutral glycoprotein synthesized in the hen oviduct and deposited in the albumen fraction of eggs. Furthermore, it has two similar domains in N and C terminal regions, each one binding one atom of transition metal (Fe^{+++}, Cu^{++}, Al^{+++}) very tightly and specifically [55]. It is implicated in the transport of iron in a soluble form to the target cells. The recognition of transferrin molecules by the target cells is mediated by membrane-bound transferrin receptors [56]. The significant structural similarities between lactoferrin and ovotransferrin justify the similarity of their biological roles. Ovotransferrin can be used as a nutritional ingredient in iron fortified products such as iron supplements, iron-fortified mixes for instant drinks, sport bars, protein supplements and iron-fortified beverages.

There is also extensive evidence of an antibacterial effect of ovotransferrin based on iron deprivation, iron being an essential growth factor for most microorganisms. The high affinity of transferrins for iron means that, in the presence of unsaturated transferrin (apotransferrin), iron will be sequestered and rendered unavailable for the growth of microorganisms. *In vivo,* ovotransferrin has been shown to have therapeutic properties against acute enteritis in infants [57].

4.3. Lysozyme

The name lysozyme was originally used to describe an enzyme which had lytic action against bacterial cells. Lysozyme is one of the oldest egg components to be utilized commercially after it was discovered by Alexander Fleming in 1922. It is a bacteriolytic enzyme commonly found in nature and is present in almost all secreted body fluids and tissues of humans and animals. It has also been isolated from some plants, bacteria and bacteriophages. Avian egg white is a rich and easily available source of lysozyme.

The lysozyme content of a laying hen's blood is 10-fold higher than in mammals because it is being transferred to the egg white. Lysozyme constitutes approximately 3.5% of hen egg white [50]. Egg white lysozyme consists of 129 amino acid residues with a molecular weight of 14.4 kDa. Because of its basic character, lysozyme binds to ovomucin, transferrin or ovalbumin in egg white [58]. In nature, lysozyme is found mainly as a monomer but it has been reported to also exist as a reversible dimer, which can be evoked by pH, concentration and/or temperature-dependent phase transition of the molecule.

It has long been believed that lysozyme's antimicrobial action could only be attributed to its catalytic effect on certain Gram-positive bacteria, by splitting the bond between N-acetylmuramic acid and N-acetyl-glucosamine of peptidoglycan in the bacterial cell wall [59]. Beside this well-known inactivation mechanism, a non enzymatic antibacterial mode of action of lysozyme was achieved by denatured form of lysozyme without enzymatic action.

Lysozyme demonstrates antimicrobial activity against a limited spectrum of bacteria and fungi [60]. However, the antimicrobial activity of lysozyme is greater for certain Gram-positive bacteria. On the other hand, Gram-negative bacteria are less susceptible to the bacteriolytic action of the enzyme [61]. The cell walls of different bacteria show varying degrees of susceptibility to digestion with hen egg white lysozyme. The walls of *Micrococcus lysodeikticus* were the most sensitive and the walls of *Staphylococci* were the less sensitive to the bacteriolytic action of lysozyme. Among Gram-negative bacteria, the walls of *Salmonella* and *Shigella* were the most sensitive whereas those of *E.coli*, *Vibrio* and *Proteus* were much less sensitive [59]. The susceptibility differences are believed to be due to the complex envelope structure of Gram-negative bacteria such as *E.coli* or *Salmonella typhimurium*. The outer membrane serves to reduce the access of lysozyme to its site of action (peptidoglycan layer).

4.3.1. Molecular modification for functional improvement

Lipophilization. A number of chemical modifications of lysozyme have been undertaken to increase its efficacy as an antimicrobial agent. The effect of lipophilization with long chain fatty acids (palmitic or stearic acid) and shorter chain saturated fatty acids (caproic, capric or myristic acid) on the bactericidal action of lysozyme was investigated [62]. Lipophilization broadened the bactericidal action of lysozyme to Gram-negative bacteria with little loss of enzymatic activity [63].

Glycosylation. It is one of the most promising techniques, involves the attachment of carbohydrate chains to lysozyme. Glycosylation produces more stable proteins with improved conformational stability, protease resistance, modulated charge effects and water-binding capacity [64]. Conjugation of lysozyme with dextran by Maillard reaction increases antimicrobial activity. In addition, emulsifying activity of the conjugate was approximately 30 times that of native lysozyme [65]. Extending the function of lysozyme by conjugation with food compounds gives a novel and potentially useful bi-functional food additive. Similarly, hen egg lysozyme conjugated with xyloglucan hydrolysates; totally conserved enzymatic activity of lysozyme and increased the emulsifying properties 5 times higher than that of the native protein [66]. An antibacterial emulsifier was prepared by conjugating a fatty acylated saccharide with lysozyme through the Maillard reaction; the conjugate exhibited considerable resistance to proteolysis and much enhanced emulsifying activity and emulsion stability. The conjugate maintained approximately 70% of the bactericidal activity of native hen egg lysozyme without significant conformational changes of the protein [67].

Combination of lipophilization and glycosylation. An egg white lysozyme, which had been modified using the Maillard-type glycosylation method prior to lipophilization with palmitic acid, was prepared [68]. The yield of lipophilized lysozyme was increased significantly by pre-glycosylation of the protein and showed strong antimicrobial activity against *Escherichia coli*. Lipophilization of lysozyme combined with glycosylation is a promising method for potential industrial applications of lysozyme due to its enhanced antimicrobial activity towards Gram-negative bacteria and improved yield.

4.3.2. Applications

The bacteriostatic and bactericidal properties of lysozyme have been used to preserve various food items, as well as in pharmacy, medicine and veterinary medicine.

Natural food preservative. Lysozyme has been used as an antimicrobial agent in various foods. In 1992, the Joint FAO/WHO Expert Committee on Food Additives declared that lysozyme was safe to be used in food [69]. The enzyme shows a number of properties important for food application. It is a heat stable protein, active in a broad range of temperatures (from 1°C to nearly 100°C), withstands boiling for 1-2 min, and stable in freeze-drying and thermal drying. Moreover, lysozyme is not inactivated by solvents and it maintains its activity when re-dissolved in water. It has optimum activity at pH 5.3 to 6.4 (*i.e.* typical for low-acidic food).

In cheese making, lysozyme has been used to prevent growth of *Clostridium tyrobutyricum,* which causes off-flavors and late blowing in some cheeses [70]. Another application of lysozyme may be the possible acceleration of cheese ripening, because lysis of starter bacteria would cause release of cytoplasmic enzymes which play a key role in proteolysis during cheese ripening [71]. Moreover, egg white lysozyme was used as an antimicrobial agent to control lactic acid bacteria in some fermented beverages [72].

Pharmaceuticals. In the pharmaceutical industry, avian egg white lysozyme can protect the body against bacterial, viral or inflammatory diseases [73]. It has been used in aerosols for the treatment of broncho pulmonary diseases, prophylactically for dental caries, for nasal tissue protection and is incorporated into various therapeutic creams for the protection and topical preparation of certain dystrophic and inflammatory lesions of the skin and soft tissues (*e.g.* burns and viral diseases).

Regardless of the direct bacteriolytic action, many other biological functions of lysozyme have recently been reported. These include anti-viral action by forming an insoluble complex with acidic viruses, enhanced antibiotic effects, anti-inflammatory and anti-histaminic actions, direct activation of immune cells and anti-tumor action [74-77].

4.4. Ovomucoid

Ovomucoid is a glycoprotein with heat stable trypsin inhibitor activity. Ovomucoid, which constitutes about 11% of the egg white protein, has a molecular weight of approximately 28 kDa and a pI of 4.1. It has nine disulfides and no free sulfydryl groups. The molecule consists of three tandem domains, each of which is homologous to pancreatic secretory trypsin inhibitor (Kazal-type). It has a putative reactive site for the inhibition of serine proteases. A large proportion of the carbohydrate present in this glycoprotein (about 25%) is joined to the polypeptide chain through asparginyl residues [78]. Ovomucoid can be heated at 100° C under acidic conditions for long periods without any apparent changes in its physical or chemical properties. Ovomucoid may play a more important role in the pathogenesis of allergic reactions to egg white than other egg white proteins [79].

4.5. Ovomucin

Ovomucin comprises 1.5-3.5% of the total egg white solids. It is a highly viscous glycoprotein with an extremely large molecular weight (8300-23000 kDa). The specific jellying property of egg white is attributed to ovomucin. It consists of two subunits; α-subunit and a β-subunit which are bound by disulfide bonds. The biological function of ovomucin is shown to inhibition of haemagglutination by viruses. Its affinity with viruses such as bovine rotavirus, hen Newcastle disease virus and human influenza virus was already proved [80]. Moreover, the β-subunit from ovomucin was shown to have a cytotoxic effect on the cultured tumor cells [81].

4.6. Avidin

Avidin is a strongly basic glycoprotein synthesized in the hen oviduct and deposited in the albumen fraction of eggs. Avidin is a tetrameric protein, composed of subunits of identical amino acid composition and sequence (15.6 kDa and 128 amino acids each). Avidin is a trace component (0.05%) of egg white, but it has been well studied because of its ability to tightly and specifically bind biotin, one of Vitamin B group. Each subunit of avidin binds to a molecule of biotin. The high affinity of avidin for biotin has been widely used as a biochemical tool in molecular biology, affinity chromatography, molecular recognition and labeling, Enzyme Linked Immuno Sorbent Assay (ELISA), histochemistry and cytochemistry [82].

4.7. Ovoglobulin

In early studies, six globulin fractions were thought to be present in egg white. They are macroglobulin, ovoglobulins G1, G2 and G3 and two other globulins. However, the two globulins were later classified as ovoinhibitors and ovoglobulin G1 was identified as lysozyme. Currently, the name ovoglobulin is given only to ovoglobulins G2 and G3, which have molecular weights of 36 and 45 kDa, respectively. The biological function of these proteins has not been clearly elucidated, but they appear to be important in the foaming capacity of egg white [5].

4.8. Ovomacroglobulin

Ovomacroglobulin is the second largest egg glycoprotein after ovomucin and its molecular weight is 760-900 kDa. Ovomacroglobulin, like ovomucin, has the ability to inhibit hemagglutination [5].

4.9. Ovoflavoprotein

Ovoflavoprotein is acidic protein with a molecular weight of 32-36 kDa, and contains a carbohydrate moiety (14%) made up of mannose, galactose and glucosamines, 7-8 phosphate groups and 8 disulfide bonds. After being transported from the blood to the egg white, most of the riboflavin (Vitamin B2) is stored in the egg white bound to an apoprotein called flavoprotein. One mole of apoprotein binds one mole of riboflavin, but this binding ability is lost when the protein is exposed to a pH below its isoelectric pH 4.2 [5]. It has antimicrobial properties due to depriving the microorganisms from its riboflavin content [50].

4.10. Ovoinhibitor

This trypsin inhibitor was discovered by Matsushima in 1958. While it is a Kazal-type inhibitor (like ovomucoid), ovoinhibitor functions as a multi-headed inhibitor and inhibits bacterial serine proteinase, fungal serine proteinase and mammalian chymotrypsin [5].

4.11. Cystatin

It is the third proteinase inhibitor in egg white (also called ficin-papain inhibitor). In contrast to ovomucin, cystatin is a small molecule (12.7 kDa) and it has no carbohydrates and a high thermal stability. The potential of their broad application in medical treatments has been reported in the literature, which includes antimicrobial and antiviral activities [83], the prevention of cerebral hemorrhage [84] and control of cancer cell metastasis [85].

4.12. Ovoglycoprotein

Ovoglycoprotein is an acidic glycoprotein with a molecular weight of 24.4 kDa. This protein contains hexoses 13.6%, glucosamine 13.8%, and N-acetylneuraminic acid 3%. The biological functions of ovoglycoprotein are still unclear [5].

5. Egg white bio peptides

5.1. Ovoalbumin peptide (Ovokinin)

Ovokinin, a vasorelaxing octapeptide derived from pepsin digest of ovalbumin, has been shown to significantly lower the systolic blood pressure of spontaneously hypertensive rats [86]. Oral vailability of ovokinin is improved after emulsification. When we eat whole egg, ovalbumin peptides will be released by the action of pepsin in the stomach, and the peptide will be instantly emulsified with egg yolk and effectively absorbed from the intestines.

A vasorelaxing peptide – ovokinin (2-7) – was isolated from chymotryptic digest of ovalbumin [87]. However, the mechanisms for the relaxation were different from ovokinin. More anti-hypertensive peptide was obtained by modifying the amino acid residues of ovokinin (2-7). The minimum effective dose of [Pro2, Phe3]-ovokinin (2-7) was about one-thirtieth of that of ovokinin (2-7). [Pro2, Phe3]-ovokinin (2-7) proved to be a potent anti-hypertensive peptide with little effect on normal blood pressure when administered orally [88].

5.2. Ovotransferrin peptides

The ovotransferrin antimicrobial peptide (OTAP-92) is a 92 amino acid cationic fragment of hen ovotransferrin located (109-299) in the N-lobe of ovotransferrin. The peptide OTAP-92 showed strong bactericidal activity against both Gram-positive *S.aureus* and Gram-negative *E.coli* strains [89]. OTAP-92 has also been shown to possess a unique structural motif similar to the insect defensins. Furthermore, this cationic antimicrobial peptide is capable of killing

Gram-negative bacteria by crossing the outer membrane by a self-promoted uptake pathway and damaging the cytoplasmic membrane by channel formation. Knowledge of the structure-function relationship may allow combinations of antimicrobial agents with different mechanisms to be designed for pharmaceutical applications. OTAP-92 may represent a novel antimicrobial agent for the food and pharmaceutical industries [55].

5.3. Ovomucoid and ovomucin peptides

A pepsin-digest of egg white ovomucoid was prepared to enhance digestibility and lower allergenicity [90]. A highly glycosylated peptide fragments (220 and 120 kDa) was separated from pronase digest of avian egg white ovomucin; It was derived from the β-subunit. Both fragments inhibited the growth of tumors [91].

5.4. Egg white peptides (Runpep®)

Egg white hydrolysate (Runpep®) has been produced by standardized technology to provide a highly nutritive source of peptides. Runpep® contains all essential amino acids with amino acid score of 100 [5, 50]. It is rich in branched chain amino acids (BCAA); these are the essential amino acids leucine, isoleucine, and valine. BCAA's are of special importance for athletes because they are metabolized in the muscle, rather than in the liver. Moreover, it is a rich source of sulpher containing amino acids as methionine and cysteine. It is in the form of small peptides (less than 3000 Da) and contains more than 91% protein providing easily absorbable amino acids source (Figure 8). An *in vitro* study showed that Runpep® has anticoagulant activity (reduce the formation of coagulum in blood vessels, reduce the risk of embolism) and helps in lowering the blood platelets aggregations [92, 93].

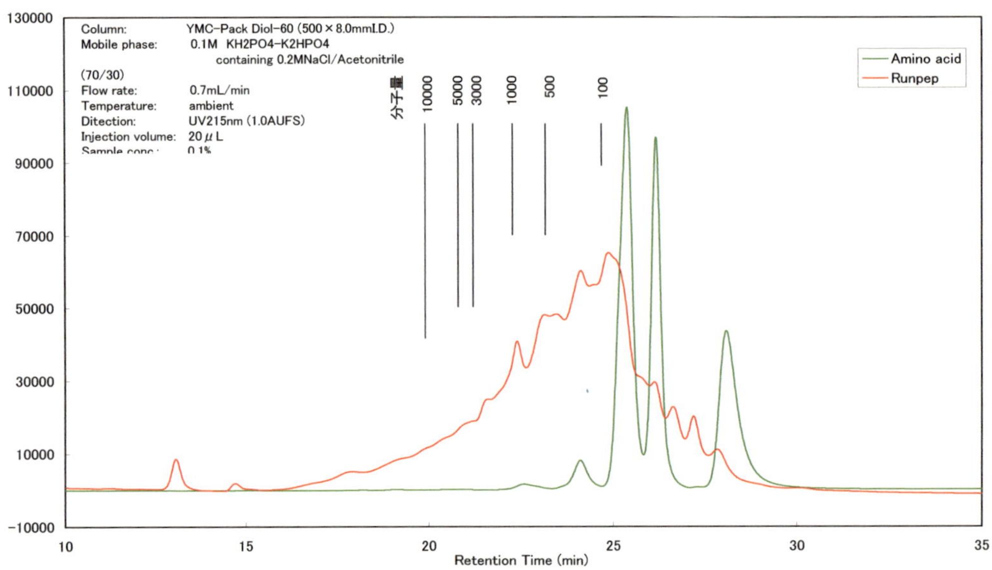

Figure 8. Runpep® profile by GPC.

In vitro, it showed dose-dependent vasorelaxant activity of both aortic and mesenteric vascular smooth muscle. *In vivo,* Runpep® exhibited dose-dependent hypotensive effect on both systolic and diastolic blood pressures with more pronounced effect on the systolic one by ingestion. Runpep® has been proved to have anti-hypertensive effect and has been shown to significantly lower the systolic blood pressure (Figure 9) of spontaneously hypertensive rats [94].

The hydrolysate of egg white with pepsin was found to exhibit a strong angiotensin I–converting enzyme (ACE) inhibitory activity in vitro [95]. Other work reports the antioxidant activity of peptides produced by pepsin hydrolysis of egg white; four peptides included in the protein sequence of ovalbumin possessed radical scavenging activity higher than that of Trolox. The combined antioxidant and ACE inhibition properties make it a very useful multifunctional preparation for the control of cardiovascular diseases, particularly hypertension. No correlation was found between antioxidant and ACE inhibitory activities of pepsin digest of egg white [96].

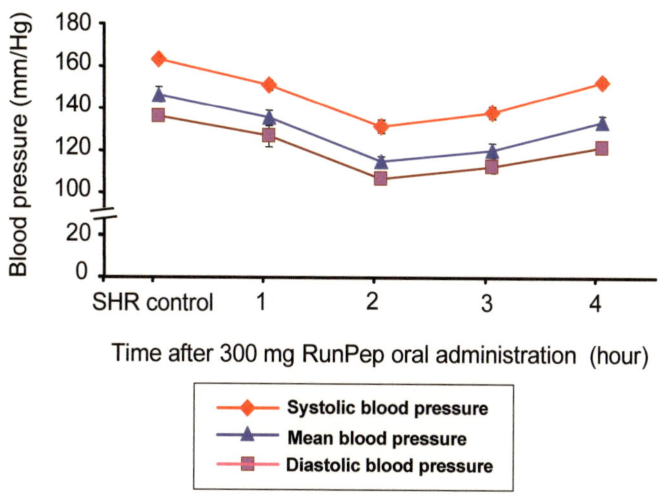

Figure 9. Antihypertensive activity of Runpep® after oral administration to SHR rats using cuff method (Sphygmomanometer).

5.5. Lysozyme peptides

Lysozyme is known to have a variety of folding topologies around the active site cleft [97]. Enzymatic hydrolysis of lysozyme is a novel technology that uses proteolytic enzymes for hydrolyzing native lysozyme to produce potent antimicrobial peptides that hidden within its folds. Lysozyme was digested by different proteolytic enzymes, such as clostripain [58, 60], pepsin, and trypsin [97, 98, 99]. All these researchers proved that the resulting peptides lost the enzymatic activities of lysozyme, but exhibited strong bactericidal activities against both Gram-negative (*E.coli, Salmonella, Pseudomonas,* and *Aeromonas*) and Gram-positive bacteria (*Listeria monocytogenes, Staph aureus, Bacillus* spp., and *Leuconostic*s spp.) as well as yeasts (*Saccharomyces*). Scanning electron microscopy clearly demonstrated that cell membrane of both Gram-negative and –positive bacteria was damaged by these peptides. Thus these peptides probably have a different mechanism of action than native lysozyme [58, 99]. Mine et al (2004) could isolation, purification, and characterization of novel antimicrobial peptides from chicken egg white lysozyme hydrolysate, obtained by peptic digestion and subsequent tryptic digestion. The hydrolysate was composed of over 20 small peptides of less than 1000 Da, and had no enzymatic activity. The water-soluble peptide mixture showed bacteriostatic activity against Gram-positive bacteria (Staphylococcus aureus 23-394) and Gram-negative bacteria (*E. coli* K-12). Two bacteriostatic peptides were purified and sequenced. One peptide, with the sequence Ile-Val-Ser-Asp-Gly-Asp-Gly-Met-Asn-Ala-Trp, inhibited Gram-negative bacteria *E. coli* K-12 and corresponded to amino acid residues 98-108, which are located in the middle part of the helix-loop-helix. Another novel antimicrobial peptide inhibited *S. aureus* and was identified as His-Gly-Leu-Asp-Asn-Tyr-Arg, corresponding to amino acid residues 15-21 of lysozyme. These peptides have broadened the antimicrobial activity of lysozyme to Gram-negative bacteria. The results obtained in this study indicate that lysozyme possesses nonenzymatic bacteriostatic domains in its primary sequence and they are released by proteolytic hydrolysis [99].

Consumers are increasingly demanding food that is free from pathogens, but with less preservatives and additives. As a response to these conflicting demands, current trends in the food industry include the investigation of alternative natural preservative in foods. Six Gram-negative bacteria (*Escherichia coli, Salmonella enteritidis* NBRC 3313, *Salmonella typhimurium, Pseudomonas fluorescens, Pseudomonas aeruginosa, Aeromonas hydrophila*) were checked for sensitivity to native hen egg white lysozyme and hydrolysate preparation of lysozyme derived peptides. Generally, lysozyme peptides preparation acts on the tested organisms and on different strains of *Bacillus* spp. with much more potency comparing to native lysozyme [100, 101]. The resulting peptides lost the enzymatic activities of lysozyme, but exhibited strong bactericidal activities against both Gram-negative and Gram-positive bacteria [97, 102]. Being natural antimicrobial, lysozyme peptides preparation will find its way as a safe shelf-life extender in the food industry.

6. Egg shell membrane bio proteins

The egg shell membrane has been thought to be beneficial in the treatment of some injuries. For example, in Japan, when Sumo wrestlers get flesh abrasions, they will often peel the egg membrane from the egg shell and cover their injuries. They believe that it facilitates their recovery. Peptides product which are stable in water have been prepared from hydrolyzed egg membrane. The effects of this egg shell membrane protein on cell growth have been studied. The growth of normal human skin fibroblasts on egg membrane protein-coated tissue cultures increased in relation to increasing egg membrane protein concentration.

The egg shell membranes contain several bacteriolytic enzymes (*e.g.* lysozyme and N-acetylglucosaminidase) and other membrane components which may alter the thermal resistance of Gram-positive and Gram-negative bacterial pathogens (*Salmonella Enteritidis, Escherichia coli* 0157:H7, *Listeria monocytogenes* and *Staphylococcus aureus*).

The presence of hydroxyproline in hydrolysates of the egg shell membrane layers has suggested that the membrane layers contain collagen [103]. This has been confirmed using biochemical and immunological tests. It has been established that about 10% of the total proteinaceous content of the membrane structure of an egg shell is collagen. The outer shell membrane contains predominately Type I collagen and the inner shell membrane contains Types I and V collagen. In addition, Type X collagen has been found in both the inner and outer shell membranes using immunohistochemical analysis. It is important to recognize the presence of collagen in egg shell membranes because of its potential value.

7. Egg shell bio proteins

Different Non-collagenous proteins have been identified in the organic matrix of the hen's egg shell. Ovocleidin-17 is a soluble matrix protein component and distributed in palisade and mammillary layers [104].

Ovocalyxin-32 has been identified as a novel 32-kDa protein. It is expressed at high levels in the uterine and isthmus regions of the oviduct, and concentrated in the eggshell. In the eggshell, ovocalyxin-32 localizes to the outer palisade layer, the vertical crystal layer, and the cuticle of the eggshell, in agreement with its demonstration by Western blotting at high levels in the uterine fluid during the termination phase of eggshell formation. Ovocalyxin-32 is therefore identified as a novel protein synthesized in the distal oviduct where hen eggshell formation occurs [105].

Osteopontin, a phosphorylated bone glycoprotein involved in formation and remodeling of the mineralised tissue, has also been demonstrated in the hen egg shell. Gene expression for this protein has been shown to be higher during the period of calcification [106].

Ovalbumin, lysozyme and ovotransferrin, as egg white proteins, have been identified in hen's egg shell. The organic matrix also contains several proteoglycan molecules [107].

Chicken eggshell powder has been proposed as an attractive source of calcium for human health to increase bone mineral density in an elderly population with osteoporosis. However, factors affecting calcium transport of eggshell calcium have not yet been evaluated. Chicken eggshell contains about 1.0% (w/w) matrix proteins in addition to a major form of calcium carbonate (95%, w/w). It was found that soluble eggshell matrix proteins remarkably enhance calcium transport using in vitro Caco-2 cell monolayers grown on a permeable support. The total calcium transport across Caco-2 monolayers showed an increase of 64% in the presence of 100 microg/well soluble eggshell matrix proteins. The active enhancer with a molecular mass of 21 kDa was isolated by reversed phase high-performance liquid chromatography and did not match to any previously identified protein. The N-terminal sequence was determined to be Met-Ala-Val-Pro-Gln-Thr-Met-Val-Gln. The possible mechanisms of eggshell matrix protein-mediated increase in calcium transport and the potential significance of eggshell calcium as a nutraceutical are discussed [108].

Experimental and clinical studies performed to date have shown a number of positive properties of eggshell powder, such as antirachitic effects in rats and humans. A positive effect was observed on bone density in animal models of postmenopausal osteoporosis in ovariectomized female rats. In vitro eggshell powder stimulates chondrocyte differentiation and cartilage growth. Clinical studies in postmenopausal women and women with senile osteoporosis showed that eggshell powder reduces pain and osteoresorption and increases mobility and bone density or arrests its loss. The bioavailability of calcium from this source, as tested in piglets, was similar or better than that of food grade purified calcium carbonate. Clinical and experimental studies showed that eggshell powder has positive effects on bone and cartilage and that it is suitable in the prevention and treatment of osteoporosis [109].

Author details

Adham M. Abdou
Food Control Department, Benha University, Moshtohor, Kaliobiya, Egypt

Mujo Kim
Research Department, Pharma Foods International Co., Ltd., Kyoto, Japan

Kenji Sato
Division of Applied Life Sciences, Kyoto Prefectural University, Kyoto, Japan

8. References

[1] Kerver JM, Park Y, Song WO. The role of eggs in American diets: health implications and benefits. In: Watson R., editor. Eggs and health promotion. Iowa: Blackwell Publishing Co. 2002; pp. 9-18.

[2] Hasler CM. The changing face of functional foods. Journal American College of Nutrition 2000; 19: 499S-506S.

[3] Watkins BA. The nutritive value of the egg. In: Stadelman WJ, Cotterill OJ, editors. Egg science and technology. New York: The Haworth Press Inc. 1995; pp. 177-194.

[4] Shin JH, Yang M, Nam SW, Kim JT, Myung NH, Bang WG, Roe IH. Use of egg yolk-derived immunoglobulin as an alternative to antibiotic treatment for control of *Helicobacter pylori* infection. Clinical Diagnostic Laboratory of Immunology 2002; 9: 1061-1066.

[5] Sugino H, Nitoda T, Juneja LR. General chemical composition of hen eggs. In: Yammamoti T, Juneja LR, Hatta H, Kim M, editors. Hen eggs: Their basic and applied science. Boca Raton: CRC Press. 1997; pp. 13-24.

[6] Shinohara K, Fukushima T, Suzuki M, Tsutsumi M, Kobori M, Kong ZL. Effect of some constituents of chicken egg yolk lipoprotein on the growth and IgM production of human-human hybridoma cells and other human-derived cells. Cytotechnology 1993; 11: 149-154.

[7] Hegenauer J, Saltmann P, Nace G. Iron (III)-phosphoprotein charts: Stoichiometricequilibrium constant for interaction of iron (III) and phosphoserine residues of phosvitin and casein. Biochemistry 1979; 18: 3865-3879.

[8] Nakamura S, Ogawa M, Nakai S, Kato A, Kitts DD. Antioxidant activity of a Maillard-type phosvitin-galactomannan conjugate with emulsifying properties and heat stability. Journal of Agriculture Food Chemistry 1998; 46: 3958-3963.

[9] Sattar Khan MA, Nakamura S, Ogawa M, Akita E, Azakami H, Kato A. Bactericidal action of egg yolk phosvitin against *Escherichia coli* under thermal stress. Journal of Agriculture Food Chemistry 2000; 48: 1503-1506.

[10] Williams T. Serum proteins and the livetins of hen's egg yolk. Biochemistry Journal 1962; 83: 346-355.

[11] Sunwoo HH, Sim JS. IgY technology for human health: Antimicrobial activities of IgY preparations against foodborne pathogens. The 225th ACS National Meeting, New Orleans, LA.: 2003.

[12] Hatta H, Tsuda K, Akachi S, Kim M, Yamamoto T. Productivity and some properties of egg yolk antibody (IgY) against human rotavirus compared with rabbit IgG. Bioscience Biotechnology and Biochemistry 1993; 57: 450-454.

[13] Sun S, Mo W, Ji Y, Lius S. Preparation and mass spectrometric study of egg yolk antibody (IgY) against rabies virus, Rapid Communications in Mass Spectrometry 2001; 15: 708-712.

[14] Woolley JA, Landon J. Comparison of antibody production to human interlukin-6 (IL-6) by sheep and chickens. Journal of Immunological Methods 1995; 187: 253-265.

[15] Shimizu M, Nagashima H, Sano K, Hashimoto K, Ozeki M, Tsuda K. Molecular stability of chicken and rabbit immunoglobulin G. Bioscience Biotechnology and Biochemistry 1992; 56: 270-274.

[16] Campbell RD, Dodds AW, Porter RR. The binding of human complement component C4 to antibody-antigen aggregates. Biochemistry Journal 1980; 189: 67-80.

[17] Fischer M, Hlinak A. The lack of binding ability of staphylococcal protein A and streptococcal protein G to egg yolk immunoglobulins of different fowl species. Berl Munch Tierarztl Wochenschr 2000; 113: 94-96.

[18] Larsson A, Karlsson-Parra A, Sjoquist J. Use of chicken antibodies in enzyme immunoassays to avoid interferences by rheumatoid factors. Clinical Biochemistry 1991; 37: 411-414.

[19] Rubinstein E, Kouns WC, Jennings LK, Boucheix C, Carroll RC. Interaction of two GPIIb/IIa monoclonal antibodies with platlet Fc receptor (Fc gamma RII). British Journal of Haematology 1991; 87: 80-86.

[20] Narat M. Production of antibodies in chickens. Food Technology and Biotechnology 2003; 41: 259-267.

[21] Zeidler G. Eggs vital to human and animal medicine. World Poultry 1998; 14: 33-34.

[22] Sarker SA, Casswall TH, Juneja LR, Hoq E, Hossain I, Fuchs GJ. Randomized, placebo-controlled, clinical trial of hyperimmunized chicken egg yolk immunoglobulin in children with rotavirus diarrhea. Journal of Pediatric Gastroenterology and Nutrition 2001; 32: 19-25.

[23] Yang H, Jin Z, Yu Q, Yang T, Wang H, Liu L. The selective recognition of IgY for digestive cancers. Chinese Journal of Biotechnology 1997; 13: 85-90.

[24] Dunn BE, Cohen H, Blaser M J. *Helicobacter pylori*. Clinical Microbiology Reviews 1997; 10: 720-741.

[25] Blaser M J. *Helicobacter pylori* and the pathogenesis of gastroduodenal inflammation. The Journal of Infectious Diseases 1990; 161: 626-633.

[26] Parsonnet J, Hansen S, Rodriguez L, Gleb AB, Warnke RA, Jellum E, Orentreich N, Vogelman JH, Friedman GD. *Helicobacter pylori* infection and gastric lymphoma. The New England Journal of Medicine 1994; 330: 1267-1271.

[27] Marshall BJ. *Helicobacter pylori*. American Journal of Gastroenterolology 1994; 89: 116-128.

[28] Chandan RC. Enhancing market value of milk by adding cultures. Journal of Dairy Science 1999; 82: 2245-2256.

[29] Yamane T, Saito Y, Takizawa S, Goshima H, Kodama Y, Horie N, Kim M. Development of anti-Helicobacter pylori urease IgY and its application for food product. Food Processing and Ingredients (in Japanese) 2003; 38: 70.

[30] Chang Y, Min S, Kim K, Han Y, Lee J. Delta [sup 13] C-urea breath test value is a useful indicator for *Helicobacter pylori* eradication in patients with functional dyspepsia. Journal of Gastroenterology and Hepatology 2003; 18: 726.

[31] Chen JP, Chang MC. Effect of anti-Helicobacter pylori urease antibody (IgY) as a food ingredient on the decrease of H. pylori in the stomach of humans infected with *H. pylori*. Taiwanese Journal of Agricultural Chemistry and Food Science (in Taiwanese) 2003; 41: 408.

[32] Horie K, Horie N, Abdou AM, Yang JO, Yun SS, Chun HN, Park CK, Kim M, Hatta H. Egg Yolk Immunoglobulin (IgY) on *Helicobacter pylori* in Humans. Journal of Dairy Science 2004; 87: 4073.

[33] Michetti P, Dorta G, Wiesel PH, Brassart D, Verdu E, Herranz M, Felley C, Porta N, Rouvet M, Blum AL, Corthésy-Theulaz I. Effect of whey-based culture supernatant of *Lactobacillus acidophilus* (johnsonii) La1 on *Helicobacter pylori* infection in humans. Digestion 1999; 60: 203.

[34] Wendakoon CN, Thomson ABR, Ozimek L. Lack of therapeutic effect of a specially designed yogurt for the suppression of *Helicobacter pylori* infection. Digestion 2002; 65: 16.

[35] Sakamoto I, Igarashi M, Kimura K, Takagi A, Miwa T, Koga Y. Suppressive effect of *Lactobacillus gasseri* OLL2716 (LG21) on *Helicobacter pylori* infection in humans. Antimicrobial Chemotherapy 2001; 47: 709.

[36] Cats A, Kuipers EJ, Bosschaert MAR, Pot RGJ, Vandenbroucke-Grauls CMJE, Kusters JG. Effect of frequent consumption of a Lactobacillus casei-containing milk drink in Helicobacter pylori-colonized subjects. Alimentary Pharmacology and Therapeutics 2003; 17: 429.

[37] Kim M, Higashiguchi S, Iwamoto Y, Yang HC, Cho HY, Hatta H. Egg yolk antibody and its application. Biotechnology and Bioprocess Engineering 2000; 5: 75.

[38] Shin JH, Roe IH, Kim HG. Production of anti- Helicobacter pylori urease-specific immunoglobulin in egg yolk using an antigenic epitope of *H. pylori* urease. Journal of Medical Microbiology 2004; 53: 31.

[39] Smith DJ. Dental caries vaccines: Prospects and concerns. Critical Reviews in Oral Biology and Medicine 2002; 13: 335.

[40] Loesche WJ. Role of *Streptococcus mutans* in human dental decay. Microbiological Reviews 1989; 50: 353.

[41] Hamada S, Koga T, Ooshima T. Virulence factors of *Streptococcus mutans* and dental caries prevention. Journal of Dental Research 1984; 63: 407.

[42] Otake S, Nishihara Y, Makimura M, Hatta H, Kim M, Yamamoto T, Hirasawa M. Protection of rats against dental carries by passive immunization with hen-egg-yolk antibody (IgY). Journal of Dental Research 1991; 70: 162.

[43] Hatta H, Tsuda K, Ozeki M, Kim M, Yamamoto T, Otake S, Hirasawa M, Katz J, Childers NK., Michalek SM. Passive immunization against dental plaque formation in humans: Effect of a mouth rinse containing egg yolk antibodies (IgY) specific to *Streptococcus mutans*. Caries Research 1997; 31: 268.

[44] Nakamura T, Umeda K, Kusonoki S, Arai J, Namiki H. Advances in IgY applications: characteristics of specific egg yolk immunoglobulin against influenza virus, Page 72 in Proc. 16thAnnual meeting Japanese Society of Bioscience Biotechnology and Agrochemistry (in Japanese), Hiroshima, Japan: 2004.

[45] Fathy G, Abdel-Raheem SM, Abdou AM, Horie M, Kim M, Suzuki H. Suppressive effect of Topical Specific Egg Yolk Immunoglobulin (IgY) on *Propionibacterium acnes* in Acne Vulgaris. Sci. J. Al-Azhar Med. Fac. (girls). 2007; 28: 2177.

[46] Jiang B, Mine M. Preparation of novel functional oligophoshopeptides from hen egg yolk phosvitin. Journal of Agricultural and Food Chemistry 2000; 48: 990.

[47] Leem KH, Kim MG, Kim HK, Oi Y, Kim M. Effect of egg yolk proteins on the longitudinal bone growth in rat, Page 200 in Proc. 16thAnnual meeting Jpn. Soc. Biosci. Biotechnol. & Agrochem. (in Japanese), Hiroshima, Japan: 2004.

[48] Leem KH, Kim MG, Kim HM, Kim M, Lee YJ, Kim HK. Effect of egg yolk proteins on the longitudinal bone growth of adolescent male rat, Bioscience Biotechnology and Biochemistry 2004; 68: 2388.
[49] Oi Y, Ji MY, Leem KH, Cho SW, Kim M. The effect of hen egg yolk peptide on bone growth, 17thAnnual meeting Japanese Society of Bioscience Biotechnology and Agrochemistry (in Japanese), Hokkaido, Japan: 2005.
[50] Ibrahim HR. Insight into the structure-function relationships of ovalbumin, ovotransferrin, and lysozyme. In: Yamamoto T, Juneja LR, Hatta H, Kim M, editors. Hen Eggs: Their basic and applied science. New York: CRC press, Inc. 1997.
[51] Mine Y, Noutomi T, Haga N. Emulsifying and structural properties of ovalbumin. Journal of Agricultural and Food Chemistry 1991; 39: 443.
[52] Nisbet AD, Saundry RH, Moir AJ, Fothergill LA, Fothergill JE. The complete amino-acid sequence of hen ovalbumin, European Journal of Biochemistry/FEBS 1981; 115: 335.
[53] Gettins PGW, Patston PA, Olson ST. Serpins: Structure, Function and Biology, LandersRG. Austin, TX: 1996.
[54] Huntington JA, Stein PE. Structure and properties of ovalbumin. Journal of Chromatography B 2001; 756: 189.
[55] Ibrahim HR. Ovotransferrin. In: Naidu AS, editor. Natural food antimicrobial systems. New York: CRS Press, Inc. 2000.
[56] Mason AB, Woodworth RC, Oliver RW, Green BN, Lin LN, Brandts JF, Savage KJ, Tam BM, MacGillivrays TA. Association of the two lobes of ovotransferrin is a prerequisite for receptor recognition: Studies with recombinant ovotransferrins. Biochemistry Journal 1996; 319: 361.
[57] Corda R, Biddau P, Corrias A, Puxeddu E. Conalbumin in the treatment of acute enteritis in the infant. Journal Tissue Reactions 1983; 1: 117.
[58] Ibrahim HR, Aoki T, Pellegrini A. Strategies for new antimicrobial proteins and peptides: lysozyme and aprotinin as model molecules. Current Pharmacological Design 2002: 8: 671.
[59] Masschalck B, Michiels CW. Antimicrobial properties of lysozyme in relation to foodborne vegetative bacteria. Critical Reviews in Microbiology 2003; 29: 191.
[60] Pellegrini A, Bramaz TN, Klauser S, Hunziker P, von Fellenberg R. Identification and isolation of bactericidal domain in chicken egg white lysozyme. Journal of Applied Microbiology 1997; 82: 372.
[61] Wild P, Gabrieli A, Schraner EM, Pellegrini A, Thomas U, Frederik PM, Stuart MC, Vonfellenberg R. Reevaluation of the effect of lysozyme on Escherichia coli employing ultrarapid freezing followed by cryoelectron microscopy or freeze substitution. Microscopy Research and Technique 1997; 39: 297.
[62] Ibrahim HR, Kato A, Kobayashi K. Antimicrobial effects of lysozyme against Gram-negative bacteria due to covalent binding of palmitic acid. Journal of Agricultural and Food Chemistry 1991; 39: 2077.

[63] Liu ST, Sugimoto T, Azakami H, Kato A. Lipophilization of lysozyme by short and middle chain fatty acids. Journal of Agricultural and Food Chemistry 2000; 48: 265.
[64] Kato A, Ibrahim HR, Watanabe H, Honma K, Kobayashi K. New approach to improve the gelling and surface functional properties of dried egg white by heating in dry state. Journal of Agricultural and Food Chemistry 1989; 37: 433.
[65] Nakamura NK, Furukawa N, Matsuoka M, Takahashi T, Yamanaka Y. Enzyme activity of lysozyme-dextran complex prepared by high-pressure treatment. Food Science and Technology International 1997; 3: 235.
[66] Nakamura S, Saito M, Goto T, Saeki H, Ogawa M, Gotoh M, Gohya Y, Hwang JK. Rapid formation of biologically active neoglycoprotein from lysozyme and xyloglucan hydroysates throughnaturally occurring Maillard reaction. Journal of Food Science and Nutrition 2000; 5: 65.
[67] Takahashi K, Lou XF, Ishii Y, Hattori M. Lysozyme-glucose stearic acid monoester conjugate formed through the Maillard reaction as an antibacterial emulsifier. Journal of Agricultural and Food Chemistry 2000; 48: 2044.
[68] Liu ST, Azakami H, Kato A. Improvement in the yield of lipophilized lysozyme by the combination with Maillard-type glycosylation. Nahrung 2000; 44: s407.
[69] Kijowski J, Lesnierowski G. Separation, polymer formation and antibacterial activity of lysozyme. Polish Journal of Food and Nutrition Sciences 1999; 8/49: 3.
[70] Stadhouders J, Stegink H, Van den Berg G, Van Ginkel W. The use of lysozyme to prevent butyric fermentation in Gouda cheese. Voedingsmiddelen technlogie 1987; 25: 17.
[71] de Roos AL, Walstra P, Geurts TJ. The association of lysozyme with casein. International Dairy Journal 1998; 8: 319.
[72] Daeschel MA, Bruslind L, Clawson J. Application of the enzyme lysozyme in brewing. Master Brewers. Association of American Technology Quart. 1999; 36: 219.
[73] Lacono VJ, Mackay BJ, Dirienzo S, Pollock JJ. Selective antibacterial properties of lysozyme for oral microorganisms. Infection and Immunity 1980; 29: 523.
[74] El-Nimr A, Hardee GE, Perrin JH. A fluorimetric investigation of the binding of drugs to lysozyme. Journal of Pharmacy and Pharmacology 1981; 33: 117.
[75] Siwicki AK, Klein P, Morand M, Kiczka W, Studnicka M. Immunostimulatory effects of dimerized lysozyme (KLP-602) on the nonspecific defense mechanisms and protection against furunculosis in salmonids. Veterinary Immunology and Immunopathology 1998; 61: 369.
[76] Sugahara T, Murakami F, Yamada Y, Sasaki T. The mode of action of lysozyme as an immunoglobulin production stimulation factor. Biochimica and Biophysica Acta 2000; 1475: 27.
[77] Sava G. Reduction of B16 melanoma metastases by oral administration of egg-white lysozyme. Cancer Chemotherapy Pharmacology 1989; 25: 221.
[78] Yamashita K, Tachibana Y, Hitoi A. Sialic acid-containing sugar chains of hen ovalbumin and ovomucoid. Carbohydrate Research 1984; 130: 271.

[79] Urisu A, Ando H, Morita Y, Wada E, Yasaki T, Yamada K, Komada K, Torii S, Goto M, Wakamatsu T. Allergenic activity of heated and ovomucoid-depleted egg white. Journal of Allergy and Clinical Immunology 1997; 100: 171.

[80] Tsuge Y, Shimoyamada M, Watanabe K. Binding of egg white proteins to viruses. Bioscience Biotechnology and Biochemistry 1996; 60: 1503.

[81] Ohami H, Ohishi H, Yokota T, Mori T, Watanabe K. Cytotoxic effect of sialoglycoprotein derived from avian egg white ovomucin on the cultured tumor cell. Medicine and Biology 1993; 126: 19.

[82] Green NM. Avidin. Advances in Protein Chemistry 1975; 29: 85.

[83] Ebina T, Tsukada K. Protease inhibitors prevent the development of human rotavirus induced diarrhea in suckling mice. Microbiology and Immunology 1991; 35: 583.

[84] Abrahamson M, Dalboge H, Olafsson I, Carlsen S, Grubb A. Efficient production of native, biologically active human cystatin C by Escherichia coli. FEBS Letters 1988; 236: 14.

[85] Kennedy AR. Anti-carcinogenic activity of protease inhibitors. In: Troll W, Kennedy AR, editors. Protease Inhibitors as Cancer Chemopreventive Agents. New York : Plenum Press. 1993.

[86] Fujita H, Sasaki R, Yoshikawa M. Potentiation of the anthypertensive activity of orally administered ovokinin, a vasorelaxing peptide derived from ovalbumin, by emulsification in egg phosphatidylcholine. Bioscience Biotechnology and Biochemistry 1995; 59: 2344.

[87] Matoba N, Usui H, Fujita H, Yoshikawa M. A novel anti-hypertensive peptide derived from ovalbumin induced nitric oxide-mediated vasorelaxation in an isolated SHR mesenteric artery. FEBS Letters 1999; 452: 181.

[88] Matoba N, Yamada Y, Usui H, Fujita H, Nakagiri R, Yoshikawa M. Designing potent derivatives of ovokinin (2-7), an anti-hypertensive peptide derived from ovalbumin. Bioscience Biotechnology and Biochemistry 2001; 65: 736.

[89] Ibrahim HR, Iwamori E, Sugimoto Y, Aoki T. Identification of a distinct antibacterial domain within the N-lobe of ovotransferrin. Biochimica et Biophysica Acta 1998; 1401: 289.

[90] Kovacs-Nolan J, Zhang JW, Hayakawa S, Mine Y. Immunochemical and structural analysis of pepsin-digested egg white ovomucoid. Journal of Agricultural and Food Chemistry 2000; 48: 6261.

[91] Watanabe K, Tsuge Y, Shimoyamada M, Ogama N, Ebina T. Antitumor effects of pronase treated fragments, glycopeptides, from ovomucin in hen egg white in a double grafted tumor system. Journal of Agricultural and Food Chemistry 1998; 46: 3033.

[92] Kittaka R, Cho HJ, Choi SH, Kang HC, Choi SA, Jung YJ, Lee TK, Kim CR, Park HJ, Kim M. Anti-thrombotic action of egg white peptides. Page 198 in Proc. 16thAnnual meeting Japanese Society of Bioscience Biotechnology and Agrochemistry (in Japanese), Hiroshima, Japan. 2004.

[93] Cho HJ, Kittaka R, Abdou AM, Kim M, Kim HS, Lee DH, Park HJ. Inhibitory effects of oligopeptides from hen egg white on both human platelet aggregation and blood coagulation. Archives in Pharmacology Research 2009; 32(6): 945.

[94] El-Mahmoudy A, Kittaka R, Okamoto K, Shimizu Y, Shiina T, Takewaki T, Kim M. Pharmacologically active peptide derived from egg white with circulatory beneficial effects. 17thAnnual meeting Japanese Society of Bioscience Biotechnology and Agrochemistry, Hokkaido, Japan. 2005.

[95] Miguel M, Recio JA, Gómez-Ruiz M, Ramos M, Amos M, López-Fandiño R. Angiotensin I–converting enzyme inhibitory activity of peptides derived from egg white proteins by enzymatic hydrolysis. Journal of Food Protection 2004; 67: 1914.

[96] Dávalos A, Miguel M, Bartolomé B, López-Fandiño R. Antioxidant activity of peptides derived from egg white proteins by enzymatic hydrolysis. Journal of Food Protection 2004; 67: 1939.

[97] Ibrahim HR, Inazaki D, Abdou AM, Aoki T, Kim M. Processing of Lysozyme at distinct loops by pepsin: A novel action for generating multiple antimicrobial peptide motifs in the newborn stomach. Biochimica et Biophysica Acta (BBA) 2005; 1726 (1): 102.

[98] Ibrahim HR, Thomas U, Pellegrini A. A helix-loop-helix peptide at the upper lip of the active site cleft of lysozyme confers potent antimicrobial activity with membrane permeabilization action. Journal of Biological Chemistry 2001; 276: 43767.

[99] Mine Y, Lauriau S, Ma F. Antimicrobial peptides released by enzymatic hydrolysis of hen egg white lysozyme. Journal of Agricultural and Food Chemistry 2004; 52: 1088.

[100] Abdou AM, Higashiguchi S, Aboueleinin A, Kim M, Ibrahim HR. Lysopep: a lysozyme peptide preparation has an exaggerated antimicrobial activity against Gram-negative bacteria, Page 226 in Proc. 16thAnnual meeting Japanese Society of Bioscience Biotechnology and Agrochemistry, Hiroshima, Japan. 2004.

[101] Abdou AM, Higashiguchi S, Aboueleinin AM, Kim M, Ibrahim HR. Antimicrobial peptides derived from hen egg lysozyme with inhibitory effect against Bacillus species. Food Control 2007; 18 (2): 173.

[102] Abdou AM, Higashiguchi S, Kim M, Ibrahim HR. A marvelous inhibitory effect of lysozyme peptides preparation against Bacillus species. Page 221 in Proc. 15thAnnual meeting Japanese Society of Bioscience Biotechnology and Agrochemistry, Yokohama, Japan. 2003.

[103] Nakano T, Ikawa NI, Ozimek L. Chemical composition of chicken eggshell and shell membranes. Poultry Science 2003; 82: 510.

[104] Hincke MT, Tsang CPW, Courtney M, Hill V, Narbaitz R. Purification and immunochemistry of a soluble matrix protein of the chicken eggshell (ovocleidin-17). Calcified Tissue International 1995; 56: 578.

[105] Gautron J, Hincke MT, Mann K, Panhéleux M, Bain M, McKee MD, Solomon SE, Nys, Y. Ovocalyxin-32, a novel chicken eggshell matrix protein. Journal of Biological Chemistry 2001; 276: 39243.

[106] Pines M, Knopov V, Bar A. Involvement of osteopontin in egg shell formation in the layingchicken. Matrix Biology 1994; 14: 765.

[107] Ferandez MS, Araya M, Arias JL. Eggshells are shaped by a precise spatio-temperal arrangement of sequentially deposited macromolecules. Matrix Biology 1997; 16: 13.

[108] Daengprok W, Garnjanagoonchorn W, Naivikul O, Pornsinlpatip P, Issigonis K, Mine Y. Chicken eggshell matrix proteins enhance calcium transport in the human intestinal epithelial cells, Caco-2. Journal of Agricultural and Food Chemistry 2003; 51: 6056.

[109] Rovensky J, Stancikova M, Masaryk P, Svik K, Istok R. Eggshell calcium in the prevention and treatment of osteoporosis. International Journal of Clinical Pharmacology 2003; 23: 83.

Chapter 6

Antihypertensive Properties of Plant Protein Derived Peptides

Anne Pihlanto and Sari Mäkinen

Additional information is available at the end of the chapter

http://dx.doi.org/10.5772/54565

1. Introduction

Cardiovascular diseases (CVD) are a major health problem in the industrialized countries, representing the main cause of death in the world. It is estimated that 17 million people globally die of CVD every year and these diseases are responsible for more than half of all deaths in Europe [1]. Therefore, primary prevention is becoming an increasing part of public health strategies aimed at reducing societal burden due to CVD-related morbidity and mortality worldwide. There are several behavioural factors such as tobacco use, ethanol consumption, unhealthy diet and physical inactivity that can lead to hypertension, hyperlipidemia, diabetes, overweight and obesity, and thereby contribute to CVD development. The World Health Organisation emphasises the importance of improved nutrition as means of controlling the expected rise in global CVD incidence over the next decades.

Angiotensin I-converting enzyme (ACE: EC 3.4.15.1) is a peptidyldipeptide hydrolase that plays an important physiological role in both the regulation of blood pressure and cardiovascular function [2] through two different reactions. First, ACE catalyzes the hydrolysis of angiotensin I, an inactive decapeptide, to angiotensin II, a powerful vasoconstrictor and salt-retaining octapeptide. Thus, ACE-inhibition has a hypotensive effect. Secondly, ACE catalyzes the inactivation of the vasodilator bradykinin that regulates different biological processes including vascular endothelial nitric oxide (NO) release [3].

Oxidative stress has a well documented role in CVD development. Oxidative stress is defined as the situation characterized by increased generation of free radicals (reactive oxygen species, ROS), resulting in increased oxidative damage of biological structures. Within the cell, physiologic levels of some ROS are involved as key intermediates in signalling pathways to maintain basal cellular functions. In contrast, when ROS are generated in the absence of a physiological stimulus, small-molecule antioxidants are depleted, or antioxidant enzymatic systems are overwhelmed. This leads to a net increase in biologically active ROS and oxidant stress ensues. In blood vessels, oxidant stress has deleterious consequences for basal vascular function. Then, the cellular mechanisms that result in vascular redox imbalance leading to an increase in oxidant stress are implicated in the pathogenesis of vascular disease [4].

The search for dietary compounds that prevent the development of CVD is deemed crucial to tackle this major health problem worldwide, and recent observational studies and clinical trials have suggested that increased protein consumption, particularly from plant sources, might reduce blood pressure and prevent CVD. Recently, interest has been emerging to identify and characterize bioactive peptides from plant and animal sources. Bioactive peptides are considered specific protein fragments that are inactive within the sequence of the parent protein. After they are released they may exert various physiological functions. The type of bioactive peptides generated from a particular protein is dependent on two factors: (a) the primary sequence of the source protein and (b) the specificity of the enzyme(s) used to generate such peptides. The hydrolysis of plant proteins has led to the production of a variety of biologically active peptides, such as opioid, antihypertensive, antioxidative, immunomodulatory or antimicrobial peptides [5, 6]. Bioactivity of peptides depend on the structure, however, the structure activity relationship is not yet fully understood for all biological activities described. This present paper focuses on the peptides beneficial to CVD derived from plant proteins.

2. Biological activities of plant proteins

Potato tuber proteins are classified into three major groups: patatins, protease inhibitors, and other proteins. Patatin is the major storage protein and an allergen for some people. The second major potato tuber storage protein is a diverse group of low molecular weight protease inhibitors [7]. The majority of the patatin and proteinase inhibitor isoforms possess enzymatic and inhibitory activities, respectively, which might be of physiological relevance. Activities are associated with the defense mechanisms of potato against pathogens and they inhibit a variety of proteases and some other enzymes, for example invertase [8]. Low molecular weight antimicrobial potato peptides from potato tubers have recently been reported to exhibit antibiotic/antifungal activity also against human pathogenic fungal and microbial strains [9]. The broad phospholipase activity of patatin has been characterized and documented rather extensively [10, 11]. The results indicate that the patatin-related enzymes are, in addition to fat metabolism, involved in the stress responses and signal transduction in the potato tubers. Patatin has also been shown to possess antioxidant or antiradical activity. Liu and colleagues [12] found that purified patatin exert antioxidant or antiradical activity in various *in vitro* tests, such as radical, scavenging activity assay and protection against hydroxyl radical-induced calf thymus DNA damage. Potato protein hydrolysates showed antioxidant activity [13] and enhanced oxidative stability of soybean oil emulsions [14]. Two peptides derived from metallocarboxypeptidase inhibitor and lipoxygenase 1 were identified. Potatoes may have a role in controlling appetite and therefore weight gain, by contributing to satiety. Gastrointestinal hormones such as cholecystokinin (CCK) are key factors in the regulation of food intake and maintaining energy homeostasis. Hill and colleagues [15] reported reduced energy intake and increased CCK release, when protease inhibitors extracted from potatoes were tested in 11 lean subjects.

The protein content of defatted rapeseed meal is high, approximately 32%, making it a potential food ingredient. Rapeseed protein has recently been demonstrated to be of high nutritional value in human subjects and substituting Cys-rich rapeseed protein, for milk protein prevented the early onset of insulin resistance, similar to those achieved by manipulating dietary fat and carbohydrates in a rat model [16, 17]. Flaxseed and its defatted meal contain high amounts of proteins, which are comparable in amino acid composition to food proteins like soy, with a preponderance of basic and branch-chain amino acids. The high Cys and Met content can boost the body's antioxidant levels, potentially stabilising DNA during cell division and reducing risk of certain forms of colon cancer [18]. Yet, there are only few studies examining rapeseed and flaxseed meals as source of bioactive peptides to enhance the value of these rapeseed and flaxseed industry by-products.

Legumes could represent valuable tools to prevent CVD, in addition to constitute an important source of dietary proteins (~18-40 %), dietary fibre, minerals and vitamins. Epidemiological studies have provided consistent evidence of the inverse relationship between legume consumption and the incidence of CVD. The majority of studies that have evaluated the hypocholesterolemic effects of legume consumption examined soybeans [19]. In addition, the meta-analysis showed that diet rich in legumes, such as a variety of beans, peas, and some seeds other than soy decreases total and low-density lipoprotein (LDL) cholesterol [20]. Different legumes have been identified as sources of ACE-inhibitory and antioxidative peptides, mainly soybean [21-24], chickpea and pea [25-28].

Cereal grains contain relatively little protein compared to legume seeds, with an average of about 10–12% of dry weight. Storage proteins account for about 50% of the total protein in mature cereal grains and have important impacts on their nutritional quality for humans and livestock and on their functional properties in food processing. Proteins can be separated into albumin, globulin, prolamin (hordein) and gluten fractions as described by [29]. The prolamins are characterized by their high content of Pro and Glu and low content of Lys and Trp. Alcohol-soluble endosperm proteins (prolamins) from some cereals (e.g. wheat, barley, and rye) give origin upon proteolytic digestion to biologically active antinutritional peptides able to adversely affect *in vivo* the intestinal mucosa of coeliac patients, whereas prolamins from other cereals (e.g. maize and rice) do not [30]. A large deal of cereal proteins originates from by-products following production of starch, malting or brewing industry. For example, α-zein protein, derived from corn starch production, is rich in Pro and hydrolysis by thermolysin liberates ACE-inhibitors [31]. Wheat proteins are also source of opioid peptides [32] and tryptic hydrolysate of rice proteins yields the immunomodulatory peptide, oryzatensin [33].

2.1. Production of peptides

ACE-inhibitory peptides have been produced by enzymatic hydrolysis and microbial fermentation of food proteins. In addition, solvent extraction has been used to isolate ACE-inhibitory activity from plant materials, such as mushrooms, broccoli and buckwheat [6]. The ACE-inhibitory potency is expressed as the IC_{50} value (inhibitor concentration leading to 50% inhibition), being used to estimate the effectiveness of different hydrolysates and peptides. The most common way to produce bioactive peptides is through enzymatic hydrolysis of whole protein molecules. The specificity of enzyme and process conditions influence the peptide composition of hydrolysates and thus their activities.

Generally, enzymatic hydrolysis is widely applied to upgrade functional features (such as emulsifying properties of hydrolysed protein) and nutritional properties of proteins [34-36]. It has been reported that additional advantage of hydrolysis can be the development of hydrophobicity since proteolysis unfolds the protein chains. The cleavage of peptide bonds enhances levels of free amino and carboxyl groups resulting in enhanced solubility. Therefore, hydrolysis can increase or decrease the hydrophobicity, which mostly depends on the nature of the precursor protein and molecular weight of the generated peptides [34]. Moreover, hydrolysis leads to production of small bioactive peptides [35] and bitterness of peptides of below 1000 Da is much less than fractions with a higher molecular mass [36]. However, it has been reported that extensive hydrolysis could adversely affect functional properties of peptides [37]. Some factors to consider in producing bioactive peptides include hydrolysis time, degree of hydrolysis of the proteins, enzyme–substrate ratio, and pre-treatment of the protein prior to hydrolysis. For example, thermal treatment of proteins can enhance enzymatic hydrolysis [38] possibly by increasing enzyme–protein interactions due to thermal-induced unfolding of the proteins.

According to the literature, enzymatic hydrolysis has been the main process for producing ACE-inhibitory and antioxidative peptides from food proteins. Use of exogenous enzymes is preferred in most cases over the autolytic process (i.e., use of endogenous enzymes present in the food source itself), due to the shorter time required to obtain similar degree of hydrolysis as well as better control of the hydrolysis to obtain more consistent molecular weight profiles and peptide composition [39-45]. Industrial food-grade proteinases such as Alcalase, Flavourzyme, and Protamex derived from microorganisms, as well as enzymes from plant (e.g. Papain) and animal sources (e.g., pepsin and trypsin), have been widely used in producing ACE-inhibitory and antioxidative peptides. The serine type endo-protease Alcalase, has produced the highest ACE-inhibitory activities *in vitro* in case of several plant proteins (Table 1).

Alcalase digests of rapeseed, canola, flaxseed, sunflower seed protein, legumes as well as and mung and chick beans showed high potency for ACE inhibition [41, 43-48]. Moreover, Alcalase digestion increased the ACE inhibition of the protein-rich by-product fraction from potato starch industry, potato tuber liquid fraction [13]. The IC_{50} -values were ranged between 0.020 and 0.64 mg protein/ml, which are similar to those reported for milk whey and casein hydrolysates [5, 41, 45-47,49]. Mäkinen and colleagues [50] reported that

Source protein	Enzyme or other process conditions	ACE inhibition	Identified peptides		*In vivo* response		Ref
		IC50 (mg/ml)	Sequences	IC50 value (μM)	Dose administration model	Response (Δ SBP mmHg)	
Sweetpotato tuber protein isolate	Thermoase PC 10F, Protease S & Proleather FG-F	0.018	ITP IIP GQY STYQT	9.5 80.8 52.3 300.4	500 mg/kg hydrolysate, oral, SHR	-30 after 8 h	[121]
Sweetpotato tuber defensin	Trypsin	0.190	GFR FK IMVAEAR GPCSR CFCTKPC MCESASSK	94.25 265.43 84.12 61.67 1.31 75.93			[118]
Sweet potato tuber Thioredoxin h	Trypsin	0.152	EVPK VVGAK FTDVDFIK MMEPMVK	1.73 1.14 0.42 1.03			[119]
Sweetpotato tuber Trypsin inhibitor	Pepsin	0.188	HDHM LR SNIP VRL TYCQ GTEKC RF VKAGE AH KIEL	276.2 746.4 228.3 208.6 2.3 275.8 392.2 141.56 523.5 849.7			[120]
Yam tuber	Powdered yam product, alcohol-insoluble-solids Water extract (15% protein) Heat treated (90°C) water extract Dioscorin, lyophilized yam powder Storage protein ion-exhange chromato-graphy Pepsin hydrolysate	ND			60 mg/kg oral, SHR 154 mg/kg oral, SHR 154 mg/kg oral, SHR 140 mg daily oral, human 40 mg/kg oral, SHR 40 mg/kg oral, SHR	-32.4 after 6 h -30.3 after 4 h -23.87 after 4 h -6.52 after 2 weeks -27.7 after 9 days -32.8 after 8 hours	[123] [86] [175] [122] [122]

Source protein	Enzyme or other process conditions	ACE inhibition	Identified peptides		*In vivo* response		Ref
		IC50 (mg/ml)	Sequences	IC50 value (μM)	Dose administratio n model	Response (Δ SBP mmHg)	
Potato tuber	Autolysed protein extract Alcalase digest of potato tuber liguid fraction	0.36 0.018					[50] [13]
Apios Americana Medikus tuber	Water extract (rich in proline)	127			200 mg/kg oral, SHR	-25 after 0.5 and 1 h	[124]
Rapeseed protein	Pepsin Subtilisin	0.16 0.16	 VWIS VW IY RIY	 30 1.6 3.7 28	500 mg/kg hydrolysate, 12.5 mg/kg 7.5 mg/kg 7.5 mg/kg 7.5 mg/kg oral, SHR	-6.8 after 4h -15.5 after 4h -12.5 after 2 h -10.8 after 2 h -9.8 after 2 h -11.3 after 4h	[125]
Rapeseed protein	Alcalase Peptide fraction from affinity purification Alcalase	0.038 0.25×10^{-3} 0.020					[45] [47]
Canola meal, defatted	Heat treatment and Alcalase	0.027	VSV FL	0.15 1.33			[46]
Flaxseed protein	Trypsin & Pronase cationic peptide fraction	0.4	QGR RW SVR GQMRQPI QQQG ASVRT DYLRSC ARDLPGQ RDLPG RGLERA TCRGLERA		200 mg/kg hydrolysate, oral, SHR	-17.9 after 2 h	[44]
Flaxseed protein	Gastrointestinal digest *in vitro*, <1kDa permeate fraction	0.040	WNI/LNA NI/LDTDI/L				[127]

Source protein	**Enzyme or other process conditions**	**ACE inhibition**	**Identified peptides**		***In vivo* response**		**Ref**
		IC_{50} (mg/ml)	**Sequences**	**IC_{50} value (μM)**	**Dose administration model**	**Response (Δ SBP mmHg)**	
Sunflower protein	Alcalase Peptide fraction from affinity purification	0.062 $1.18x10^{-3}$					[61]
Pea protein	Gastrointestinal digest *in vitro*	0.070			50 mg/kg intravenous, SHR	-44, transient and sharp reduction	[144]
Pea protein isolate	Thermolysin, 3kDa MWCO permeate	ND			200 mg/kg oral, SHR 25-30 g/day oral, Han:SPDR-cy rat 3g/day oral, Human, consumption	-19 after 4h -30 at weeks 7 and 8 -6 after 3 weeks	[143]
Chick pea protein	Alcalase		Peptide fractions	0.103-0.117 mg/ml			[41]
Chick pea legumin	Alcalase		Peptide peaks	0.011-0.021 mg/ml			[43]
Common beans Pinto beans Green lentils	Heat treatment and *in vitro* gastro-intestinal digestion	0.78-0.83 0.15-0.69 0.008-0.89					[131]
Red and green lentil protein	*In vitro* gastro-intestinal digestion	0.053- 0.190					[64]
Mung bean protein	Alcalase, 6kDa MWCO permeate	0.64	KDYRL VTPALR KLPAGTLF	26.5 82.4 13.4	600 mg/kg oral, SHR	-30.8 after 6 h	[48]
Mung bean sprout	Raw sprout extract Dried sprout extract Pepsin, trypsin and chymotrypsin				600 mg/kg intragastric, SHR	-41 after 6 h -24 after 3 h -29 after 3 h	[142]
Soy protein	Alcalase	0.065	DLP DG	4.8 12.3			[39]

Source protein	**Enzyme or other process conditions**	**ACE inhibition**	**Identified peptides**		***In vivo* response**		**Ref**
		IC_{50} (mg/ml)	**Sequences**	**IC_{50} value (μM)**	**Dose administration model**	**Response (Δ SBP mmHg)**	
Soybean	Fermentation	0.45	AW GW AY SY GY AF VP AI VG	μg/ml 0.03 0.05 0.07 0.10 0.19 0.48 0.69 1.1	Diet contained 10% v/w of fermented soy product oral, SHR	-20 after 11 week	[184]
Rye	Sourdough *Lactobacillus reuteri* TMW 1.106 and added protease		VPP IPP LQP LLP	9 5 2 57			[149]
Rice	Alcalase	0.14	TQVY	18.2	600 mg/kg hydrolysate 30 mg peptide/kg SHR, oral	-25.6 after 6 h -40 after 6 h	[153]
Maize, α-zein	Thermolysin		LRP LSP LQP	0.29 1.7 2.0	Peptide (LRP) 30 mg/kg SHR, intravenous	-15 2 min after injection	[31]
Wheat bran	Autolysis	0.08	LQP IQP LRP VY IY TF	peptide-fraction 0.14 mg/ml	10 mg/ml oral, SHR	-45 after 2 h	[185] [186]

Table 1. ACE-inhibitory activities *in vitro* and antihypertensive effect *in vivo* of plant protein–derived hydrolysates and peptides

autolysis of protein isolates from the potato tuber tissue enhances ACE-inhibition which may be due to the native proteolytic activity of potato tuber proteins.

Peña-Ramos and Xiong [51] used different enzymes to produce hydrolysates from native and heated soy protein isolates. They reported that using different enzymes resulted in the formation of a mixture of peptides with different degrees of hydrolysis and accordingly different ranges of antioxidant activity. It has been found that antioxidant activity of Alcalase derived hydrolysates is higher than that of other hydrolysates [52-54]. It is also reported that peptides produced by Alcalase have diverse biological activities, including antioxidant activity [55]. In comparison to other proteases, it provided higher yields of antioxidative peptides and develops shorter peptides. Udenigwe and colleagues [55] observed that release of radical scavenging peptides depends on the specificity of protease used in hydrolysis. In another study, flaxseed protein was treated with thermolysin followed by pronase to produce antioxidant peptides [42]. Moreover, pepsin, pancreatin, neutrase and esperase have been used to produce antioxidative hydrolysates and peptides [13, 42, 51, 56, 57].

Simulated gastrointestinal enzymatic process has also been used to mimic normal human digestion of proteins to evaluate the possibility of releasing potent bioactive peptides after normal consumption of food proteins. The combination of pepsin-trypsin-chymotrypsin or pepsin-pancreatin has been used to simulate the gastrointestinal degradation of proteins in humans [58]. Pepsin treatment alone cannot effectively elicit ACE-inhibitory peptides from buckwheat protein, while this enzyme followed by chymotrypsin and trypsin lead to a significant increase in ACE-inhibitory activity [59]. In some studies, plant protein hydrolysates generated during pepsin digestion have potent ACE-inhibitory peptides. Lower ACE-inhibitory activity was found after subsequent digestion with pancreatin, suggesting that the active components were hydrolyzed [60, 61]. For the pea proteins, high ACE-inhibitory activity is reached at the early stage of pepsin hydrolysis and the level is maintained during the small intestine phase using trypsin-chymotrypsin treatment [62]. While digestion of red lentils with trypsin showed moderate ACE inhibition (IC_{50} value of 0.44 mg/ml), addition of pepsin and chymotrypsin clearly improved it (IC_{50} of 0.09 mg/ml) [63, 64].

Careful choice of suitable enzymes and digestion conditions such as optimal temperature, degree of hydrolysis and enzyme-substrate ratio, as well as the control of hydrolysis time, are crucial for obtaining protein hydrolysates with desirable functional and bioactive properties. Hydrolysis can be performed by conventional batch hydrolysis or by continuous hydrolysis using ultrafiltration membranes. The traditional batch method has several disadvantages, such as the relatively high cost of the enzymes and their inefficiency compared to a continuous process, as noted in numerous studies [65, 66]. The hydrolysis process is feasible to scale-up production of peptides from laboratory scale to pilot and industrial plant scales with conserved peptide profiles and bioactivity of the resulting products [67]. The crude protein hydrolysate may be further processed, for example by passage through ultrafiltration membranes, in order to obtain a more uniform product with the desired range of molecular mass [68]. Alcalase hydrolysate from soy isolate was ultrafiltrated with 1-30 kDa membranes and ACE-inhibitory activities were analysed. The IC_{50}-values for 1 kDa and 10 kDa permeates were almost the same, 0.080 and 0.078 mg/ml, respectively, but recovery yield of 10 kDa permeate was much higher than that of 1 kDa permeate [69]. Ultrafiltration membrane reactors have been shown to improve the efficiency of enzyme-catalysed bioconversion, to increase product yields, and to be easily scaled up. Furthermore, ultrafiltration membrane reactors yield a consistently uniform product with desired molecular mass characteristics [65, 68]. Low molecular mass cut-off membranes are useful for concentrating bioactive peptides from the higher molecular mass components remaining, including undigested polypeptide chains and enzymes. Other techniques such as nanofiltration, ion-exchange membranes, or column chromatographic methods can be used in further concentration and purification of the peptides [70].

A number of studies have shown that antihypertensive peptides are liberated during fermentation of milk. Fermented milk products prepared using different strains of lactic acid bacteria have been found to exert antihypertensive and antioxidative activities [71, 72]. Moreover, several antioxidative and ACE-inhibitory peptide sequences have been identified from fermented milk [73]. Only few experimental investigations to produce these compounds by fermentation of plant proteins have been reported. Fermentation of rapeseed and flaxseed proteins with *Lactobacillus helveticus* and *Bacillus subtilis* yields to products containing compounds with ACE-inhibitory and inhibition of lipid peroxidation capacities [74]. Fermented soybean products such as natto, tempeh, and douche also contain antioxidative and ACE-inhibitory peptides due to the action of fungal proteases. The results have indicated that the processing techniques have an impact on the ACE-inhibitory activities of soy products. Different fermented soybean foods showed IC_{50} values of 0.51 mg/ml for tempeh, 1.77 mg/ml for tofuyo, 3.44 and 0.71-17.80 mg/ml for soy sauce, 5.35 and 1.27 mg/ml for miso paste, and 0.16, 0.19 and 0.27 mg/ml for natto [24,75, 76]. Commercial Chinese style soy paste exhibited ACE-inhibitory activities with the lowest and the highest IC_{50} values of 0.012 and 3.241 mg/ml, respectively [77]. Tsai and colleagues [78] fermented soy milk with lactic acid bacteria (*Lb. casei, Lb., acidophilus, Lb. bulcaricus, Streptococcus thermophilus* and *Bifidobacterium longum*) and IC_{50}-value was 2.89 mg/ml after 30 h fermentation. When a protease (Prozyme 6) was added after 5 h fermentation, much lower IC_{50}-value (0.66 mg/ml) was obtained. Natto has shown to have radical scavenging activity and inhibitory effect on the oxidation of rat plasma LDL *in vitro* [79]. The aqueous extracts of Douchi showed radical scavenging activities and chelating ability of ferrous ions. The radical scavenging activities were higher than that of Trolox, an analogue of vitamin E used as a standard [80].

2.2. Structure-activity relationship

ACE-inhibitory peptides are generally short sequences often carrying polar amino acid residues like Pro. This is in agreement with the results of Natesh and co-workers [81] who showed that the active site of ACE cannot accommodate large peptide molecules. The C-terminal tripeptide strongly influences the binding of substrate or a competitive inhibitor to ACE. Many ACE-inhibitory peptides identified from different food sources, structure–activity studies indicated that C-terminal tripeptide residues play a predominant role in competitive binding to the active site of ACE. It has been reported that this enzyme prefers substrates or inhibitors containing hydrophobic (aromatic or branched side chains) amino acid residues at each of the three C-terminal positions. The most effective ACE-inhibitory peptides identified contain Tyr, Phe, Trp, and/or Pro at the C-terminal. In addition, structure–activity data suggests that the positive charge of Lys (ε–amino group) and Arg (guanidine group) as the C-terminal residue may contribute to the inhibitory potency [82-84]. Quantitative structure–activity relationship (QSAR) modelling was used to develop statistical computer models potentially capable of identifying ACE-inhibitory peptides based on structure–activity data [85]. A relationship was found between hydrophobicity and positively charged amino acid in C-terminal position, size of amino acid next to C-terminal position and ACE inhibition of peptides up to six amino acids in length. Moreover, no relationship between N-terminal structure and inhibition activity was found.

The exact mechanism underlying the antioxidant activity of peptides has not fully been understood, as various studies have displayed that they are inhibitors of lipid peroxidation, scavengers of free radicals and chelators of transition metal ions [86, 87]. In addition, it has been reported that antioxidative peptides keep cells safe from damage by ROS through the induction of genes [88]. Antioxidative properties of the peptides are more related to their composition, structure, and hydrophobicity. Tyr, Trp, Met, Lys, Cys, and His are examples of amino acids that cause antioxidant activity. Amino acids with aromatic residues can donate protons to electron deficient radicals. This property improves the radical scavenging properties of the amino acid residues [89, 90]. It is proposed that the antioxidative activity of His containing peptides is in relation with the hydrogen donating, lipid peroxyl radical trapping and/or the metal ion chelating ability of the imidazole group [87, 91]. On the other hand, SH group in Cys has an independently crucial antioxidant action due to its direct interaction with radicals [92]. In addition to the amino acid composition, their correct position in peptide sequence plays an important role in antioxidative properties of peptides. Chen and colleagues [93] designed 28 synthetic peptides following the structure of an antioxidative peptide (Leu-Leu-Pro-His-His) from digestion of soybean protein conglycinin. According to the results, Pro-His-His sequence displayed the greatest antioxidative activity among all tested peptides. The antioxidant activity of a peptide was more dependent on His-His segment in the Leu-Leu-Pro-His-His-domain and its activity was decreased by removing a His residue from the C-terminus. Moreover, substitution of L-His by D-His in a peptide leads reduction of activity [93]. They concluded that the correct position of imidazole group is the key factor influencing the antioxidant activity. Saito and co-workers [94] also studied antioxidative activity of peptides created in two tripeptide libraries. According to their results, for the 114 peptides containing either His or Tyr residues, tripeptides containing two Tyr residues showed higher activity in the linoleic acid peroxidation system than tripeptides containing two His residues. Further, Tyr-His-Tyr showed strong synergistic effects with phenolic antioxidants. It has been shown that certain amino acids can exert higher antioxidative properties when they are incorporated in dipeptides [95] and some peptide bond or its structural conformation can reduce the antioxidant activity of the constituent amino acids [96].

2.3. *In vitro* and *in vivo* activity

The search for *in vitro* ACE-inhibitors is the most common strategy followed in the selection of potential antihypertensive peptides derived from food proteins. *In vitro* ACE-inhibitory activity is generally measured by monitoring the conversion of an appropriate substrate by ACE in the presence and absence of inhibitors. There are several methods, and those based on spectrophotometric [97-99] and high-performance liquid chromatography (HPLC) assays are most commonly utilized [100-102]. Hippuryl-His-Leu (HHL) is one of the oldest and most used methods for determining the ACE activity or inhibition [97, 99-101]. The broadly used spectrophotometric method of Cushman and Cheung [97] is based on the hydrolysis of HHL by ACE to hippuric acid and His-Leu, and the extent of hippuric acid released is measured after its extraction with ethyl acetate. The liberated hippuric acid can also be measured by chromatographic assays avoiding the extraction step [102]. The method based on the substrate 2-furanacryloyl-phenylalanylglycyl-glycine (FAPGG) was developed by Holmquist and colleagues [103] and can be applied in microtiter plates [98]. Substrates such as o-aminobenzoylglycyl-p-nitrophenylalanylproline are designed to perform in fluorometric assays [104, 105].

Specific assays have not yet been developed or standardized to measure the antioxidative capacity of peptides or peptide mixtures. Therefore, assays that are commonly used for measuring antioxidative capacity of non-peptidic antioxidants have been used in the literature to measure the antioxidative capacity of peptides as well. Due to the complexity of oxidative processes occurring in food or biological systems as well as the different antioxidative mechanisms by which various compounds may act, finding one method that can characterize the overall antioxidative potential of food is not an easy task. Nevertheless, methods such as the Trolox equivalent antioxidant capacity (TEAC) assay, oxygen radical absorbance capacity (ORAC) assay, and the total radical-trapping antioxidant parameter (TRAP) assay have been widely reported in the literature for measuring antioxidative capacity of food and biological samples [106, 107]. Commonly used assays include measuring the inhibition of lipid peroxidation in a linoleic acid model system and the capacity to scavenge the 2,2'-azino-bis(3-ethylbenzothiazoline-6-sulphonic acid)/2,2-Diphenyl-1-picrylhydrazyl (ABTS/DPPH) radicals.

In vitro cultured cell model systems allow for rapid, inexpensive screening of compounds for their bioavailability, metabolism, as well as bioactivity, compared to expensive and time-consuming animal studies and human clinical trials. Endothelial cells are currently used as *in vitro* model systems for various physiological and pathological processes, especially in angiogenesis research. Endothelial dysfunction, an initiator of atherosclerosis, is manifested by altered NO and endothelin-1 (ET-1) homeostasis. ET-1 is one of the most potent endogenous vasoconstrictors and impaired NO release precedes the development of atherosclerosis [108]. NO is produced by the enzyme endothelial NO synthase (eNOS) which converts L-Arg in the presence of O_2 and NADPH into L-citruline and NO. There are several assays available to analyze NO, ET-1, or expression of eNOS after exposure to compounds of interest. Use of cell culture models for antioxidant research is particularly important since the studies to date have demonstrated that the mechanism of the action of antioxidants in human health promotion go beyond the antioxidant activity of scavenging free radicals [109]. During experiments, intracellular oxidation of cells can be induced by using a peroxy radical generator or by using hydrogen peroxide (H_2O_2) [110]. The 20,70-dichlorofluorescein diacetate (DCFH-DA) probe can be used to measure the extent of intracellular radical formation with and without added antioxidative compound in order to assess the cellular antioxidant activity (CAA) [111].

The antihypertensive effects can be assessed by *in vivo* experiments using spontaneously hypertensive rats (SHR) that constitute an accepted animal model to study human essential hypertension [88]. A great number of studies have addressed the effects of both short-term and long-term administration of potential antihypertensive milk protein-derived peptides using this animal model [88, 112]. *In vitro* measurements of antioxidant capacity of compounds in interest cannot be directly related to their capacity *in vivo*. Biomarkers of lipid and protein peroxides as well as DNA damage can be assessed to monitor changes in oxidative stress *in vivo*. Only few studies so far have been done on plant derived peptides in animal models or human clinical trials.

2.4. ACE-inhibitory and antioxidant peptides

2.4.1. Potato and other root crops

During the last decade the *in vitro* capacity of tuber protein -derived peptides to inhibit ACE have gained increasing interest (Table 1). Hsu and colleagues [113] reported that yam (*Dioscorea alata*) tuber dioscorin possess high ACE-inhibitory capacity and the digestion with pepsin increased the efficacy further. Moderate ACE-inhibition *in vitro* has been reported for purified yam (*Dioscorea batatas*) tuber mucilage [114] and for an enzymatic digest as well as for an autolysate of yam (*Dioscorea opposita*) tuber extract [115, 116]. However, the potential impact of other compounds, such as phenolic compounds and sugars, on the observed ACE inhibition should be taken into consideration. It has been shown that naturally occurring phenolic compounds, such as flavonoids and proanthocyanidins, have inhibition activity towards ACE [117].

Sweet potato proteins defensin and thioredoxin h2 which were overproduced in *Esherichia coli* showed moderate ACE-inhibitory capacity (IC_{50} of 0.190 and 0.152 mg/ml, respectively) and both proteins showed mixed type inhibitor against ACE using FAPGG as substrate. Hydrolysis with trypsin increased the capacity. Several peptides contained in the hydrolysate with IC_{50} values from 1.31 to 265.43 μM were analyzed [118, 119]. Trypsin inhibitor from the root storage protein of sweet potato, inhibited ACE in a dose-dependent manner (50-200 μg/ml, with 31.9-53.2% inhibition), and the IC_{50} value was 187.96 μg/ml. After digestion with pepsin the ACE-inhibition increased and peptides were designed by simulating the pepsin cutting sites of sporamin A. Finally, ten new ACE-inhibitory peptides showed IC_{50}-values from 2.3 to 849.7 μM [120]. Sweet potato protein isolate digested with 16 different proteases showed variability in digestibility from 44.7 to 97.3% and IC_{50} values from 0.16 to 1.08 mg/ml. Based on these results four most potent enzymes (Thermoase PC 10F, Protease S, Proleather FG-F and Orientase 22BF) were selected and combined effect of enzymes were tested. Combination of Thermoase PC 10F, Protease S and Proleather FG-F produced potent ACE-inhibition (IC_{50} of 0.137 mg/ml). Finally, four different peptides derived from sweet potato storage protein, sporamin, were identified with IC_{50} values from 9.5 to 300.4 μM [121]. The lowest IC_{50} values were obtained for synthetic tripeptides, Ile-Thr-Pro (9.5 μM), Gly-Gln-Tyr (52.3 μM) and Ile-Ile-Pro (80.8 μM).

The protein-rich by-product fraction from potato (*Solanum tuberosum*) starch industry, potato tuber liquid, has been found to be a valuable source of ACE-inhibitory peptides [13]. The ACE-inhibitory activity of potato tuber liquid was moderate and enzymatic digestion was needed to enhance the activity to high level. Alcalase digest showed the highest ACE-inhibitory activity and the digest was chromatographically separated to highly active peptide fractions. The potato liquid Alcalase hydrolysate produced the highest radical scavenging potency even though no statistically significant differences were found among hydrolysates produced by Alcalase, Neutrase and Esperase. Mäkinen and colleagues [47] reported that autolysis of protein isolates from the potato tuber tissue enhances ACE inhibition which may be due to the native proteolytic activity of potato tuber proteins. The results indicated a relevant role of potato tuber storage proteins in the production of ACE-inhibitory peptides during the autolysis. Enrichment of recombinant potato tuber protein to the autolysis enhanced the production of activity significantly, which suggests possibility to enhance potato tuber ACE-inhibitory potential by means of biotechnological tools. Anyhow, more research is needed to characterize and identify the ACE-inhibitory potato peptides and to evaluate the *in vivo* antihypertensive potential.

Recently, antihypertensive effects of some tuber plant –derived protein digests have been evaluated *in vivo* using SHR animal model, although no tuber protein –derived peptides in pure form have been reported. The proteins tested in the *in vivo* trials have concerned the antihypertensive effects of the main storage proteins of the tubers and peptides derived from these proteins. Among the tuber proteins, the *in vivo* antihypertensive effects of yam (*Dioscorea alata*) tuber proteins are the most studied. Lin and co-workers [122] purified storage proteins, dioscorins from yam tubers, that were digested with pepsin and evaluated for their antihypertensive effects in SHR. The maximum effect after single oral administration was observed after 4 h with the dioscorin isolate and after 8 h with the peptic hydrolysate, while the antihypertensive effect of the peptic digest was more pronounced (-33.7 mmHg Mean arterial pressure, MAP) and less transient than that of the dioscorin isolate (-21.5 mmHg MAP). The long-term antihypertensive effect of the dioscorin isolate was tested for 25 days with daily oral administration and the greatest reductions in systolic blood pressure (SBP) and diastolic blood pressure (DBP) were observed on the ninth day. Liu and co-workers [123] tested the antihypertensive effects of different yam tuber products on SHR. The yam tuber alcohol-insoluble solids and water extract before and after heat treatment were observed to decrease the blood pressure after single oral administration. The most pronounced effect with the lowest dose was found with the alcohol-insoluble-solids product, which contained the yam tuber dioscorins. Iwai and Matsue [124] reported moderate antihypertensive effects of an edible tuber *Apios Americana* Medikus in SHR. The animals ingested water extract of the tubers that was rich in Pro. The antihypertensive effect was suggested to be due to Pro-rich peptides, which were released during digestion. Sweet potato protein digest made with combination of three proteases (Thermoase PC10F, Protease S and Proleather FG-F) showed a dose dependent decrease in SBP after single oral administration in SHR [121].

2.4.2. *Oil seed plants-derived peptides*

According to the published data, enzymatic hydrolysis is required to release antihypertensive peptides from oil seed plant proteins. Only few peptide sequences have been identified from oil seed plants. Four ACE-inhibitory peptides were isolated from the rapeseed subtilisin digest of which Ile-Tyr and Arg-Ile-Tyr can be found in the primary structure of napin and Val-Trp and Val-Trp-Ile-Ser exist in the primary structure of cruciferin and ribosomal protein, respectively. Among the peptides isolated, Ile-Tyr and Val-Trp can be considered true ACE-inhibitors because IC_{50} values for these peptides before and after pre-incubation with ACE were found to be the same. Val-Trp-Ile-Ser is a pro-drug type ACE-inhibitor, as pre-incubation with ACE of Val-Trp-Ile-Ser intensified inhibitory activity of this peptide [125]. In addition, to these rapeseed peptides, two peptide sequences with high ACE-inhibitory capacity were identified from canola meal hydrolysed with Alcalase, Val-Ser-Val and Phe-Leu, located in the primary structure of canola napin and cruciferin proteins [46]. Low-molecular weight cationic peptide fractions from flaxseed protein hydrolysed by Alcalase or thermolysin showed concentration dependent ACE-inhibition (IC_{50} 0.0275-0.151 mg/ml) [126]. The Alcalase cationic peptide and thermolysin hydrolysate showed mixed type inhibition of ACE activity. Several peptides were detected in a cationic peptide fraction of a trypsin & Pronase digest of flaxseed, which showed moderate ACE inhibition *in vitro* and antihypertensive effects in SHR [44]. A pentapeptide Trp-Asn-Ile-Leu-Asn-Ile-Leu and a hexapeptide Asn-Ile-Leu-Asp-Thr-Asp-Ile-Leu were identified from flaxseed protein digested with an *in vitro* digestion model [127]. Anyhow, ACE-inhibitory activity of these flaxseed derived peptides has not been evaluated individually and thus, the ACE-inhibitory capacity of these peptides in a pure form is not clear [44, 126, 127].

Despite the high *in vitro* ACE-inhibitory potency of Alcalase digests of oil seed proteins, their antihypertensive effects *in vivo* have not been evaluated yet. Subtilisin digest of rapeseed and tryptic digest of flaxseed have shown antihypertensive properties in SHR. Marczak and co-workers [125] studied the Subtilisin and pepsin digests of rapeseed in SHR. Subtilisin digest of rapeseed protein showed dose dependent antihypertensive effect after oral administration to SHR and its effect was significant even at a single dose of 0.15 g/kg. The Subtilisin digest was subjected to hydrolysis with different proteases to simulate gastrointestinal digestion *in vitro* and the ACE-inhibitory activity was changed only slightly indicating that ACE-inhibitory peptides present in the Subtilisin digest are relatively resistant. The antihypertensive activities of Val-Trp, Val-Trp-Ile-Ser, Ile-Tyr and Arg-Ile-Tyr were tested following oral administration to SHR. The maximum hypotensive activity of Val-Trp, Val-Trp-Ile-Ser and Ile-Tyr occurred 2 h after administration, whereas Arg-Ile-Tyr (rapakinin) had the maximum effect 4 h after administration. All peptides displayed dose-dependent antihypertensive effect. Hypotensive activity of the peptides was compared after oral administration to young (19-20 weeks) and old (28-30 weeks) SHR. Usually the hypotensive effect of ACE inhibitors in old SHR is lower than in young SHR. The hypotensive effects of Val-Trp, Val-Trp-Ile-Ser and Ile-Tyr were lower in old rats, but in the case of rapakinin the effect was similar in old and young rats. The authors suggested that another mechanism besides ACE inhibition may be involved in hypotensive effect of rapakinin. Recently, the hypotensive effect of rapakinin was found to be mediated mainly by the prostaglandin IP (PGI2-IP) receptor followed by CCK1 receptor-dependent vasorelaxation [128]. In addition to hypotensive effects, the napin-derived peptide rapakinin has been reported to possess multifunctional properties. Rapakinin dose-dependently decreased food intake and gastric emptying after oral administration at a dose of 150 mg/kg in mice [129] and recently, Yamada and colleagues [130] reported anti-opioid activity related to the hypotensive effects for rapakinin.

Peptide fraction from the enzymatic digest of flaxseed protein was recently reported to possess hypotensive activity in SHR [44]. An Arg-rich peptide fraction was produced from flaxseed protein using trypsin and pronase and by subsequent concentration with combined electrodialysis and ultrafiltration. The hypotensive effect of the flaxseed derived Arg-rich peptide fraction was tested on SHR and the effects were compared to the effects of inherent flaxseed protein isolate and amino acid form of Arg. The maximum hypotensive effect of the cationic peptide fraction was observed already 2 h after oral administration while the amino acid form of Arg showed lower hypotensive activity, -10.3 mmHg. On the other hand, the flaxseed protein isolate exhibited a slow-acting hypotensive effect with maximum of -18.4 mmHg (SBP) at 6 h after the administration. The hypotensive effect of the Arg-rich peptide fraction was longer-lasting when compared to the free amino acid form of Arg and the authors suggested that this might be related to more efficient absorption of peptides and the ability of peptides to translocate directly into the cells, obviating the need for transporters. The observed hypotensive activity of flaxseed protein and peptide fraction could be due to vasodilatory activity of NO synthesized from the Arg through the L-Arg-NO pathway in the vascular endothelium, or ACE- and renin-inhibition observed *in vitro* by the cationic peptide fraction.

2.4.3. Legume derived peptides

Several enzymes have been used to produce pulse protein hydrolysates having bioactive properties. It has been suggested that hydrolysates of chickpea legume and mung bean obtained by Alcalase treatments are good sources of ACE-inhibitory peptides [43, 48]. Potential ACE-inhibitory potencies of common dry beans, dry pinto beans and green lentils increased during *in vitro* gastrointestinal digestion have been reported, with IC_{50} values of 0.78–0.83, 0.15–0.69 and 0.008–0.89 mg protein/ml, respectively [131]. In addition, 15 min heat treatment effectively increased the ACE-inhibitory activity of the stomach digest [43, 131]. Digestion simulating the physiological conditions of pea proteins sufficed to achieve the highest ACE-inhibitory activity with IC_{50} value of 0.076 mg/ml [132]. Furthermore, it has been suggested that red lentil protein hydrolysates have ACE-inhibitory properties. The ACE-inhibitory property of the tryptic hydrolysates varied as a function of the protein fraction with the total lentil protein hydrolysate having the lowest IC_{50} (0.440 ± 0.004 mg/ml). This indicates that lentil varieties having higher amounts of legumin and albumin proteins may have higher ACE-inhibitory properties [63]. Pedroche and co-workers [41] hydrolysed chickpea protein isolate with Alcalase to produce a bioactive hydrolysate having ACE-inhibitory properties with an IC_{50} value of 0.190 mg/ml. Four peptide-fractions with average molecular weight of 900 Da, representing peptides with six to eight amino acid residues were isolated. The IC_{50} values were 0.103-0.117 mg/ml and two of the peptides showed competitive and two showed uncompetitive mechanism [41]. In addition, six ACE-inhibitory peptides with IC_{50} values ranging from 0.011 to 0.012 mg/ml have been isolated from chickpea hydrolysed by Alcalase. All these peptides contained Met and were rich in other hydrophobic amino acids [43]. Two lentil varieties were hydrolysed with different enzymes and IC_{50} values ranged between 0.053 and 0.190 mg/ml. Furthermore, the inhibition mechanism investigated using Lineweaver–Burk plots revealed a non-competitive inhibition of ACE with inhibitor constants (Ki) between 0.16 and 0.46 mg/ml [40]. Three dipeptides, Ile-Arg, Lys-Phe and Glu-Phe were isolated and identified from Alcalase hydrolysate of pea protein isolate. The peptides showed strong inhibitions (IC_{50} values <25 mM) of ACE and renin [133].

Enzymatic hydrolysate of soy protein showed a moderate ACE-inhibitory activity (0.034 mg/ml) [39] as compared with values reported ranging from 0.021 to 1.73 mg/ml [5, 6, 49, 70]. Lo and co-workers [134] applied the dynamic model for the *in vitro* digestion of isolated soy proteins. They concluded that ACE-inhibitory activity was dependent on the digestion time and the heat treatment of soy protein. Pepsin hydrolysis of isolated soy protein produced peptides with a higher ACE-inhibitory activity due to the increased digestion time. Hydrolysis by pancreatin produced soy peptides with higher ACE-inhibitory activity as compared to pepsin hydrolysates, but decresed inhibitory activity appeared after longer digestion time. These results suggest that at the longer digestion time pancreatin may have hydrolysed the peptides from pepsin digestion that had strong ACE-inhibitory activity, and then turned them into peptides with lower ACE-inhibitory activity.

Pea protein hydrolysates by thermolysin contained low molecular weight (<3 kDa) peptides with various antioxidant activities that were dependent on the amounts of hydrophobic and aromatic amino acid constituents [135] (Table 2). Peptide fractions with the least cationic property had significantly stronger scavenging activity against DPPH and H_2O_2. Generally, the scavenging of $O_2^{\cdot -}$ and H_2O_2 was negatively related with the cationic property of the peptide fractions [136]. Chick pea hydrolysate with antioxidant activities was prepared from chickpea protein isolates by Alcalase. This hydrolysate was separated with Sephadex G-25. Four fractions were obtained, and fraction IV had the highest antioxidant activities assayed by free radical scavenging effects [137]. The active peptide was identified as Asn-Arg-Tyr-His-Glu. This peptide quenched the free radical sources DPPH, hydroxyl, and superoxide free radicals. Furthermore, the inhibition of the peptide on lipid peroxidation was greater than that of α-tocopherol [57].

Different hydrolysis conditions of soy protein isolates have resulted in peptide mixtures with different antioxidant properties. Native and heated soy protein isolate hydrolysed with different enzymes resulted in different degree of hydrolysis ranging from 1.7-20.6 with antioxidant activity ranging from 28% to 65% [138]. Zhang and coworkers [139] used three microbial proteases to produce hydrolysates with degree of hydrolysis values from 13.4% to 26.1% and with different oxygen radical absorbance capacity (ORAC), DPPH-radical

Source of proteins or hydrolysates	*In vitro* methods used in measuring antioxidant capacity	Enzymes or other process conditions used	Antioxidative peptides identified	Ref
Potato liquid fraction	$ABTS^{+-}$ scavenging	**Alcalase**[1], Esperase, Neutrase	Not specified	[13]
Potato protein concentration	$ABTS^{+-}$- scavenging Emulsion oxidative stability	Alcalase	SSEFTY IYLGQ	[14]
Soybean proteins	Liposome oxidizing system	**Chymotrypsin**, Pepsin, Papain, **Flavourzyme,** Alcalase, Protamex	Not specified	[51]
Soybean protein b-conglycin	Linoleic acid peroxidation system	Protease M, Protease N, Protease P, **Protease S**	VNPHDHQN LVNPHDQN LLPHH LLPHHADADY VIPAGYP LQSGDALRVPSGTTYY	[141]
Yellow pea seed protein	Radical (DPPH, O_2^-, H_2O_2) scavenging and inhibition of linoleic acid oxidation	Thermolysin	NRYHE	[135]
Barley glutelin	Radical scavenging capacity (DPPH/O_2^-/OH^-), Fe^{2+-} chelating effect and reducing power	Alcalase	QKPFPQQPPF PQIPEEF LRTLPMSVNVPL	[54]
Wheat gluten	Linoleic acid peroxidation system	Pepsin	LQPGQGQQG AQIPQQ	[56]
Rice endosperm protein	O_2^-, OH and DPPH radical sacavenging capacity, Linoleic acid peroxidation system	Alcalase, Chymotrypsin, **Neutrase,** Papain, Flavorase	FRDEHKK KHNRGDEF	[151]

1 The enzyme indicated in bold is the most effective of the enzymes to produce antioxidative activity/peptides

Table 2. Antioxidative capacity of plant protein-derived hydrolysates and peptides

scavenging activities as well transition metal chelating activities. Chen and coworkers [140] isolated 6 antioxidative peptide fragments from the digests of β-conglycinin, a main soybean protein component, by using protease S from *Bacillus* sp. The antioxidant activity of the soybean hydrolysates, based on a linoleic acid oxidation system study, was attributed to the Leu-Leu-Pro-His-His peptide sequence [89, 93]. A potent antioxidant peptide, with inhibitory activity of lipid peroxidation, was isolated from soy protein isolate hydrolysed by

Alcalase. Purification of peptide by ultrafiltration and chromatographic techniques enhanced the specific activity 67.9-fold compared to hydrolysate. The final potent antioxidant peptide contained hydrophobic amino acids and among them, Phe was especially abundant [141].

Pea and mung bean protein digests have been reported to possess antihypertensive activity in SHR [48, 142]. Mung bean protein hydrolysate prepared with alcalase decreased significantly SBP (-30.8 mmHg) of SHR 6 h after single oral administration at a dose of 600 mg/ml. The blood pressure-lowering effect continued for at least 8 h, and the blood pressure returned to initial levels at 12 h after administration [48]. Single administration of mung bean raw sprout extract (at dose of 600 mg/kg) reduced significantly SBP (-40 mmHg) 6 h after administration. Plasma ACE activities in the treated rats also decreased (0.007 Unit/ml). Long-term intervention (1 month) test showed that blood pressure in the treated animals fluctuated according to the treatments. While raw sprout extract showed effective results after 1 week of intervention, dried sprout extracts did not have significant effects until 2 weeks [142]. Pea protein hydrolysate was made by thermolysin action followed by membrane filtration. Oral administration of the pea protein hydrolysate, containing <3 kDa peptides, to SHR at doses of 100 and 200 mg/kg body weight led to a lowering of SBP, with a maximum reduction of 19 mmHg at 4 h. In contrast, orally administered unhydrolysed pea protein isolate had no blood pressure reducing effect in SHR, suggesting that thermolysin hydrolysis may have been responsible for releasing bioactive peptides from the native protein [143]. Pea protein peptides from *in vitro* gastrointestinal digestion were observed to absorb poorly with *in vitro* model and the hypotensive effect was tested with intravenous administration [144].

2.4.4. Cereals

The seed storage proteins of wheat, barley, rye, and oats contain known ACE-inhibitory di- and tripeptides in their primary structures. Barley and barley by-products extract possesses a biological activity such as free radical scavenging activity, tyrosinase, xanthin oxidase, and ACE-inhibition effect [145]. Hydrolysates of barley prolamin fraction exhibited the highest antioxidant and ACE-inhibitory activity compared to other protein fractions and protein isolate. Moreover, positive correlations were obtained between antioxidant and ACE-inhibitory activity and the degree of hydrolysis of hydrolysed protein fractions and protein isolate [146].

The computer analysis of amino acid sequences of wheat gliadins made by means of BIOPEP database [147] showed the presence of fragments that are homological with the sequences regarded as antihypertensive peptides. They were: Leu-Gln-Pro (α-, β- and γ-gliadins), Pro-Tyr-Pro (α-, β- and γ-gliadins), Ile-Pro-Pro (α-and β-gliadins), Leu-Pro-Pro (γ-gliadins) and Leu-Val-Leu (γ-gliadins). Bioinformatic analysis of cereal proteins sequences revealed that particularly four tripeptides with known ACE-inhibitory activity, Leu-Gln-Pro, Val-Pro-Pro, Ile-Pro-Pro, and Leu-Leu-Pro, are frequently encrypted in the primary structure of rye secalin, wheat gluten, and barley hordein. Sourdoughs fermented with different strains showed different concentrations of Leu-Gln-Pro and Leu-Leu-Pro. These differences corresponded to strain-specific differences in endopeptidase (PepO) and aminopeptidase (PepN) activities. The highest levels of peptides Val-Pro-Pro, Ile-Pro-Pro, Leu-Gln-Pro, and Leu-Leu-Pro, 0.23, 0.71, 1.09, and 0.09 mmol/ kg dry matter (DM), respectively, were observed in rye malt: gluten sourdoughs fermented with *Lactobacillus reuteri* TMW 1.106 and added protease [148]. Several clinical trials with hypertensive humans show a moderate but relatively consistent reduction of blood pressure upon consumption of the fermented milk products containing Val-Pro-Pro and Ile-Pro-Pro [88, 112]. Cheung and co-workers [149] used *in silico* approach to evaluate the potential of using oats as a protein source for generation of ACE-inhibitory peptides, and to screen for candidate enzymes to hydrolyse the oat protein for this purpose. It was found that thermolysin under high enzyme to substrate ratio (3%) and short time (20 min) conditions produced strong and stable ACE-inhibitory activity.

Barley glutelin possess high hydrophobic amino acid content and enzymatic release by Alcalase produced peptides that had antioxidant capacity. Large size peptides possessed stronger DPPH scavenging activity and reducing power, whereas small-sized peptides were more effective in Fe^{2+} and hydroxyl radical scavenging activity [54]. Pepsin hydrolysis of by-product of the wheat starch industry has shown antioxidant properties. Especially ultrafiltration produced fraction showed strong inhibition of the autoxidation of linoleic acid and scavenging activity of DPPH, superoxide and hydroxyl free radicals. The molecular weight distribution ranged from 0.1-1.7 kDa and high content of total hydrophobic amino acid was found in the active fraction [150]. Rice endosperm protein was, respectively, digested by five different protease treatments (Alcalase, chymotrypsin, Neutrase, Papain and Flavorase), and Neutrase produced the most desirable quality of antioxidant peptides. Two different peptides showing strong antioxidant activities were isolated from the hydrolysate using consecutive chromatographic methods. Especially, Phe-Arg-Asp-Glu-His-Lys-Lys significantly inhibited lipid peroxidation in a linoleic acid emulsion system more effectively than α-tocopherol [151].

The Alcalase-generated rice hydrolysate showed ACE-inhibitory activity with an IC_{50} value of 0.14 mg/ml. A potent ACE-inhibitory peptide with the amino acid sequence of Thr-Gln-Val-Tyr (IC_{50} of 18.2 μM) was isolated and identified from the hydrolysate. Single oral administration of the hydrolysate (600 mg/kg) and Thr-Gln-Val-Tyr (30 mg/kg) showed significantly decreased blood pressure in SHR, -25.6 and -40 mmHg SBP, respectively, after 6h [152]. Three strong ACE-inhibitors with the Leu-Arg-Pro, Leu-Ser-Pro and Leu-Gln-Pro sequences were isolated from maize α-zein hydrolysed with thermolysin. After intravenous administration of these peptides (30 mg/kg body weight), SBP was found to decrease up to a maximum of 15 mmHg [31]. Moreover, a tripeptide (Ile-Val-Tyr) isolated from wheat germ hydrolysate reduced MAP of 19.2 mmHg at dose of 5 mg/ml in SHR [153].

3. Bioavailability

Bioavailability is a major issue when establishing correspondence between *in vitro* and *in vivo* activities of bioactive peptides. The capacity to reach target organ in an active conformation determines the physiological effect of bioactive peptides. Various processes take place after oral administration of a bioactive peptide and need to be considered on the final activity. It's highly likely that antihypertensive reported peptide sequences are subjected to alteration before the final activity *in vivo* after the various steps, such as attack of gastrointestinal enzymes and brush border peptidases, absorption through the intestinal barrier, attack of intracellular peptidases in the transcellular absorption and plasma enzymes after the peptides have entered the circulation [58, 154]. Therefore, the different aspects of bioavailability of antihypertensive peptide sequences have attracted a growing interest in the last years. The possibility of modification or breakdown of peptides during the gastrointestinal digestion is one of the most important factors to be considered when evaluating potential food-derived peptides for promotion of human health. Various models have been implemented to simulate gastrointestinal digestion; static and dynamic models which both differ in enzymes applied and reaction conditions, such as agitation and duration. For instance, authors in references [155] and [47] utilized human digestive liquids to model digestion *in vitro* whereas several reports have concerned implementation of porcine enzyme mixtures [e.g. 60, 127, 132]. In addition to studying the resistance of antihypertensive peptide sequences against the digestive enzymes, the models have been utilized in order to produce bioactive peptides, plant derived ACE-inhibitory peptides among them. For instance pea, lentil, bean and chickpea proteins have been reported to release ACE-inhibitory peptides during *in vitro* digestion [40, 127, 132]. The digestive characteristics of commercial proteases mixtures are known to differ from those of human origin [155]. Zhu and co-workers [156] reported that the antioxidative activity of a zein hydrolysate, which had previously shown antioxidant activity in aqueous solutions and in food systems, was either decreased or improved during the course of *in vitro* digestion, depending on the enzymes encountered and the duration of hydrolysis. Thus, direct comparison of the results between the different models is difficult. A consensus concerning the basic parameters would be relevant in order to harmonize the various *in vitro* digestion models.

Study of intestinal absorption *in vitro* is another common aim when elucidating the bioavailability. It has been indicated that a small portion of bioactive peptides can pass the intestine barrier and although it is usually too small to be considered nutritionally important, it can present the biological effects in tissue level [157, 158]. Molecular size and structural properties, such as hydrophobicity, affect the major transport route for peptides [158]. Research findings indicate that peptides with 2–6 amino acids are absorbed more readily in comparison to protein and free amino acids. As the molecular weight of peptides increases, their chance to pass the intestinal barrier decreases. Peptides are transported by active transcellular transport or by passive process [159]. The absorption studies are commonly performed with the monolayer of intestinal cell lines, such as Caco-2 cells, simulating intestinal epithelium, and analysis of peptides and metabolites in serum after *in vivo* and clinical studies. Foltz *et al.* [160] investigated the transport of Ile-Pro-Pro and Val-Pro-Pro by using three different absorption models and demonstrated that these tri-peptides are transported in small amounts intact across the barrier of the intestinal epithelium. In another study, the absolute bioavailability of the tri-peptides in pigs was below 0.1%, with an extremely short elimination half-life ranging from 5 to 20 min [161]. In humans, maximal plasma concentration did not exceed picomolar concentration [162]. Studies concerning absorption of plant derived peptides are rare, but milk-derived peptide Leu-His-Leu-Pro-Leu-Pro is an interesting example of a peptide with evaluation of bioavailability. This peptide resisted gastrointestinal simulation, but cellular peptidases digested the peptide to His-Leu-Pro-Leu-Pro before crossing Caco-2 cell monolayer [163, 164]. The degradation product, His-Leu-Pro-Leu-Pro, has been demonstrated to absorb in human intestine as it has been identified in human plasma after oral administration [165].

Fujita and colleagues [166] established a bioavailability factor in relation to antihypertensive activity and ACE inhibition mechanism. The classification is based on inhibitor type and substrate type, the possible conversion of peptides by ACE into peptides with weaker activity and pro-drug type inhibitors, or possible conversion of peptides into true inhibitors by ACE or gastrointestinal proteases. A delayed antihypertensive effect is characteristic for pro-drug type peptides as they need to degrade further to reach the final active form [167, 168]. For instance, flaxseed protein showed pro-drug type characteristics compared to hydrolysed cationic peptide fraction [126]. The protein fraction showed a delayed hypotensive effect in SHR comparable to captopril (3 mg/kg body weight) and the effect was more sustained than the effect of the digested peptide fraction. The slow-acting character of the protein fraction was expected since the digestion of the proteins. Anyhow, more research is needed to identify the active peptide sequences released in the digestive tract and to evaluate the bioavailability of these peptides.

It can be deduced due to the incomplete bioavailability of peptide following oral ingestion, a peptide with pronounced antioxidant activity *in vitro* may exert little or no activity *in vivo*. However, bypass routes which increase the chance of peptide absorption can diminish the problem and it is possible that *in vivo* antioxidant activity can be higher than *in vitro* activity. In such cases, bioactive peptides may display their biological functions by mechanisms other than what is applied in experiment. In addition, it has been suggested that the strong *in vivo* activity can be due to increased activity of peptides following their breakdown by gastrointestinal proteases [88].

The improvement and optimization of bioavailability of antihypertensive peptides have gained a great interest during the last decade. The improvement of limited absorption and stability of peptides has been a goal when evaluating their effectiveness. For example, some carriers interact with the peptide molecule to create an insoluble entity at low pH, which later dissolves and facilitates intestinal uptake, by enhancing peptide transport over the non-polar biological membrane [169]. Bioavailability of bioactive tri-peptides (Val-Pro-Pro, Ile-Pro-Pro, Leu-Pro-Pro) was improved by administering them with a meal containing fibre, as compared to a meal containing no fibre. High methylated citrus pectin was used as a fibre [170]. Among drug delivery systems, emulsions have been used to enhance oral bioavailability or promoting absorption through mucosal surfaces of peptides and proteins [169]. Individually, various components of emulsions have been considered as candidates for improving bioavailability of peptides. Anyhow, it seems that no general strategy for improving bioavailability of antihypertensive peptides exists and due to the number of processes involved and different characteristics of peptides depending on the sequence each case must be studied. Many strategies are currently demonstrated for enhancing bioavailability [171], among them microencapsulation for controlled release of the active compounds, stabilization of the active molecules to improve transportation through the intestinal barrier and provide resistance against degradation, and development of highly stabile peptide analogues [172-174].

4. Health benefits

ACE-inhibitory peptides have been studied extensively in the past two decades and ACE-inhibition is the main mechanism concerning bioactive peptides with proven antihypertensive effects. ACE is a constituent enzyme of the Renin-Angiotensin-Aldosterone System (RAAS), which is a crucial regulator in human physiology. It controls blood pressure, fluid and electrolyte balance and affects the heart, vasculature and kidney [2]. Among the food-derived ACE-inhibitory peptides milk-derived peptides are the most extensively studied. The relevance of vegetable proteins as a source of antihypertensive peptides is increasing and several *in vivo* studies performed in SHR have demonstrated that plant-derived ACE-inhibitory protein hydrolysates and peptides significantly reduce blood pressure, either after oral or intravenous administration. For instance, a clinical randomized, placebo-controlled crossover study was performed in order to elucidate further the antihypertensive potential of yam tuber dioscorins [175]. The dioscorin meal or placebo was intervened as a morning drink daily for five weeks, followed by a washout stage for one week and the trial was then crossed over for five weeks. The SBP and DBP values were decreased after the five weeks of dioscorin meal intervention. The clinical trial as well as the animal trials with dioscorin intervention suggests that the gastrointestinal digestion may produce antihypertensive peptides from the yam tuber dioscorins.

Furthermore, related to the RAAS, plant derived ACE-inhibitory peptides have been reported to possess inhibition activity against renin, the first and rate-determining enzyme in RAAS [2]. The inhibition of renin is being suggested as a major alternative in hypertension prevention. The first direct renin-inhibitor, aliskiren, is currently under phase III trials to evaluate its potential as an antihypertensive drug [176]. Thermolysin digest of pea protein decreased remarkably the renal expression of renin mRNA levels *in vivo* and lowered plasma levels of angiotensin II, thus the reduction in blood pressure in SHR and human subjects was likely due to the effects on the renal angiotensin system [143]. Pea-derived peptides Ile-Arg, Lys-Phe and Glu-Phe showed strong inhibitions *in vitro* studies of ACE and renin [133] as well as ACE-inhibitory peptide fractions from flaxseed protein hydrolysates possessed inhibition also against renin [126, 177, 178].

Opioid receptors are involved in various physiological phenomenons, e.g. in the regulation of blood pressure and circulation, and these receptors are related to the antihypertensive properties of some food derived peptides. Other vasodilatory substances, such as ET-1, have also been suggested to be involved in the antihypertensive effects of food-derived peptides [173, 178, 179]. However, peptide sequences derived specifically from plant proteins inducing endothelial NO liberation have not been reported this far. A cationic Arg-rich peptide fraction from flaxseed, which possessed hypotensive effects in SHR, was suggested to mediate blood pressure through vasodilatory activity of NO synthesized from Arg. The observed effect might also be due to ACE- and renin-inhibition and in-depth research is needed to measure the renin and ACE protein levels and activities in SHR tissues and plasma and to specify the prior mechanism of antihypertensive action [44].

Calmodulin (CaM) –dependent cyclic nucleotide phosphodiesterase (CaMPDE) regulates a large variety of cellular functions and excessive levels of CaM and CaMPDE play important roles in many physiological conditions, symptoms of cardiovascular disease among them. Recently, food derived peptides capable to inhibit CaMPDE have been reported, flaxseed and pea protein derived peptides among them [125, 133, 178]. Oxidative stress is a crucial causative factor for the initiation and progression of hypertension and CVD. Increased production of ROS, such as H2O2 and superoxide anion, reduced NO synthesis, and decreased bioavailability of antioxidants have been demonstrated in experimental and human hypertension. Diet rich in antioxidants can reduce blood pressure and thus, antioxidant properties of food-derived peptides may also affect on blood pressure regulation [180, 181]. Several food derived peptides have been reported to possess dual (ACE-inhibition and antioxidant) activity, among them plant protein derived hydrolysates of flaxseed [44, 55,182], rapeseed [47], potato [13] and yam [115, 116].

New mechanisms of action of antihypertensive peptides have been demonstrated in the recent years. The antihypertensive effect of a rapeseed derived tri-peptide rapakinin, was suggested to be mediated through other mechanism than ACE-inhibition [125]. Later on, different mechanisms were considered and the vasorelaxing activity of rapakinin was not blocked by eNOS inhibitor, while antagonists of IP and CCK1 receptor blocked the vasorelaxing effect of rapakinin significantly [128]. The results demonstrated that rapakinin relaxes the mesenteric artery of SHR through the PGI2-IP receptor followed by CCK pathway and the antihypertensive activity is mediated mainly by the PGI2-IP CCK-CCK1 receptor-dependent vasorelaxation. Moreover, inhibition of platelet-activating factor acetylhydrolase (PAF-AH) is suggested to play a crucial role in the hypertension prevention. PAF-AH is a circulatory enzyme secreted by inflammatory cells and it is involved in atherosclerosis. The discovery and application of natural PAF-AH into health promoting foods open up considerable potential [183].

5. General conclusions

The interest on foods possessing health-promoting or disease-preventing properties has been increasing. So far most of the studies on antihypertensive peptides have been done on milk protein-derived peptides. In fact, much work has been done with dietary antihypertensive peptides and evidence of their effect in animal and clinical studies. However, it has been highlighted that there is a huge potential for obtaining antihypertensive peptides from protein sources other than milk. Much work has been done on plant protein hydrolysates and their activity *in vitro*. So far only limited number of peptides has been identified from plant proteins. In addition very little is known on the activity of these hydrolysates and peptides in animal or in humans. These findings open up an interesting field aiming to revaluation of plant derived protein-rich by-products formed in food industry processes in remarkable amounts.

Certain aspects, such as identification of the active form of the peptides in the organism and the different mechanisms of action that contribute in the antihypertensive effect still need to be further investigated. Recent advances on specific analytical techniques enable to follow small amounts of the peptides or derivatives in complex matrices and biological fluids. This will allow performing the kinetic studies in model animals and humans. Similarly, identifying novel and more complex biomarkers of exposure and activity by advances in new disciplines such as nutrigenomic and nutrigenetic will open new ways to follow bioactivity in the organism. There is still poor knowledge on the resistance of peptides to gastric degradation, and low bioavailability of peptides has been observed. This reinforces the need of various strategies to improve the oral bioavailability of peptides.

More emphasis has been put on the legal regulation of the health claims attached to the products. Systematic approaches for review and assessment of scientific data have been developed by authorities around the world. The scientific evidence on the beneficial effects of the product should be enough detailed, extensive and conclusive for the use of a health claim in the functional food product labeling and marketing. First, it is necessary to identify and quantify the active sequences in the product. It is mandatory to monitor the hydrolytic or fermentative industrial production process as the antihypertensive peptides are only minor constituents in highly complex food matrices. Second, the antihypertensive effect in humans as well as the minimal dose needed to show the effect has to be proven in extensive investigations to fulfill the requirements of the legislation concerning functional foods. Besides being based on generally accepted scientific evidence, the claims should be well understood by the average consumer. Japan is the pioneer in the area of regulation of the health claims concerning food products. The concept of Foods for Specified Health Use (FOSHU) was established in 1991. In EU, the European Regulation on nutrition and health claims was established in January 2007 and the regulations are governed by European Food Safety Authority (EFSA).

Author details

Anne Pihlanto and Sari Mäkinen
MTT, Biotechnology and Food Research, Jokioinen, Finland

6. References

[1] World Health Organisation (2011) Cardiovascular Diseases (CVD's). Fact sheet N°317.

[2] Lavoie JL, Sigmund CD (2003) Minireview: Overview of the renin-angiotensin system - An endocrine and paracrine system. Endocrinol. 144: 2179-2189.

[3] Landmesser U, Spiekermann S, Dikalov S, Tatge H, Wilke R, Kohler C, Harrison DG, Hornig B, Drexler H (2002) Vascular oxidative stress and endothelial dysfunction in

patients with chronic heart failure - Role of xanthine-oxidase and extracellular superoxide dismutase. Circulation 106: 3073–3078.

[4] Yao EH, Yu Y, Fukuda N (2006) Oxidative stress on progenitor and stem cells in cardiovascular diseases. Curr. pharm. biotechnol. 7: 101-108.

[5] Pihlanto A, Korhonen H (2003) Bioactive peptides and proteins. Adv. food res. 47: 175-276.

[6] Guang C, Phillips RD (2009) Plant food-derived angiotensin I converting enzyme peptides. J. agric. food chem. 57: 5113-5120.

[7] Pihlanto A, Mäkinen S, Mattila P (2012) Potential health-promoting properties of potato-derived proteins, peptides and phenolic Compounds. In: Caprara C, editor. Potatoes: Production, Consumption and Health Benefits. Nova Publishers pp. 173-194.

[8] Heibges A, Salamini F, Gebhardt C (2003) Functional comparison of homologous members of three groups of Kunitz-type enzyme inhibitors from potato tubers (Solanum tuberosum L.). Mol. gen. genet. 269: 535-541.

[9] Kim J-Y, Park C-S, Hwang I, Cheong H, Nah J-W, Hahm K-S, Park Y (2009) Protease inhibitors from plants with antimicrobial activity. Int. j. mol. sci. 10: 2860–2872.

[10] Scherer GFE, Ryu SB, Wang XM, Matos AR, Heitz T (2010) Patatin-related phospholipase A: nomenclature, subfamilies and functions in plants. Trends plant sci. 15: 693-700.

[11] Bauw G, Nielsen HV, Emmersen J, Nielsen KL, Jorgensen M, Welinder KG (2006) Patatins, Kunitz protease inhibitors and other major proteins in tuber of potato cv. Kuras. FEBS J. 273: 3569-3584.

[12] Liu YW, Han CH, Lee MH, Hsu FL, Hou WC (2003) Patatin, the tuber storage protein of potato (*Solanum tuberosum L.*), exhibits antioxidant activity *in vitro*. J. agric. food chem. 51: 4389-4393.

[13] Pihlanto A, Akkanen S, Korhonen HJ (2008). ACE-inhibitory and antioxidant properties of potato (Solanum tuberosum). Food chem. 109: 104-112.

[14] Cheng Y, Xiong YL, Chen J (2010) Fractionation, separation, and identification of antioxidative peptides in potato protein hydrolysate that enhance oxidative stability of soybean oil emulsions. J. food sci.75: C760-C765.

[15] Hill AJ, Peikin SR, Ryan CA, Blundell JE (1990) Oral administration of proteinase inhibitor II from potatoes reduces energy intake in man. Physiol. behave. 48: 241-246.

[16] Bos C, Airinei G, Mariotti F, Benamouzig R, Bérot S, Evrard J, Fénart E, Tomé D, Gaudichon C (2007) The poor digestibility of rapeseed protein is balanced by its very high metabolic utilization in humans. J. nutr. 137: 594-600.

[17] Mariotti F, Hermier D, Sarrat C, Magné J, Fenart E, Evrard J, Tome D, Huneau JF (2008) Rapeseed protein inhibits the initiation of insulin resistance by a high-saturated fat, high-sucrose diet in rats. Br. j. nutr. 100: 984-991.

[18] Ooman DB (2001) Flaxseed as a functional food source. J. sci. food agric. 81: 889-894.

[19] Bazzano LA, He J, Ogden LG, Loria C, Vupputuri S, Myers L, Whelton PK (2001) Legume consumption and risk of coronary heart disease in US men and women. Arch. intern. med. 161: 2573-2578.

[20] Reynolds K, Chin A, Lees KA, Nguyen A, Bujnowski D, He J (2006) A meta-analysis of the effect of soy protein supplementation on serum lipids. Am j. cardiol. 98: 633-640.

[21] Gibbs BF, Zougman A, Masse R, Mulligan C (2004) Production and characterization of bioactive peptides from soy hydrolysate and soy-fermented. Food res. int. 37: 123-131.

[22] Zhang JH, Tatsumi E, Ding CH, Li LT (2006) Angiotensin I-converting enzyme inhibitory peptides in Douchi, a Chinese traditional fermented soybean product. Food chem. 98: 551-557.

[23] Fan J, Hu X, Tan S, Zhang Y, Tatsumi E, Li L (2009) Isolation and characterisation of a novel angiotensin I-converting enzyme-inhibitory peptide derived from douchi, a traditional Chinese fermented soybean food. J. sci. food agric. 89: 603-608.

[24] Ibe S, Yoshida K, Kumada K, Tsurushin S, Furusho T, Otobe K (2009) Antihypertensive effects of natto, a traditional Japanese fermented food, in spontaneously hypertensive rats. Food sci. technol. res. 15: 199-202.

[25] Roy F, Boye JI, Simpson BK (2010) Bioactive proteins and peptides in pulse crops: Pea, chickpea and lentil. Food res. int. 43: 432-442.

[26] Aluko RE (2008) Determination of nutritional and bioactive properties of peptides in enzymatic pea, chickpea, and mung bean protein hydrolysates. J. AOAC int. 91: 947-956.

[27] Ndiaye F, Tri Vuong, Duarte J, Aluko RE, Matar C (2012) Anti-oxidant, anti-inflammatory and immunomodulating properties of an enzymatic protein hydrolysate from yellow field pea seeds. Eur. j. nutr. 51: 29-37.

[28] Arcan I, Yemenicioglu A (2010) Effects of controlled pepsin hydrolysis on antioxidant potential and fractional changes of chickpea proteins. Food res. int. 43: 140-147.

[29] Shewry PR, Halford NG (2002) Cereal seed storage proteins: structures, properties and role in grain utilization. J. exp. bot. 53: 947–958.

[30] Silano M, De Vincenzi M (1999) Bioactive antinutritional peptides derived from cereal prolamins: a review. Nahrung 43:175-184.

[31] Miyoshi S, Ishikawa H, Kaneko T, Fukui F, Tanaka H, Maruyama S (1991) Structures and activity of angiotensin-converting enzyme inhibitors in an alpha-zein hydrolysate. Agric. biol. chem. 55: 1313-1318.

[32] Kong X, Zhou H, Hua Y, Qian H (2008). Preparation of wheat gluten hydrolysates with high opioid activity. Eur. food res. technol. 227: 511-517.

[33] Takahashi M, Moriguchi S, Yoshikawa M, Sakasaki R (1994) Isolation and characterization of oryzatensin - a novel bioactive peptide with ileum-contracting and immunomodulating activities derived from rice albumin. Biochem. mol. biol. int. 33: 1151-1158.

[34] Calderón de la Barca AM, RuizSalazar RA, JaraMarini ME (2000) Enzymatic hydrolysis of soy protein to improve its amino acid composition and functional properties. J. food sci. 65: 246–253.

[35] Moure A, Domínguez H, Parajó JC (2005) Fractionation and enzymatic hydrolysis of soluble protein present in waste liquors from soy processing. J. agric. food chem. 53: 7600–7608.

[36] Yim MH, Lee JH (2000) Functional properties of fractionated soy protein isolates by proteases from Meju. Food sci. biotechnol. 9: 253–257.

[37] Kristinsson HG, Rasco BA (2000) Biochemical and functional properties of Atlantic salmon (Salmo salar) muscle hydrolyzed with various alkaline proteases. J. agric. food chem. 48: 657–666.

[38] Inouye K, Nakano K, Asaoka K, Yasukawa K (2009) Effects of thermal treatment on the coagulation of soy proteins induced by subtilisin Carlsberg. J agric. food chem. 57:717–23.

[39] Wu J, Ding X (2002) Characterization of inhibition and stability of soy-protein-derived angiotensin I-converting enzyme inhibitory peptides. Food res. int. 35: 367-375.

[40] Barbana C, Boye JI (2010) Angiotensin I-converting enzyme inhibitory activity of chickpea and pea protein hydrolysates. Food res. int. 43: 1642–1649.

[41] Pedroche J, Yust MM, Giron-Calle J, Alaiz M, Millan F, Vioque J (2002) Utilisation of chickpea protein isolates for production of peptides with angiotensin I-converting enzyme (ACE)-inhibitory activity. J. sci. food. agric. 82: 960-965.

[42] Udenigwe CC, Aluko RE (2010) Antioxidant and angiotensin converting enzyme-inhibitory properties of a flaxseed protein-derived high fischer ratio peptide mixture. J. agric. food chem. 58: 4762-4768.

[43] Yust MM, Pedroche J, Giron-Calle J, Alaiz M, Francisco M, Vioque J (2003) Production of ace inhibitory peptides by digestion of chickpea legumin with alcalase. Food chem. 81: 363–369.

[44] Udenigwe CC, Adebiyi AP, Doyen A, Li H, Bazinet L, Aluko RE (2012) Low molecular weight flaxseed protein-derived arginine-containing peptides reduced blood pressure of spontaneously hypertensive rats faster than amino acid form of arginine and native flaxseed protein. Food chem. 132: 468-475.

[45] Megias C, Pedroche J, Yust MDM, Alaiz M, Giron-Calle J, Millan F, Vioque J (2006) Affinity purification of angiotensin converting enzyme inhibitory peptides using immobilized ACE. J. agric. food chem. 54: 7120-7124.

[46] Wu J, Aluko RE, Muir AD (2008) Purification of angiotensin I-converting enzyme-inhibitory peptides from the enzymatic hydrolysate of defatted canola meal. Food chem. 111: 942-950.

[47] Mäkinen S, Johansson T, Vegarud G, Pihlava J-M, Pihlanto A (2012) Angiotensin I-converting enzyme inhibitory and antioxidant properties of rapeseed hydrolysates. J. funct. foods 4: 575-583.

[48] Li G, Shi Y, Liu H, Le G (2006) Antihypertensive effect of alcalase generated mung bean protein hydrolysates in spontaneously hypertensive rats. Eur. food res. technol. 222: 733–736.

[49] FitzGerald R, Meisel H (2000) Milk protein-derived peptide inhibitors of angiotensin-I-converting enzyme. Br. j. nutr. 84: S33-S37.

[50] Makinen S, Kelloniemi J, Pihlanto A, Makinen K, Korhonen H, Hopia A, Valkonen JPT (2008) Inhibition of angiotensin converting enzyme I caused by autolysis of potato

proteins by enzymatic activities confined to different parts of the potato tuber. J. agric. food chem. 56: 9875-9883.
[51] Pêna-Ramos EA, Xiong YL (2002) Antioxidant activity of soy protein hydrolyzates in a liposomial system. J. food sci. 67: 2952–2956
[52] Li Y, Jiang B, Zhang T, Mu W, Liu J (2008) Antioxidant and free radical scavenging activities of chickpea protein hydrolysate (CPH). Food chem. 106: 444–450.
[53] Dryakova A, Pihlanto A, Marnila P, Curda L, Korhonen HJT (2010) Antioxidant properties of whey protein hydrolysates as measured by three methods. Eur. food res. technol. 230: 865-874.
[54] Xia Y, Bamdad F, Gänzle M, Chen L (2012) Fractionation and characterization of antioxidant peptides derived from barley glutelin by enzymatic hydrolysis. Food chem. 134: 1509-1518.
[55] Udenigwe CC, Lu Y, Han C, Hou W, Aluko RE (2009) Flaxseed protein-derived peptide fractions: Antioxidant properties and inhibition of lipopolysaccharide-induced nitric oxide production in murine macrophages. Food chem. 116: 277-284.
[56] Suetsuna K, Chen JR (2002) Isolation and characterization of peptides with antioxidant activity derived from wheat gluten. Food sci. tehnol. res. 8: 227–230.
[57] Zhang T, Li Y, Miao M, Jiang B (2011) Purification and characterisation of a new antioxidant peptide from chickpea (Cicer arietium L.) protein hydrolysates. Food chem. 128: 28-33.
[58] Vermeirssen V, Van Camp J, Verstraete A, Verstraete W (2004) Bioavailability of angiotensin I converting enzyme inhibitory peptides. Br. j. nutr. 92: 357.
[59] Li C, Matsui T, Matsumoto K, Yamasaki R, Kawasaki T (2002) Latent production of angiotensin I-converting enzyme inhibitors from buckwheat protein. J. pept. sci. 8: 267–274.
[60] Lo WMY, Li-Chan ECY (2005) Angiotensin I converting enzyme inhibitory peptides from *in vitro* pepsin-pancreatin digestion of soy protein. J. agric. food chem. 53: 3369-3376.
[61] Megías C, Yust MM, Pedroche J, Lquari H, Girón-Calle J, Alaiz M, Millán F, Vioque J (2004) Purification of an ACE inhibitory peptide after hydrolysis of sunflower (*Helianthus annuus* L.) protein isolates. J. agric. food chem. 52: 1928–1932.
[62] Vermeirssen V, van der Bent A, Van Camp J, van Amerongen A, Verstraete W (2004) A quantitative *in silico* analysis calculates the angiotensin I converting enzyme (ACE) inhibitory activity in pea and whey protein digests. Biochimie 86: 231-239.
[63] Boye JI, Roufik S, Pesta N, Barbana C (2010) Angiotensin I-converting enzyme inhibitory properties and SDS-PAGE of red lentil protein hydrolysates. LWT – Food sci. technol. 43: 987-991.
[64] Barbana C, Boye JI (2011) Angiotensin I-converting enzyme inhibitory properties of lentil protein hydrolysates: Determination of the kinetics of inhibition. Food chem. 127: 94-101.
[65] Mannheim A, Cheryan M (1990) Continuous hydrolysis of milk protein in a membrane reactor. J. food sci. 55: 381-385.

[66] Wang Y-K, He H-L, Wang G-F, Wu H, Zhou B-C, Chen X-L, Zhang YZ (2010) Oyster (Crassostrea gigas) hydrolysates produced on a plant scale have antitumor activity and immunostimulating effects in BALB/c mice. Mar. drugs 8:255–68.

[67] Chiang WE, Cordle CT, Thomas RL (1995) Casein hydrolysate produced using a formed-in-place membrane reactor. J. food sci. 60: 1349-1352.

[68] Guérard F (2007). Enzymatic methods for marine by-products recovery. In: Shahidi F editor. Maximizing the value of marine by-products. Campridge: Woodward Publishing Limited pp. 107–143.

[69] Chiang WD, Tsou MJ, Tsai ZY, Tsai TC (2006) Angiotensin I-converting enzyme inhibitor derived from soy protein hydrolysate and produced by using membrane reactor. Food chem. 98: 725–732.

[70] Korhonen H, Pihlanto A (2003) Food-derived bioactive peptides - Opportunities for designing future foods. Curr. pharm. design 9: 1297-1308.

[71] Virtanen T, Pihlanto A, Akkanen S, Korhonen H (2007) Development of antioxidant activity in milk whey during fermentation with lactic acid bacteria. J. appl. microbiol. 102: 106–115.

[72] Pihlanto A, Virtanen T, Korhonen H (2010) Angiotensin I converting enzyme (ACE) inhibitory activity and antihypertensive effect of fermented milk. Int. dairy j. 20: 3-10.

[73] Hernández-Ledesma B, Miralles B, Amigo L, Ramos M, Recio, I (2005) Identification of antioxidant and ACE-inhibitory peptides in fermented milk. J. sci. food agric. 85: 1041-1048.

[74] Pihlanto A, Johansson T, Mäkinen S (2012) Inhibition of angiotensin I-converting enzyme and lipid peroxidation by fermented rapeseed and flaxseed meal. Eng. life sci. DOI: 10.1002/elsc.201100137.

[75] Okamoto A, hanagata H, Matsumoto E, kawamura Y, Koizumi Y, Yanagida F (1995) Angiotensin I-converting enzyme inhibitory peptides isolated from tofuyo fermented soybean food. Biosci. biotech. biochem 59: 1147-1149.

[76] Kuba M, Tanaka K, Tawata S, Takeda Y, Yasuda M (2003) Angiotensin I-converting enzyme inhibitory peptides isolated from tofuyo fermented soybean food. Biosci. biotech. biochem. 67: 1278-1283.

[77] Li F-J, Yin L-J, Cheng Y-Q, Saito M, Yamaki K, Li L-T (2010) Angiotensin I-converting enzyme inhibitory activities of extracts from commercial chinese style fermented soypaste. JARQ 44:167-172.

[78] Tsai JS, Lin YS, Pan BS, Chen TJ (2006) Antihypertensive peptides and γ-aminobutyric acid from prozyme 6 facilitated lactic acid bacteria fermentation of soymilk. Process biochem. 41: 1282-1288.

[79] Iwai K, Nakaya N, Kawasaki Y, Matsue H (2002) Inhibitory effect of natto, a kind of fermented soybeans, on LDL oxidation in vitro. J. agric. food chem. 50: 3592-3596.

[80] Wang D, Wang L, Zhu F, Zhu J, Chen XD, Zou L, Saito, M, Li L (2008) In vitro and in vivo studies on the antioxidant activities of the aqueous extracts of Douchi (a traditional Chinese salt-fermented soybean food). Food chem. 107:1421-1428.

[81] Natesh R, Schwager SLU, Sturrock ED, Acharya KR (2003) Crystal structure of the human angiotensin-converting enzyme lisonopril complex. Nature 421: 551–554.

[82] Ondetti MA, Rubin B, Cushman DW (1977) Design of specific inhibitors of angiotensin-converting enzyme: new class of orally active antihypertensive agents. Science 196: 441-444.

[83] Cheung HS, Wang FL, Ondetti MA, Sabo EF, Cushman DW (1980) Binding of peptide substrates and inhibitors of angiotensin-converting enzyme: importance of the COOH-terminal dipeptide sequence. J biol. chem. 255: 401-407.

[84] Ariyoshi Y (1993) Angiotensin-converting enzyme inhibitors derived from food proteins. Trends food sci. technol. 4:139-144.

[85] Pripp AH, Isaksson T, Stepaniak L, Sorhaug T (2004) Quantitative structure–activity relationship modelling of ACE-inhibitory peptides derived from milk proteins. Eur. food res. technol. 219: 579-583.

[86] Moure A, Domínguez H, Parajó JC (2006) Antioxidant properties of ultrafiltration recovered soy protein fractions from industrial effluents and their hydrolysates. Process biochem. 41: 447–456.

[87] Rajapakse N, Mendis E, Jung WK, Je JY, Kim SK (2005) Purification of a radical scavenging peptide from fermented mussel sauce and its antioxidant properties. Food res. int. 38: 175–182.

[88] Erdmann K, Cheung BWY, Schröder H (2008) The possible roles of food derived bioactive peptides in reducing the risk of cardiovascular disease. J. nutr. biol. 19: 643–654.

[89] Chen HM, Muramoto K, Yamauchi F, Fujimoto K, Nokihara K (1998) Antioxidative properties of histidine-containing peptides designed from peptide fragments found in the digests of a soybean protein. J. agric. food chem. 46: 49–53.

[90] Pihlanto A (2006) Antioxidative peptides derived from milk proteins. Int. dairy j. 16: 1306–1314.

[91] Chan KM, Decker EA (1994) Endogenous muscle antioxidants. Crit. rev. food. sci. 34: 403–426.

[92] Qian ZJ, Jung WK, Kim SK (2008) Free radical scavenging activity of a novel antioxidative peptide purified from hydrolysate of bullfrog skin, Rana catesbeiana Shaw. bioresource technol. 99: 1690–1698.

[93] Chen HM, Muramoto K, Yamauchi F, Nokihara K (1996) Antioxidant activity of designed peptides based on the antioxidative peptide isolated from digests of a soybean protein. J. agric. food chem. 44: 2619–2623.

[94] Saito K, Jin DH, Ogawa T, Muramoto K, Hatakeyama E, Yasuhara T, Nokihara K (2003) Antioxidative properties of tripeptide libraries prepared by the combinatorial chemistry. J. agric. food chem. 51: 3668–3674.

[95] Nagasawa T, Yonekura T, Nishizawa N, Kitts DD (2001) *In vitro* and *in vivo* inhibition of muscle lipid and protein oxidation by carnosine. Mol. cell. biochem. 225: 29–34.

[96] Hernández-Ledesma B, Davalos A, Bartolome B, Amigo L (2005) Preparation of antioxidant enzymatic hydrolysates from a-lactalbumin and b-lactoglobulin. Identification of active peptides by HPLC–MS/MS. J. agric. food chem. 53: 588–593.

[97] Cushman DW, Cheung HS (1971) Spectrophotometric assay and properties of the angiotensin converting enzyme of rabbit lung. Biochem. pharmacol. 20: 1637.

[98] Vermeirssen V, Van Camp J, Verstraete W (2002) Optimisation and validation of an angiotensin-converting enzyme inhibition assay for the screening of bioactive peptides. J. biochem. biophys. methods 51: 75.

[99] Li GH, Liu H, Shi Y-H, Le GW (2005) Direct spectrophotometric measurement of angiotensin I-converting enzyme inhibitory activity for screening bioactive peptides. J. pharm. biomed. anal. 37: 219.

[100] Shalaby SM, Zakora M, Otte J (2006) Performance of two commonly used angiotensin-converting enzyme inhibition assays using synthetic peptide substrates. J. dairy res. 73: 178.

[101] Doig MT, Smiley JW (1993) Direct injection assay of angiotensin-converting enzyme by high-performance liquid-chromatography using a shielded hydrophobic phase column. J. chrom. biomed. appl. 613: 145.

[102] Hyun C, Shin H (2000) Utilization of bovine blood plasma proteins for the production of angiotensin I converting enzyme inhibitory peptides, Process biochem. 36: 65-71.

[103] Holmquist B, Bunning P, Riordan J (1979) Continuous spectrophotometric assay for angiotensin converting enzyme. Anal. biochem. 95: 540-548.

[104] Carmel A, Yaron A (1978) An intramolecularly quenched fuorescent tripeptide as a fuorogenic substrate of angiotensin-I-converting enzyme and of bacterial dipeptidyl carboxypeptidase. Eur. j. biochem. 87: 265.

[105] Sentandreu MA, Toldrá F (2006) A rapid, simple and sensitive fuorescence method for the assay of angiotensin-I converting enzyme. Food chem. 97: 546.

[106] Cao G, Prior RL (1998) Comparison of different analytical methods for assessing total antioxidant capacity of human serum. Clin. chem. 44: 1309–1315.

[107] Re R, Pellegrini N, Proteggente A, Pannala A, Yang M, Rice- Evans C (1999) Antioxidant activity applying an improved ABTS radical cation decolorization assay. Free radical bio. med. 26: 1231–1237.

[108] Verma S, Buchanan MR, Anderson TJ (2003) Endothelial function testing as a biomarker of vascular disease. Circulation 108: 2054-9.

[109] Liu RH, Finley J (2005) Potential cell culture models for antioxidant research. J. agric. food chem. 53: 4311–4314.

[110] Elisia I, Kitts DD (2008) Anthocyanins inhibit peroxy radical-induced apoptosis in Caco-2 cells. Mol. cell. biochem. 312: 139–145.

[111] Wolfe KL, Liu RH (2007) Cellular antioxidant activity (CAA) assay for assessing antioxidants, foods, and dietary supplements. J. agric. food chem.. 55: 8896–8907.

[112] Fitzgerald RJ, Murray BA, Walsh DJ (2004) Hypotensive peptides from milk proteins. J. nutr. 134: 980-988.

[113] Hsu F, Lin Y, Lee M, Lin C, Hou W (2002) Both dioscorin, the tuber storage protein of yam (Dioscorea alata cv. Tainong No. 1), and its peptic hydrolysates exhibited angiotensin converting enzyme inhibitory activities. J. agric. food chem. 50: 6109-6113.

[114] Lee M, Lin Y, Lin Y, Hsu F, Hou W (2003) The mucilage of yam (Dioscorea batatas Decne) tuber exhibited angiotensin converting enzyme inhibitory activities. Bot. bull. acad. sinica 44: 267-273.

[115] Nagai T, Suzuki N, Nagashima T (2007) Autolysate and enzymatic hydrolysates from yam (Dioscorea opposita Thunb.) tuber mucilage tororo have antioxidant and angiotensin I-converting enzyme inhibitory activities. J. food agric. environ. 5: 39-43.

[116] Nagai T, Suzuki N, Tanoue Y, Kai N, Nagashima T (2007) Antioxidant and antihypertensive activities of autolysate and enzymatic hydrolysates from yam (Dioscorea opposita Thunb.) ichyoimo tubers. J. food agric. environ. 5: 64-68.

[117] Lacaille-Dubois MA, Franck U, Wagner H (2001) Search for potential angiotensin converting enzyme (ACE)-inhibitors from plants. Phytomedicine 8: 47–52.

[118] Huang G, Lu T, Chiu C, Chen H, Wu C, Lin Y, Hsieh W, Liao J, Sheu M, Lin Y (2011) Sweet potato storage root defensin and its tryptic hydrolysates exhibited angiotensin converting enzyme inhibitory activity *in vitro*. Bot. stud. 52: 257-264.

[119] Huang G, Chen H, Susumu K, Wu J, Hou W, Wu C, Sheu M, Huang S, Lin Y (2011) Sweet potato storage root thioredoxin h2 and their peptic hydrolysates exhibited angiotensin converting enzyme inhibitory activity *in vitro*. Bot. stud. 52: 15-22

[120] Huang G, Ho Y, Chen H, Chang Y, Huang S, Hung H, Lin Y (2008) Sweet potato storage root trypsin inhibitor and their peptic hydrolysates exhibited angiotensin converting enzyme inhibitory activity *in vitro*. Bot. stud. 49: 101-108.

[121] Ishiguro K, Sameshima Y, Kume T, Ikeda K, Matsumoto J, Yoshimoto M (2012) Hypotensive effect of a sweet potato protein digest in spontaneously hypertensive rats and purification of angiotensin I-converting enzyme inhibitory peptides. Food chem. 131: 774-779.

[122] Lin C, Lin S, Lin Y, Hou W (2006) Effects of tuber storage protein of yam (Dioscorea alata cv. Tainong No. 1) and its peptic hydrolyzates on spontaneously hypertensive rats. J. sci. food agric. 86: 1489-1494.

[123] Liu Y, Lin Y, Liu D, Han C, Chen C, Fan M, Hou W (2009) Effects of different types of yam (Dioscorea alata) products on the blood pressure of spontaneously hypertensive Rats. Biosci. biotech. biochem. 73: 1371-1376.

[124] Iwai K, Matsue H (2007) Ingestion of Apios americana Medikus tuber suppresses blood pressure and improves plasma lipids in spontaneously hypertensive rats. Nutr. res. 27: 218-224.

[125] Marczak E, Usui H, Fujita H, Yang Y, Yokoo M, Lipkowski A, Yoshikawa M (2003) New antihypertensive peptides isolated from rapeseed. Peptides 24: 791-798.

[126] Udenigwe CC, Lin YS, Hou WC, Aluko RE (2009) Kinetics of the inhibition of renin and angiotensin I-converting enzyme by flaxseed protein hydrolysate fractions. J. funct. foods 1: 199-207.

[127] Marambe HK, Shand PJ, Wanasundara, JPD (2011) Release of angiotensin I-converting enzyme inhibitory peptides from flaxseed (Linum usitatissimum L.) protein under simulated gastrointestinal digestion. J. agric. food chem. 59: 9596-9604.

[128] Yamada Y, Iwasaki M, Usui H, Ohinata K, Marczak ED, Lipkowski, AW, Yoshikawa M (2010) Rapakinin, an anti-hypertensive peptide derived from rapeseed protein, dilates mesenteric artery of spontaneously hypertensive rats via the prostaglandin IP receptor followed by CCK1 receptor. Peptides 31: 909-914.

[129] Marczak, E. D., Ohinata, K., Lipkowski, A. W., & Yoshikawa, M. (2006). Arg-Ile-Tyr (RIY) derived from rapeseed protein decreases food intake and gastric emptying after oral administration in mice. Peptides 27: 2065-2068.

[130] Yamada Y, Ohinata K, Lipkowski AW, Yoshikawa M (2011) Rapakinin, Arg-Ile-Tyr, derived from rapeseed napin, shows anti-opioid activity via the prostaglandin IP receptor followed by the cholecystokinin CCK2 receptor in mice. Peptides 32: 281-285.

[131] Akıllıoglu HG, Karakaya, S (2009) Effects of heat treatment and *in vitro* digestion on the angiotensin converting enzyme inhibitory activity of some legume species. Eur. food res. technol. 229: 915–921.

[132] Vermeirssen V, Van Camp J, Decroos K, Van Wijmelbeke L, Verstraete W (2003) The impact of fermentation and *in vitro* digestion on the formation of angiotensin-I-converting enzyme Inhibitory activity from pea and whey protein. J. dairy. sci. 86: 429-438.

[133] Li H, Aluko RF (2010) Identification and inhibitory properties of multifunctional peptides from pea protein hydrolysate. J. agric. food chem. 58: 11471-11476.

[134] Lo WMY, Farnworth ER, Li-Chan ECY (2006) Angiotensin I-converting enzyme inhibitory activity of soy protein digests in a dynamic model system simulating the upper gastrointestinal tract. J. food. sci. 71: S231–S237.

[135] Pownall TL, Udenigwe CC, Aluko RE (2010) Amino acid composition and antioxidant properties of pea seed (Pisum sativum L.) enzymatic protein hydrolysate fractions. J. agric. food chem. 58: 4712–4718.

[136] Pownall TL, Udenigwe CC, Aluko RE (2011) Effects of cationic property on the *in vitro* antioxidant activities of pea protein hydrolysate fractions. Food res. int. 44: 1069-1074.

[137] Li Y, Jiang B, Zhang T, Mu W, Liu J (2008) Antioxidant and free radical-scavenging activities of chickpea protein hydrolysate (CPH). Food chem. 106: 444-450.

[138] Liu JR, Chen MJ, Lin CW (2005) Antimutagenic and antioxidant properties of milk-kefir and soymilk-kefir. J. agric. food chem 53: 2467-2474.

[139] Zhang L, Lia J, Zhoub K (2010) Chelating and radical scavenging activities of soy protein hydrolysates prepared from microbial proteases and their effect on meat lipid peroxidation. Bioresource technology, 101: 2084-2089.

[140] Chen HM, Muramoto K, Yamauchi F. 1995. Structural analysis of antioxidant peptides from soybean b-conglycinin. J Agric Food Chem 43:574-578.

[141] Park SY, Lee JS, Baek HH, Lee HG (2010) Purification and characterization of antioxidant peptides from soy protein hydrolysate. J. food biochem.34:120-134

[142] Hsu GW, Lu Y, Chang S, Hsu S (2011) Antihypertensive effect of mung bean sprout extracts in spontaneously hypertensive rats. J. food biochem. 35: 278-288.

[143] Li H, Prairie N, Udenigwe CC, Adebiyi AP, Tappia PS, Aukema HM, Jones PJH, Aluko RE (2011) Blood pressure lowering effect of a pea protein hydrolysate in hypertensive rats and humans. J. agric. food chem. 59: 9854-9860.

[144] Vermeirssen V, Augustijns P, Morel N, Van Camp J, Opsomer A, Verstraete W (2005) *In vitro* intestinal transport and antihypertensive activity of ACE inhibitory pea and whey digests. Int. j. food sci. nutr. 56: 415-430.

[145] Lee NY, Kim Y, Choi I, Cho S, Hyun J, Choi J, Park K, Kim K, Lee M (2010) Biological activity of barley (Hordeum vulgare L.) and barley byproduct extracts. Food sci. biotechnol. 19: 785-791.

[146] Aludatt MH, Ereifej K, Abou-zaitoun A, Al-Rababah M, Almajwal A, Rababeh T, Yang W (2012) Oxidant, anti-diabetic and anti-hypertensive effects of extracted phenolics and hydrolyzed peptides from barley protein fraction. Int. j. food prop. 15: 781-795.

[147] Iwaniak A, Dziuba B (2009) Motifs with potential physiological activity in food proteins –Biopep database. Acta Scientiarum Polonorum: Technologia Alimentaria 8: 59-85.

[148] Ying Hu, Stromeck A, Loponen J, Lopes-Lutz D, Schieber A, Gänzle MG (2011) LC-MS/MS Quantification of bioactive angiotensin I-converting enzyme inhibitory peptides in rye malt sourdoughs. *J. agric. food chem. 59:* 11983–11989.

[149] Cheung IWY, Nakayama S, Hsu MNK, Samaranayaka AGP, Li-Chan ECY (2009) Angiotensin-I converting enzyme inhibitory activity of hydrolysates from oat (Avena sativa) proteins by *in silico* and *in vitro* analyses. J agric. food chem. 57: 9234–9242.

[150] Kong X, Zhou H, Hua Y (2008) Preparation and antioxidant activity of wheat gluten hydrolysates (WGHs) using ultrafiltration membranes. J sci. food agric. 88: 920–926.

[151] Zhang J, Zhang H, Wang L, Guo X, Wang X, Yao H (2010) Isolation and identification of antioxidative peptides from rice endosperm protein enzymatic hydrolysate by consecutive chromatography and MALDI-TOF/TOF MS/MS. Food chem. 119: 226-234.

[152] Li GH, Qu MR, Wan JZ, You JM (2007) Antihypertensive effect of rice protein hydrolysate with *in vitro* angiotensin I-converting enzyme inhibitory activity in spontaneously hypertensive rats. Asia pa. j. clin. nutr. 16: 275-280.

[153] Matsui T, Li CH, Tanaka T, Maki T, Osajima Y, Matsumoto K (2000) Depressor effect of wheat germ hydrolysate and its novel angiotensin I–converting enzyme inhibitory peptide, Ile-Val-Tyr, and the metabolism in rat and human plasma. Biol. pharm. bull. 23: 427–431.

[154] De Leo F, Panarese S, Gallerani R, Ceci LR (2009) Angiotensin converting enzyme (ACE) inhibitory peptides: production and implementation of functional food. Curr. pharm. design 15: 3622-3643.

[155] Eriksen EK, Holm H, Jensen E, Aaboe R, Devold TG, Jacobsen M, Vegarud GE (2010) Different digestion of caprine whey proteins by human and porcine gastrointestinal enzymes. Br. j. nutr. 104: 374-381.

[156] Zhu L, Chen J, Tang X, Xiong YL (2008) Reducing, radical scavenging, and chelation properties of *in vitro* digests of alcalase-treated zein hydrolysate. J. agric. food chem. 56: 2714-2721.

[157] Gardner MLG (1988) Gastrointestinal absorption of intact proteins. Annu. rev. nutr. 8: 329–350.

[158] Gardner MLG (1998) Transmucosal passage on intact peptides in mammalian metabolism. In: Grimble GK, Backwell FRG, editors. Tissue utilization and clinical targeting. London: Portland Press Ltd.

[159] Shimizu M, Tsunogai M, Arai S (1997) Transepithelial transport of oligopeptides in the human intestinal cell, Caco2. Peptides 18: 681–687.

[160] Foltz M, Cerstiaens A, van Meensel A, Mols R, van der Pijl PC, Duchateau GSMJE, Augustijns P (2008) The angiotensin converting enzyme inhibitory tripeptides Ile-Pro-Pro and Val-Pro-Pro show increasing permeabilities with increasing physiological relevance of absorption models. Peptides 29: 1312–1320.

[161] van der Pijl PC, Kies AK, Ten Have GA, Duchateau GS, Deutz NE (2008) Pharmacokinetics of proline-rich tripeptides in the pig. Peptides 29: 2196–2202.

[162] Foltz M, Meynen EE, Bianco V, van Platerink C, Koning TMMG, Kloek J (2007) Angiotensin converting enzyme inhibitory peptides from a lactotripeptide-enriched milk beverage are absorbed intact into the circulation. J. nutr. 137: 953–958.

[163] Quiros A, Davalos A, Lasuncion MA, Ramos M, Recio I (2008) Bioavailability of the antihypertensive peptide LHLPLP: Transepithelial flux of HLPLP. Int. dairy j. 18: 279-286.

[164] Quiros A, del Mar Contreras M, Ramos M, Amigo L, Recio I (2009) Stability to gastrointestinal enzymes and structure-activity relationship of beta-casein-peptides with antihypertensive properties. Peptides 30: 1848-1853.

[165] van Platerink C, Janssen H, Horsten R, Haverkamp J (2006) Quantification of ACE inhibiting peptides in human plasma using high performance liquid chromatography-mass spectrometry. J. chromatogr. B 830: 151-157.

[166] Fujita H, Yokoyama K, Yoshikawa M (2000) Classification and antihypertensive activity of angiotensin I-converting enzyme inhibitory peptides derived from food proteins. J. food sci. 65: 564-569.

[167] Zhao Y, Li B, Dong S, Liu Z, Zhao X, Wang J, Zeng M (2009) A novel ACE inhibitory peptide isolated from Acaudina molpadioidea hydrolysate. Peptides 30: 1028-1033.

[168] Muguruma M, Ahhmed AM, Katayama K, Kawahara S, Maruyama M, Nakamura T (2009) Identification of pro-drug type ACE inhibitory peptide sourced from porcine myosin B: Evaluation of its antihypertensive effects *in vivo*. Food chem. 114: 516-522.

[169] Shaji J, Patole V (2008) Protein and peptide drug delivery: Oral approaches. J. pharm. sci. 70: 269-277.

[170] Kies AK, Van Der Pijl P (2012) Peptide availability. USA Patent Application 20120040895.
[171] Nestor JJ Jr. (2009) The medicinal chemistry of peptides. Curr. med. chem. 16: 4399-4418.
[172] Yamada Y, Matoba N, Usui H, Onishi K, Yoshikawa M (2002) Design of a highly potent anti-hypertensive peptide based on ovokinin(2-7). Biosci. biotechn. biochem. 66: 1213-1217.
[173] Matoba N, Yamada Y, Usui H, Nakagiri R, Yoshikawa H (2001) Designing potent derivatives of ovokinin(2-7), an anti-hypertensive peptide derived from ovalbumin. Biosci. biotechn. biochem. 65: 736-739.
[174] Gomez-Guillen MC, Gimenez B, Lopez-Caballero ME, Montero MP (2011) Functional and bioactive properties of collagen and gelatin from alternative sources: A review. Food hydrocolloids 25: 1813-1827.
[175] Liu D, Liang H, Han C, Lin S, Chen C, Fan M, Hou W (2009) Feeding trial of instant food containing lyophilised yam powder in hypertensive subjects. J. sci. food agric. 89: 138-143.
[176] Peach MJ (1997) Renin-angiotensin system: biochemistry and mechanism of action. Physiol. rev. 57: 313-370.
[177] Gradman A, Schmieder R, Lins R, Nussberger J, Chiang Y, Bedigian, M (2005) Aliskiren, a novel orally effective renin inhibitor, provides dose-dependent antihypertensive efficacy and placebo-like tolerability in hypertensive patients. Circulation 111: 1012-1018.
[178] Udenigwe CC, Aluko RE (2012) Multifunctional cationic peptide fractions from flaxseed protein hydrolysates. Plant food hum. nutr. 67: 1-9.
[179] Erdmann K, Grosser N, Schipporeit K, Schroeder H (2006) The ACE inhibitory dipeptide Met-Tyr diminishes free radical formation in human endothelial cells via induction of heme oxygenase-1 and ferritin. J. nutr. 136: 2148-2152.
[180] Touyz R (2004) Reactive oxygen species, vascular oxidative stress, and redox signaling in hypertension - What is the clinical significance? Hypertension 44: 248-252.
[181] Akpaffiong M, Taylor A (1998) Antihypertensive and vasodilator actions of antioxidants in spontaneously hypertensive rats. Am. j. hyp. 11: 1450-1460.
[182] Marambe PWMLHK, Shand PJ, Wanasundara JPD (2008) An in-vitro investigation of selected biological activities of hydrolysed flaxseed (Linum usitatissimum L.) proteins. J. am. oil chem. soc. 85: 1155-1164.
[183] Fitzgerald C, Gallagher E, Tasdemir D, Hayes M (2011) Heart health peptides from macroalgae and their potential use in functional foods. J. agric. food chem. 59: 6829-6836.
[184] Nakahara T, Sano A, Yamaguchi H, Sugimoto K, Chikata H, Kinoshita E, Uchida R (2010) Antihypertensive effect of peptide-enriched soy sauce-like seasoning and identification of its angiotensin I-converting enzyme inhibitory substances. J. agric. food chem.58: 821-827.

[185] Nogata Y, Nagamine T, Sekiya K (2011) Antihypertensive effect of angiotensin I-converting enzyme inhibitory peptides derived from wheat bran in spontaneously hypertensive rats. J. jpn. soc. food sci. technol. 58: 67-70.

[186] Nogata Y, Nagamine T, Yanaka M, Ohta, H (2009) Angiotensin I converting enzyme inhibitory peptides produced by autolysis reactions from wheat bran. J. agric. food chem. 57: 6618-6622.

Chapter 7

Vigna Unguiculata as Source of Angiotensin-I Converting Enzyme Inhibitory and Antioxidant Peptides

Maira R. Segura-Campos, Luis A. Chel-Guerrero and David A. Betancur-Ancona

Additional information is available at the end of the chapter

http://dx.doi.org/10.5772/53045

1. Introduction

The frequency of lifestyle-related diseases is steadily increasing, particularly of hypertension, a risk factor for cardiovascular diseases such as coronary heart disease, peripheral arterial disease and stroke. Indeed, cardiovascular diseases are the primary cause of morbidity and mortality in Western countries, with hypertension affecting about 20% of the world's adult population [1]. Blood pressure is controlled by various regulatory factors in the body, including angiotensin I-converting enzyme (ACE-I). ACE-I (peptidyldipeptidaseA, kininase II, EC 3.4.15.1) is a zinc dipeptidylcarboxypeptidase. This membrane-bound exopeptidase is found on the plasma membranes of various cell types, including vascular endothelial cells, microvillar brush border epithelial cells and neuroepithelial cells. It is thought to be physiologically important. The primary activity of ACE-I is to cleave broad specificity free carboxyl group oligopeptides. Substrates containing Pro at the P1' position and Asp or Glu at P2' are resistant to ACE-I. However, on certain substrates ACE-I can also function as an endopeptidase or a tripeptidylcarboxypeptidase. With ACE-I, endopeptidase activity is observed on substrates having amidated carboxyl groups where the enzyme can cleave a C-terminal dipeptide amide and/or a C-terminal tripeptide amide [2]. ACE-I is responsible for converting angiotensin I (Ang I) to the powerful vasoconstrictor angiotensin II (Ang II) and inactivating the vasodilator peptide bradykinin (BK) by removal of C-terminal dipeptides [3]. In a functional sense, therefore, the enzymatic actions of ACE-I potentially cause increased vasoconstriction and decreased vasodilation. ACE-I has attracted interest for development of orally-active ACE-I inhibitors to treat hypertension due to its central role in vasoactive peptide metabolism. Inhibition of ACE-I prevents conversion of Ang I into Ang II, making it becomes one of the most effective methods for suppressing increases in blood pressure [4]. Since discovery of ACE-I inhibitors in snake venom, extensive research has been done on synthesizing ACE-I inhibitors such as captopril, enalapril, alacepril and lisinopril, all currently in use for treatment of hypertension and heart failure in humans. However, these synthetic drugs occasionally produce side effects such as cough, taste alterations and skin rashes. Interest has consequently increased in natural ACE-I inhibitors as safer and lower cost alternatives to synthetic ones [5].

Antioxidant deficiency also has been implicated in the occurrence of hypertension and other degenerative diseases. Reactive oxygen species (ROS) such as the superoxide anion radical (O^{-}_{2}), hydrogen peroxide and hydroxyl radicals (•OH) are physiological metabolites formed as result of respiration in aerobic organisms. ROS are very unstable, and react rapidly with other substances including DNA, membrane lipids and proteins. Oxidative stress is produced by an imbalance between oxidizing species and natural antioxidants in the body, and has been associated with aging, cell apoptosis and severe diseases such as cancer, Parkinson, Alzheimer, and cardiovascular disorders [6]. Epidemiological studies have demonstrated an inverse association among intake of antioxidants from fruits and vegetables, and morbidity and mortality due to coronary heart diseases and cancer. In response, researchers are searching for natural antioxidants in food that may protect the body from free radicals and retard the evolution of many chronic diseases [7].

In recent years, food proteins have gained increasing value due to the rapidly expanding knowledge about physiologically active peptides. Peptides from various dietary sources have been shown to have clearly positive effects on health by functioning as antihypertensives, antioxidants, anticarcinogens, antimicrobials and anticariogenics, among others. These properties have led to their labeling as functional or biologically-active (i.e. bioactive) peptides. Bioactive peptides may be encrypted within the amino acid sequence of a larger protein. These peptides usually consist of 3-20 amino acids and are released from the original protein after degradation [8]. The most common way to produce bioactive peptides is through enzymatic hydrolysis of whole protein molecules. After enzymatic processing, amino acid sequences that were inactive in the core of the source protein are released and can exercise special properties. Many of the known bioactive peptides have been produced using gastrointestinal enzymes, usually pepsin and trypsin. Other digestive enzymes and different enzyme combinations of proteinases-including Alcalase®, chymotrypsin, pancreatin, pepsin and thermolysin have also been utilized to generate bioactive peptides from various proteins [9].

Continued population growth worldwide, consequent food resource shortages in developing countries, and the health risks associated with excessive animal protein (and saturated fats) intake has led researchers to search for new sources of proteins from non-conventional raw materials. Legumes are cultivated worldwide and constitute an excellent protein source (protein content =20-30%). Cowpea (*Vigna unguiculata*) is a major legume crop worldwide, particularly in tropical and subtropical areas such as southeast Mexico. It serves as a major dietary protein source in both human and animal diets, and its protein content makes it a good raw material for preparation of protein extracts and hydrolysates. Protein extract has been used as substrates for production of hydrolysates with functional and/or nutritional properties better than the original extract [10]. Bioactive peptides with ACE-I inhibitory activity has been isolated from protein hydrolysates from a number of animal and vegetable sources. Vegetable origin proteins are of particular interest, and legumes are especially promising due to their high protein content and diverse physiological activities in the human organism. Extensive hydrolysis of *V. unguiculata* protein concentrates with commercial and digestives enzymes could therefore produce a number of peptides with a myriad of potential applications; for example, as natural-source therapeutic agents in medical treatments and/or as an ingredient in functional foods. Taking this into account, the aim of the present study was to modify enzymatically protein concentrates of *V. unguiculata,* evaluate the ACE-I inhibitory and antioxidant potential of the hydrolysates and relate the biological activity to their amino acid compositions.

2. Material and methods

2.1. Materials

V. unguiculata seeds were obtained from the February 2007 harvest in Yucatan state, Mexico. Reagents were of analytical grade and purchased from J.T. Baker (Phillipsburg, NJ, USA), Sigma Chemical Co. (St. Louis, MO, USA), Merck (Darmstadt, Germany) and Bio-Rad Laboratories, Inc. (Hercules, CA, USA). Alcalase® 2.4L FG and Flavourzyme® 500MG enzymes were purchased from Novo Laboratories (Copenhagen, Denmark).

2.2. Protein isolates

A single extraction was performed with 6 kg of cowpea seeds. Impurities and damaged seeds were removed, and sound seeds milled in a Mykros impact mill (Industrial Machinery, Monterrey, Mexico) until passing through a 20-mesh screen (0.85 mm) followed by milling in a Cyclotec 1093 (Tecator, Sweden) mill until passing through a 60-mesh screen (0.24 mm). The resulting flour was processed using the wet fractionation method of Betancur-Ancona [11]. Briefly, whole flour was suspended in distilled water at a 1:6 (w/v) ratio, pH was adjusted to 11.0 with 1 M NaOH, and the dispersion was stirred for 1 h at 0.178 x *g* with a mechanical agitator (Caframo Rz-1, Heidolph Schwabach, Germany). This suspension was wet-milled with a Kitchen-Aid® food processor, and the fiber solids were separated from the starch and protein mix by straining through 80- and 150-mesh sieves and washing the residue five times with distilled water. The protein-starch suspension was allowed to sediment for 30 min at room temperature to recover the starch and protein fractions. The pH of the separated solubilized proteins was adjusted to the isoelectric point (4.5) with 1 N HCl. The suspension was then centrifuged at 1317 x *g* for 12 min (Mistral 3000i, Curtin Matheson Sci.), the supernatants were discarded, and the precipitates were freeze-dried until use.

2.3. Enzymatic hydrolysis

Hydrolysis of the protein extract was done using a totally randomized design with the treatments being the enzymatic system applied: Alcalase® 2.4L FG; Flavourzyme® 500MG; or a sequential system using pepsin from porcine gastric mucosa (Sigma, P7000-100G) and pancreatin from porcine pancreas (Sigma, P3292-100G). The response variable was degree of hydrolysis (DH).

Hydrolysis was done under controlled conditions (temperature, pH and stirring) in a 1000 mL reaction vessel equipped with a stirrer, thermometer and a pH electrode. Hydrolysis with Alcalase® and Flavourzyme® was done according to Pedroche et al. [12]. Protein extracts were suspended in distilled water to produce a 4% (w/v) protein solution. This solution was equilibrated at optimum temperature and pH for each protease before adding the respective enzyme. Protease was then added to the solution at a ratio of 0.3 UA/g for Alcalase® and 50 UAPL/g for Flavourzyme®. Hydrolysis conditions were 90 min at 50°C for both enzymes, and pH 8.0 for Alcalase® and pH 7.0 for Flavourzyme®. The pH was kept constant by adding 1.0 M NaOH during hydrolysis. Hydrolysis with the sequential pepsin-pancreatin system was done with a pH-stat method for 90 min: pre-digestion with pepsin for 45 min followed by incubation with pancreatin for 45 min. Hydrolysis parameters were substrate concentration 4%; enzyme/substrate ratio 1:10; pH 2 for pepsin; pH 7.5 for pancreatin; and 37°C [13, 14]. In all three treatments the reaction was stopped by heating to 80°C for 20 min, followed by centrifugation at 9,880 x *g* for 20 min to remove the insoluble portion.

2.4. Degree of hydrolysis

DH was calculated by calculating free amino groups with o-phthaldialdehyde [15]: $DH= h/h_{tot} *100$, where h_{tot} is the total number of peptide bonds per protein equivalent, and h is the number of hydrolyzed bonds. The h_{tot} factor is dependent on raw material amino acid composition.

2.5. Hydrolysate fractionation by ultrafiltration

Following Cho et al. [16], the hydrolysate was fractionated by ultrafiltration using a high performance ultrafiltration cell (Model 2000, Millipore, Inc., Marlborough, MA, USA). Five fractions were prepared using four molecular weight cut-off membranes: 1 kDa, 3 kDa, 5 kDa and 10 kDa. Soluble fractions prepared by centrifugation were passed through the membranes stating with the largest Molecular Weight Cut off (MWCO) membrane cartridge (10 kDa). The retentate and permeate were collected separately, and the retentate recirculated into the feed until maximum permeate yield was reached, as indicated by a decreased permeate flow rate. The permeate from the 10 kDa membrane was then filtered through the 5 kDa membrane with recirculation until maximum permeate yield was reached. The 5 kDa permeate was then processed with the 3 kDa membrane and the 3 kDa permeate with the 1 kDa membrane. This process minimized contamination of the larger molecular weight fractions with smaller molecular weight fractions while producing enough retentates and permeates for the following analyses. The five ultrafiltered peptide fractions were designated as >10 kDa (10 kDa retentate); 5-10 kDa (10 kDa permeate-5 kDa retentate); 3-5 kDa (5 kDa permeate- 3 kDa retentate); 1-3 kDa (3 kDa permeate-1 kDa retentate); and <1 kDa (1 kDa permeate).

2.6. ACE-I inhibitory activity

ACE-I inhibitory activity in the hydrolysate and its purified peptide fractions was analyzed following Hayakari et al. [17]. ACE-I hydrolyzes hippuryl-L-histidyl-L-leucine (HHL) to yield hippuric acid and His-Leu. This method relies on the colorimetric reaction of hippuric acid with 2,4,6-trichloro-s-triazine (TT) in a 0.5 mL incubation mixture containing 40 μmol potassium phosphate buffer (pH 8.3), 300 μmol sodium chloride, 40 μmol 3% HHL in potassium phosphate buffer (pH 8.3) and 100 mU/mL ACE-I. The mixture was incubated at 37ºC for 45 min and then reaction terminated by addition of TT (3% v/v) in dioxane and 3 ml of 0.2 M potassium phosphate buffer (pH 8.3). After centrifuging the reaction mixture at 10,000 x *g* for 10 min, enzymatic activity was determined in the supernatant by measuring absorbance at 382 nm. All runs were performed in triplicate. ACE-I inhibitory activity was quantified by a regression analysis of ACE-I inhibitory activity (%) versus peptide concentration and defined as an IC_{50} value, that is, the peptide concentration in (μg protein/mL) required to produce 50% ACE-I inhibition under the described conditions.

2.7. Antioxidant activity by ABTS assay

2,2'azino-bis (3-ethylbenzothiazoline-6-sulfonic acid) (ABTS) radical cation ($ABTS^{\bullet+}$) was produced by reacting ABTS with potassium persulfate following Pukalskas et al. [18]. To prepare the stock solution, ABTS was dissolved at a 2 mM concentration in 50 mL phosphate-buffered saline (PBS) prepared from 4.0908 g NaCl, 0.1347 g KH_2PO_4, 0.7098 g Na_2HPO_4, and 0.0749 g KCl dissolved in 500 mL ultrapure water. If pH was lower than 7.4, it was adjusted with NaOH. A 70 mM $K_2S_4O_8$ solution in ultrapure water was prepared. $ABTS^{\bullet+}$ radical was produced by reacting 10 mL of ABTS stock solution with 40 μL $K_2S_4O_8$ solution and allowing the mixture to stand in darkness at room temperature for 16-17 h before use. The radical was stable in this form for more than 2 days when stored in darkness at room temperature.

Antioxidant compound content in the hydrolysates and their UF peptide fractions was analyzed by diluting the $ABTS^{\bullet+}$ solution with PBS to an absorbance of 0.800 ± 0.030 AU at 734 nm. After adding 990 μL of diluted $ABTS^{\bullet+}$ solution ($A_{734\ nm}$= 0.800 ± 0.030) to 10 μL antioxidant compound or Trolox standard (final concentration 0.5-3.5 mM) in PBS, absorbance was read at ambient temperature exactly 6 min after initial mixing. All analyses were run in triplicate. The percentage decrease in absorbance at 734 nm was calculated and plotted as a function of the concentration of Trolox for the standard reference data. The radical scavenging activity of the tasted samples, expressed as inhibition percentage, was calculated by the following formula:

$$\% \text{ Inhibition} = [(A_B - A_A)/A_B] \times 100$$

Where A_B was the absorbance of the blank sample (t=0), and A_A was the absorbance of sample with antioxidant after 6 min.

Trolox equivalent antioxidant coefficient (TEAC) was quantified by a regression analysis of % Inhibition versus Trolox concentration using the following formula:

$$TEAC = (\%I_M - b)/m$$

Where b was the intersection and m was the slope.

2.8. G-50 gel filtration chromatography

After filtration through 10, 5, 3 and 1 kDa membranes in a high performance ultrafiltration cell, 10 mL of the fraction with highest ACE-I inhibitory activity was injected into a Sephadex G-50 gel filtration column (3 cm x 79 cm) at a flow rate of 25 mL/h of 50 mM ammonium bicarbonate (pH 9.1). The resulting fractions were collected for to assay ACE-I inhibitory activity. Peptide molecular masses were determined by reference to a calibration curve created by running molecular mass markers on the Sephadex G-50 under conditions identical to those used for the test samples. Molecular mass standards were thyroglobulin (670 kDa), bovine gamma globulin (158 kDa), equine myoglobin (17 kDa), vitamin B12 (1.35 kDa) and Thr-Gln (0.25 kDa). Fractions selected for further purification of peptides were pooled and lyophilized before RP-HPLC.

2.9. HPLC C18 chromatography

The fractions isolated with the Sephadex G-50 column were redissolved in deionized water and injected into a preparative HPLC (Agilent, Model 1110, Agilent Technologies, Inc. Santa Clara, CA, USA) reverse-phasecolumn (C18 Hi-Pore RP-318, 250 mm x10 mm, Bio-Rad). The injection volume was 100 μL, and the sample concentration was 20 mg/mL. Elution was achieved by a linear gradient of acetonitrile in water (0-30% in 50 min) containing 0.1% trifluoroacetic acid at a flow rate of 4 mL/min and 30°C [13]. Elution was monitored at 215 nm, and the resulting fractions were collected for assay of ACE-I inhibitory activity as described above.

2.10. Amino acid composition

Protein amino acid composition was determined for the hydrolysate, and the peptides were purified by ultrafiltration, gel filtration chromatography and HPLC [19]. Samples (2-4 mg protein) were treated with 4 mL of 6 mol equi/L HCl, placed in hydrolysis tubes and gassed with nitrogen at 110°C for 24 h. They were then dried in a rotavapor and suspended in 1 mol/L sodium borate buffer at pH 9.0. Amino acid derivatization was performed at 50°C using diethyl ethoxymethylenemalonate. Amino acids were separated using HPLC with a reversed-phase column (300 x 3.9 mm, Nova Pack C18, 4 mm; Waters), and a binary gradient system with 25 mmol/L sodium acetate containing (A) 0.02 g/L sodium azide at pH 6.0, and (B) acetonitrile as solvent. The flow rate was 0.9 mL/min, and the elution gradient was: time 0.0–3.0 min, linear gradient A:B (91:9) to A-B (86:14); time 3.0–13.0 min, elution with A-B (86–14); time 13.0–30.0 min, linear gradient A-B (86:14) to A-B (69:31); time 30.0–35.0 min, elution with A-B (69:31).

2.11. Statistical analysis

All results were analyzed in triplicate using descriptive statistics with a central tendency and dispersion measures. One-way ANOVAs were performed to evaluate protein isolate hydrolysis data and *in vitro* ACE-I inhibitory activity. A Duncan's multiple range test was used to determine differences between treatments. All analyses were performed according to Montgomery [20] and processed using the Statgraphics Plus version 5.1 software.

3. Results and discussion

3.1. Protein extract hydrolysis

Alcalase®, Flavourzyme® and pepsin-pancreatin were used to produce extensively hydrolyzed *V. unguiculata* protein extracts. Degree of hydrolysis (DH) differed ($P<0.05$) between the enzymatic systems with values of 53.0%, 58.8%, and 35.7% for Alcalase® hydrolysate (AH), Flavourzyme® hydrolysate (FH) and Pepsin-pancreatin hydrolysate (PPH), respectively.

The AH had a 53.0% DH, which is lower than reported by Vioque et al. [21] for rapeseed protein hydrolysates (60% DH) produced with a mixture of Alcalase® and Flavourzyme® during 180 min. However, this DH was higher than that reported for mung bean protein hydrolysates (20.0% DH) produced with Alcalase® for 10 h[22] and for *V. unguiculata* hydrolysates (32.3% DH) produced with Alcalase® for 60 min [10].The variation in DH observed here is probably the result of protease specificity since Alcalase® is an industrial alkaline protease produced from *Bacillus licheniformis,* the main enzyme component of which (serine endopeptidase subtilisin Carlsberg) presents broad specificity and hydrolyzes most peptide bonds, with a preference for those containing aromatic amino acid residues [23]. ACE-I prefers substrates or competitive inhibitors containing hydrophobic amino acid (aromatic or branched lateral chain) residues (Hong et al., 2005). Alcalase® is therefore very suitable for production of bioactive peptides, such as those with ACE-I inhibitory activity. According to Pedroche et al. [24] the controlled liberation of biologically active peptides from protein by enzymatic hydrolysis is one of the most promising trends concerning medical applications of the protein hydrolysates with DH higher than 10% while hydrolysates with a low DH (lower than 10%) are used for the improvement of functional properties of flours or protein isolates. Therefore, the results suggest (DH=53.0%) that *V. unguiculata* protein is an appropriate substrate for producing these bioactive peptides when hydrolyzed with Alcalase®.

Hydrolysis with Flavourzyme® produced a *V. unguiculata* hydrolysate with 58.8% DH, somewhat higher than obtained with the Alcalase® system. A similar discrepancy has been reported for chickpea protein hydrolysates (27.0% DH) produced with Flavourzyme® for 180 min [25]. DH was higher with Flavourzyme® since it is a protease complex produced by *Aspergillus orizae,* which contains endoproteinases and exopeptidases. The fungal protease complex Flavourzyme® has a broader specificity; thus, when combined with its exopeptidase activity high DH values can be achieved, perhaps as much as 50% giving mostly dipeptides in the hydrolysate.

Sequential hydrolysis with pepsin-pancreatin produced cowpea protein hydrolysates with the lowest DH (35.74%) of the three studied enzymatic systems. This DH was similar to

37.0% at 360 min reported for sunflower protein hydrolysates obtained with the same system [13], but higher than the reported values for soy protein hydrolysates produced with pancreatin for 60 (11.0%) and 180 min (17.0%) [26]. The *V. unguiculata* PPH represents a pool of peptides resembling those generated during digestion of *V. unguiculata* proteins in the organism. This coincides with the behavior of extensively hydrolyzed sunflower protein reported by Megías et al. [13].Pepsin is the main proteolytic enzyme generated in the stomach during food digestion, while pancreatin includes proteases such as trypsin, chymotrypsin and elastase, which are released by the pancreas in the small intestine. The resulting peptides are therefore resistant to pepsin and pancreatin, suggesting that they might be absorbed by digestive epithelial cells in the small intestine, probably might be bioavailable and exercise their biological activity.

3.2. ACE-I inhibitory activity

ACE-I inhibitory activity of AH, FH and PPH was measured and calculated as IC_{50}. Biological activity was highest in the PPH, as indicated by a lower IC_{50} value (1397.9 μg/ml) compared to the AH (2564.7 μg/ml) and FH (2634.4 μg/ml). In other words, more ACE-I inhibitory active peptides were produced using the pepsin-pancreatin treatment.

ACE-I inhibitory activity has been reported for enzymatic hydrolysates from different protein sources with IC_{50} values ranging from 0.2 to 246.7 μg/mL [22]. Many of these hydrolysates have been shown to have antihypertensive activities in spontaneously hypertensive rats (SHR). In the present study, the IC_{50} value for the hydrolysate prepared with pepsin-pancreatin at 90 min incubation is within the concentration range likely to mediate an antihypertensive effect. Therefore, it is to be expected that *V. unguiculata* protein derived ACE-I inhibitory peptides would have antihypertensive activity. However, further investigations are necessary to examine whether the peptide mixture may exert antihypertensive activity *in vivo* because the inhibitory potencies of the peptides on ACE-I activity do not always correlate with their antihypertensive activities found in SHR[22].

Ultrafiltration of AH, FH and PPH produced peptide fractions with increased biological activity (Fig. 1). ACE-I inhibitory activity of peptide fractions ranged from 24.3-123 μg/ml in the AH, from 0.04 to 170.6 μg/ml in the FH and from 44.7 to 112 μg/ml in the PPH. ACE-I inhibitory activity was significantly ($P<0.05$) dependent on peptide fraction molecular weight, with the lowest activity being in the >10kDa fractions and the highest in the <1kDa fractions for all hydrolysates. Similar ACE-I inhibitory activity behavior was reported by Je et al.[5]for five peptide fractions from pepsin-hydrolyzed Alaska pollack frame protein run through an ultrafiltration membrane-bioreactor system with MWCOs of 30, 10, 5, 3 and 1 kDa. Higher ACE-I inhibitory activity (%) in lower molecular weight fractions was also reported by Xue-Ying et al. [27] for yak casein hydrolysate fractions separated using 6 and 10 kDa MWCOs: >10kDa (23.1%), 6-10kDa (29.2%) and <6 kDa (85.4%).

Of the three hydrolysates tested in the present study, FH, which had the highest DH, also exhibited the highest ACE-I inhibitory activity in the <1kDa fraction. The biological activity of this fraction provides it potential commercial applications as a 'health-enhancing ingredient' in functional food production.

In terms of ACE-I inhibitory activity, the pepsin-pancreatin system produced hydrolysate with the highest activity, while the Flavourzyme® system produced peptide fractions with the highest activity. This also confirms that *V. unguiculata* is a good protein source for bioactive peptide extraction by gastrointestinal or commercial proteases.

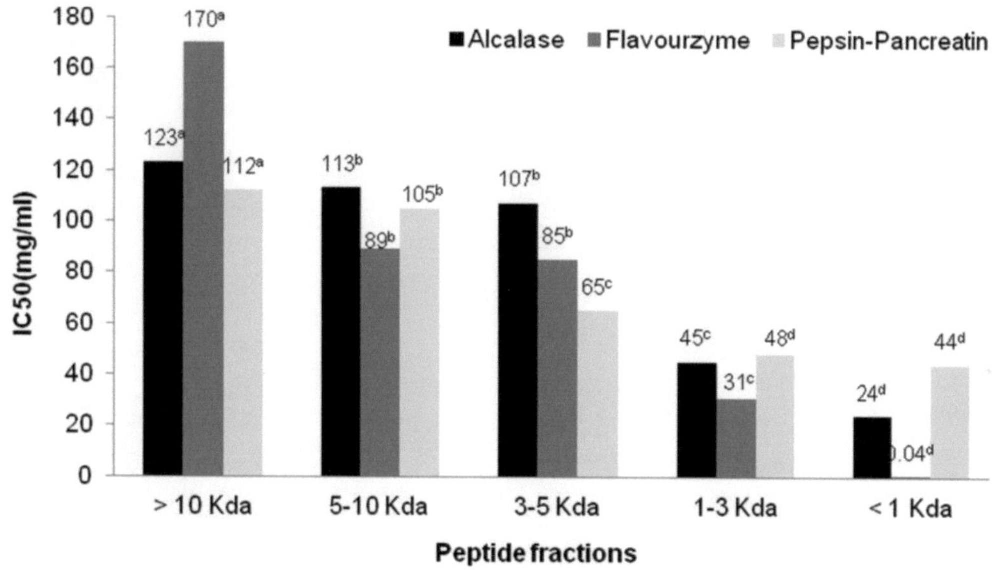

Figure 1. IC_{50} values of peptide fractions obtained by ultrafiltration from *V. unguiculata* protein hydrolysates.

3.3. Antioxidant activity by ABTS assay

Antioxidant activity of the protein hydrolysates and their corresponding UF peptide fractions was quantified and calculated as TEAC values (mM/mg protein). Antioxidant activity was not significantly different ($P>0.05$) between the three hydrolysis systems (14.7 for AH, 14.5 for FH and 14.3 mM/mg protein for PPH). Ultrafiltration improved antioxidant activity, which was dependent ($P<0.05$) on fraction molecular weight (Fig. 2).TEAC values were 303.2-1457 mM/mg protein for the AH, 357.4-10211 mM/mg protein for the FH, and 267.1-2830.4 mM/mg protein for the PPH. These values, and consequently the fraction's antioxidant activities, were higher ($P<0.05$) as fraction molecular weight decreased. According to Dávalos et al. [28] this behavior among the *V. unguiculata* peptide fractions may reflect the enhanced accessibility of small peptides to the redox reaction system, for the prescence of critical amino acid residues.

Antioxidant activity in the *V. unguiculata* protein hydrolysates and their UF peptide fractions was measured with an ABTS assay, which quantifies an antioxidant's (i.e. hydrogen or electron donor) suppression of the radical cation $ABTS^{\bullet+}$ based on single-electron reduction of the relatively stable radical cation $ABTS^{\bullet+}$ formed previously by an oxidation reaction. When added to PBS medium (pH 7.2) containing $ABTS^{\bullet+}$, the proteins in the hydrolysates and peptide fractions very probably acted as electron donors, transforming this radical cation (maximum absorbance at 734nm) into the non-radical ABTS. The higher antioxidant activity of the UF peptide fractions versus their source hydrolysates is related to unique properties provided by their amino acid composition. The fractions' increased ability to decrease free radical reactivity is linked to the greater exposure of their amino acids, which leads to increased peptide/free radical reactions.

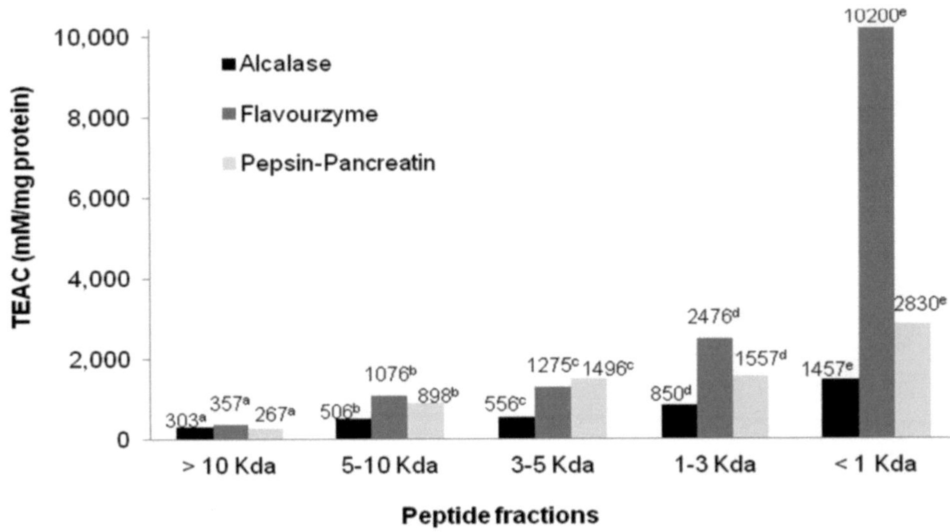

Figure 2. Antioxidant activity of peptide fractions obtained by ultrafiltration from *V. unguiculata* protein hydrolysates.

Overall, the <1 kDa peptide fraction from the FH had the highest TEAC values and was shown to undergo single-electron transfer reactions in the $ABTS^{\bullet+}$ reduction assay, demonstrating its antioxidant capacity. This is an extremely attractive property since oxidants are known to be involved in many human diseases and aging processes. Oxidants are associated with the chronic damage of ageing, and destructive oxidants and oxygen-free radicals can be extremely toxic to tissues by promoting tissue necrosis and cell damage. Some authors claim that proteins possess antioxidant properties; for instance, [29] reported that protein insufficiency aggravates enhanced lipid peroxidation and reduces antioxidative enzyme activities in rats, while Larson et al. [30] observed that proteins affect lipid metabolism in laboratory animals. The biological effect exhibited by the FH<1kDa fraction apparently reinforces the claim that proteins possess antioxidant properties. This makes the FH<1kDa fraction a potential "antioxidant" ingredient in functional food production.

3.4. Gel filtration chromatography

Because it exhibited the highest ACE-I inhibitory (IC_{50} value of 0.04 μg/mL) and antioxidant activity (TEAC value of 10211 mM/mg protein), the <1 kDa fraction from FH was selected for further fractionation. Gel filtration chromatography was used to generate a molecular weight profile of this fraction (Fig. 3).

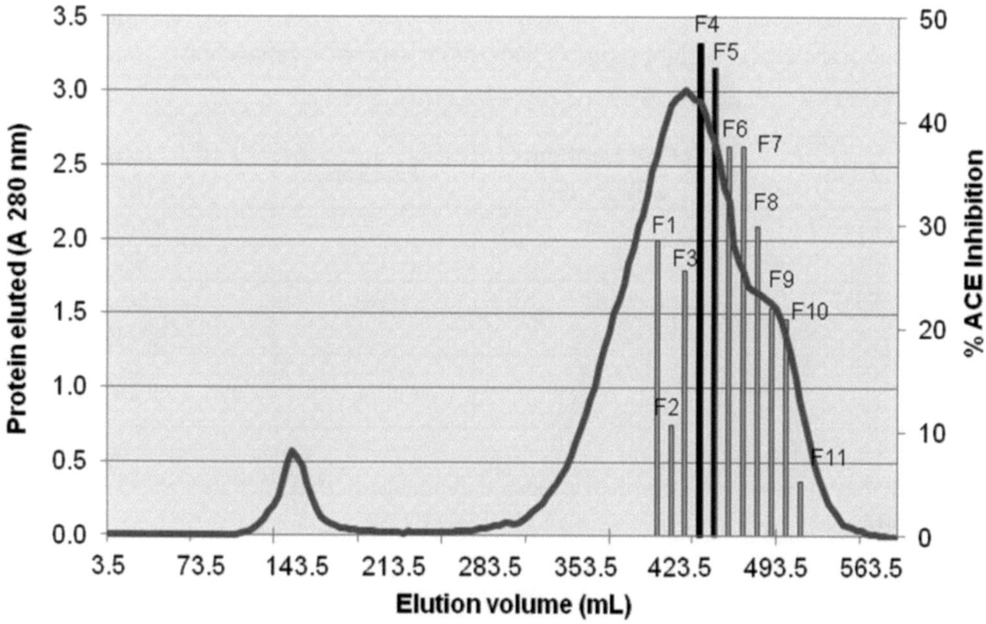

Figure 3. Elution profile of the <1 kDa ultrafiltration fraction of the cowpea *V. unguiculata* protein hydrolysate with Flavorzyme® purified in a Sephadex G-50 gel filtration column.

The profile was typical of a protein hydrolysate formed by a pool of peptides, with gradually decreasing molecular masses. Elution volumes between 406 and 518 mL included free amino acids and peptides with molecular masses ranging from 3.6 to 0.4 kDa. This range was fractionated into eleven fractions (1 to 11) and ACE-I inhibitory activity determined for each. Fractions with elution volumes smaller than 406 mL and greater than 518 mL were not analyzed because they largely included peptides with high molecular weights, as well as free amino acids. ACE-I inhibitory activity (%) in the eleven fractions ranged from 5.29 to 47.43% and differed ($P<0.05$) between fractions (Table 1). The highest ACE-I inhibitory activity was observed in fractions F4 (47.43%; 437.5-444.5 mL elution volume) and F5 (45.14%, 448-455 mL elution volume), which were not statistically different ($P<0.05$). Their molecular masses were approximately 1.8 kDa (indicative of 7 amino acid residues) and 1.5 kDa (indicative of 10 amino acid residues), respectively. The IC_{50} value for F4 (14.19 μg/mL) was similar than those of *Fagopirum esculentum* peptide fractions purified by Sephadex LH-20 gel filtration (15.1 μg/mL) [31], but lower than those of gel filtration (Sephadex G-25) peptide fractions from tuna broth hydrolysate (210 to 25,260 μg/mL) [32], from *Fagopyrum esculentum* Moench (Sephadex C-25 = 25,715.1 μg/mL; Sephadex G-10 = 21,315.1 μg/mL), and from the peptic hydrolysate of *Acetes chinensis* (Sephadex C-15 = 770 – 1590 μg/mL) [33]. A similar behavior pattern was observed with Konjac peptides purified by series connection of Sephadex G-25 and Sephadex G-15 columns that resulted in purified peptides with molecular weights of 1500 and 1000 Da and IC_{50} values of 120 μg/mL and 88 μg/mL, respectively [34]. Ji-Eun et al. [35] reported similar results to separate peptide fractions below 3 kDa through size exclusion chromatography (IC_{50}= 500 μg/mL) from textured and fermented vegetable protein (IC_{50} = 2190 μg/mL). They purified peptide fraction with a molecular weight range of 500-999 Da with IC_{50} values of 94 μg/mL and that represented peptides of approximately 7 amino acids residues.

Fraction	ACE-I inhibitory activity (%)	Fraction	ACE-I inhibitory activity (%)
F1	28.47^{C}	F7	37.61^{d}
F2	10.67^{a}	F8	29.9^{c}
F3	25.61^{bc}	F9	22.01^{b}
F4	47.43^{e}	F10	20.92^{b}
F5	45.14^{e}	F11	5.29^{a}
F6	37.58^{d}		

Table 1. ACE-I inhibition percentage of peptide fractions purified in a Sephadex G-50 gel filtration column.$^{a\text{-}e}$Different superscripts letters indicate statistical difference (P <0.05). Data are the mean of three replicates.

3.5. Reverse-phase HPLC chromatography of pooled fractions

Fractions F4 and F5 from the gel filtration chromatography treatment were pooled and analyzed using RP-HPLC to produce a chromatographic profile from mass-transfer between stationary and mobile phases. Mixture components were separated by dissolving fractions F4 and F5 in acetonitrile and forcing them through a chromatographic column under high pressure. The mixture resolved into its components in the column, separating F4 and F5 based on differences in hydrophobicity. In this process, the components of both fractions passed over stationary-phase particles containing pores large enough for them to enter, and in which interactions with the hydrophobic surface removed them from the flowing mobile-phase stream. The strength and nature of the interaction between the sample particles and stationary phase depended on hydrophobic and polar interactions. As the eluent organic solvent concentration increased, it reached a critical value for each analyte and desorbed it from the hydrophobic stationary-phase surface to allowing it to elute from the column into the flowing mobile phase. Because this elution depended on the precise distribution of hydrophobic residues in each specie, each analyte eluted from the column at a characteristic time and the resulting peaks were used to qualitatively analyze both fractions' components. Within each gel filtration fraction, the eluates were divided into four major fractions: F4-1, F4-2, F4-3, F4-4; F5-1, F5-2, F5-3, F5-4. The peptides were relatively pure, although a small shoulder still appeared behind the peaks in the chromatogram.

Enough material from each fraction was collected in successive analyses to determine ACE-I inhibitory activity (Fig. 4). Fraction F4 had a larger (P<0.05) ACE-I inhibitory activity range (33.83 to 75.42%) than F5 (32.31 to 49.71%). Overall, F4-2 (75.42%) had the highest ACE-I inhibitory activity (IC_{50} = 0.4704 μg/mL). This value is lower than the 6.3 μg/ml reported for a peptide from *F. esculentum* Moench purified by RP-HPLC [31] and within the 8.1 to 91.6 μg/mL range reported for fractions from caprine kefir water-soluble extract purified by preparative RP-HPLC [36]. The same behavior was observed when comparing the results of

RP-HPLC fractions obtained of *Mustelus mustelus* intestines with alkaline proteases (130- 783 μg/ml) [37] and with peptide fractions obtained from the peptic hydrolysate of the freshwater rotifer *Brachionus calyciflonus* (40.01 μg/ml) [38].

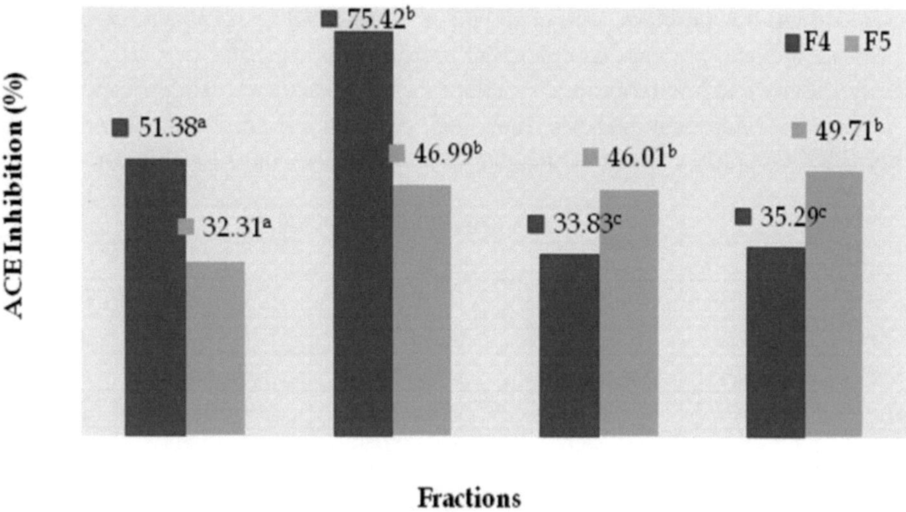

Figure 4. ACE inhibition percentage of RP-HPLC fractions obtained from fractions F4 and F5 produced after hydrolysis of *V. unguiculata* with Flavourzyme®.[a-c]Different letters in the same gel filtration chromatography fraction indicate statistical difference ($P<0.05$). Data are the mean of three replicates.

3.6. Amino acid composition

Fractions with the highest ACE-I inhibitory activity were analyzed to produce an amino acid profile. During hydrolysis, asparagine and glutamine partially converted to aspartic acid and glutamic acid, respectively; the data for asparagine and/or aspartic acid were therefore reported as Asx while those for glutamine and/or glutamic acid were reported as Glx. The higher ACE-I inhibitory activity exhibited by the <1 kDa fraction (IC_{50}=0.04 μg/mL) compared to FH (IC_{50}=2634.4 μg/mL) was probably due to its higher concentration of neutral amino acids, such as Ser (3.03%) and Thr (11.36%), hydrophilics such as His (12.5%) or hydrophobics such as Ala (7.84%), Pro (23.80%), Val (10%), Met (66.66%), Ile (10%), Leu (21.6%), Phe (18.6%) and Trp (16.6%) (Table 2).

Compared to the <1 kDa fraction amino acid profile, the G-50 gel filtration chromatography fractions had higher Asx, Glx and Arg concentrations. Hydrophobic amino acid content decreased by 20.27% in F4 (34.56g/100g) and 21.19% in F5 (34.16g/100g) compared to the <1 kDa fraction (43.35 g/100g), while hydrophilic residues increased by 19.25% (48 g/100g) in F4 and 22.48% (50 g/100g) in F5. The ACE-I inhibitory activity observed in F4 (47.43%) and F5 (45.14%) could therefore be the result of their higher Arg, Asx or Asp concentrations. These residues are known to play an important role in the antihypertensive activity of peptides from white and red wines and from French flor-sherry wine [38]. The F4 and F5 fractions may also have the added benefit of low bitterness. Many of the ACE-I-inhibitory peptides isolated from food sources are composed of multiple food components and hydrophobic and/or aromatic amino acid residues. However, practical use of these food protein hydrolysates is complicated by formation of peptides, which impart a bitter taste, the result of the formation of low molecular weight peptides containing mostly hydrophobic amino acids. To address this problem, Kim et al. [40] recommended use of Flavourzyme®, a fungal endoprotease and exo-protease complex that produces hydrolysates or peptides with ACE-I inhibitory activity and low bitterness. Fractions F4 and F5 are promising prospects for use in new product development because they had clear ACE-I inhibitory activity and low hydrophobic amino acid content, which may ensure that they have low bitterness.

Amino acid	Composition (g/100g)*				
	PC	**FH**	**F<1 kDa**	**F4**	**F5**
Asx	10.8 ± 0.003	9.6 ± 0.387	9.4 ± 0.448	14 ± 0.274	16.7 ± 0.152
Glx	19 ± 0.016	18.6 ± 0.013	14.9 ± 0.171	17.9 ± 0.182	18.3 ± 0.17
Ser	6.5 ± 0.041	6.4 ± 0.016	6.6 ± 0.046	5.8 ± 0.302	5.0 ± 0.126
His	2.9 ± 0.028	2.8 ± 0.001	3.2 ± 0.024	2.9 ± 0.008	2.7 ± 0.061
Gly	4.4 ± 0.005	4.5 ± 0.006	4.1 ± 0.035	4.3 ± 0.011	3.0 ± 0.177
Thr	4.3 ± 0.009	3.9 ± 0.009	4.4 ± 0.032	4.0 ± 0.040	3.4 ± 0.036
Arg	7.8 ± 0.067	7.9 ± 0.025	6.0 ± 0.073	7.5 ± 0.026	6.9 ± 0.039
Ala	4.4 ± 0.008	4.7 ± 0.002	5.1 ± 0.106	4.3 ± 0.016	3.6 ± 0.625
Pro	2.7 ± 0.078	1.6 ± 0.038	2.1 ± 0.01	0 ± 0	0 ± 0
Tyr	2.6 ± 0.098	2.7 ± 0.006	2.6 ± 0.026	2.2 ± 0.018	1.2 ± 0.078
Val	5.4 ± 0.052	6.3 ± 0.015	7.0 ± 0.029	5.3 ± 0.004	4.2 ± 0.211
Met	0.1 ± 0.109	0.3 ± 0.001	0.9 ± 0.013	0.7 ± 0.006	2.2 ± 0.292
Cys	0.4 ± 0.002	0.4 ± 0.006	0.2 ± 0.002	1.2 ± 0.01	3.2 ± 0.1
Ile	4.4 ± 0.019	5.4 ± 0.207	6.0 ± 0.073	5.2 ± 0.015	4.7 ± 0.236
Leu	9.2 ± 0.115	9.8 ± 0.175	12.5 ± 0.029	11.6 ± 0.092	12.6 ± 0.229
Phe	6.9 ± 0.008	7.0 ± 0.004	8.6 ± 0.056	6.7 ± 0.046	5.8 ± 0.043
Lys	7.3 ± 0.089	7.1 ± 0	5.3 ± 0.072	5.7 ± 0.042	5.3 ± 0.263
Trp	0.7 ± 0.064	1.0 ± 0.038	1.2 ± 0.041	0.8 ± 0.241	1.0 ± 0.535

*Data are the mean of three replicates

Table 2. Amino acid contents of the cowpea *V. unguiculata* protein concentrate (PC), Flavourzyme hydrolysate (FH), <1 kDa ultrafiltered fraction and F4 and F5 gel filtration chromatography fractions.

Comparison of amino acid composition and properties between the cowpea *V. unguiculata* hydrolysate and its peptides isolated by ultrafiltration and G-50 gel chromatography showed the most active fraction to be the <1 kDa fraction, which had abundant aromatic (e.g., Phe) and cyclic amino acids (e.g., Pro). Amino acids such as Phe, with large bulky chains and hydrophobic side chains, are preferred in both positions of a dipeptide for their high steric properties and low lipophilicity, while amino acids such as Pro are preferred in the carboxyl terminus of active tripeptides for their low lipophilicity, and high steric and electronic properties [41]. Aromatic and basic amino acids are important to the ACE-I inhibitory activity of peptides. For instance, similar observations have been made for an orientase hydrolysate with notable ACE-I inhibition attributed to its high basic and aromatic amino acids contents [33]. Cheung et al. [43] reported a peptide with strong, competitive ACE-I inhibition in which aromatic amino acid residues at its C-terminal and basic or hydrophobic ones at its N-terminal played an essential role. Based on the above, the higher ACE-I inhibitory potential of the <1 kDa fraction from the cowpea hydrolysate can be attributed to the steric properties of its aromatic amino acids and the lipophilicity and electronic properties of its cyclic amino acids.

The most active fraction among those purified from F4 by preparative RP-HPLC was F4-2, which had much higher neutral amino acid content (80.6 g/100g) than F4-1 (46.6 g/100g), F4-3(6.1 g/100g) and F4-4 (28.6 g/100g) (Table 3). Of the neutral amino acids, Tyr was higher in F4-2 (71.3 g/100g) compared to F4-3 (1.5 g/100g), F4-4 (3.9 g/100g) and F4-1 (3.9 g/100g). This supports the importance of aromatic residues in a peptides' biological potential, probably due to their high steric properties and low lipophilicity. The Tyr amino acid has been reported in peptides from milk (e.g., Tyr-Pro-Tyr-Tyr) isolated by a combination of lactic acid bacteria fermentation and Flavourzyme® hydrolysis. These peptides are bioavailable and exhibit *in vitro* (90.9 μM) and *in vivo* ACE-I inhibitory activity, the latter in the form of reduced hypertension in SHR (15.9 mmHg reduction in systolic blood pressure) [43].

F4-2 was also unique in containing Met (0.5 g/100g) and Trp (0.2 g/100g), as well as higher concentrations of Ile (3.2 g/100g) and Leu (10.2 g/100g) than the other fractions eluted from F4, all of which could have significantly increased its relative ACE-I inhibition activity. This would coincide with previous reports of inhibitory activity in peptides with these amino acids. The residues Val, Leu and Ile are preferred in the amino terminal position in active tripeptides for their low lipophilicity and steric properties or side chain bulk/ molecular size [40]. Clearly, the structure of ACE-I inhibitory peptides influences their activity, as shown by Cheung et al. [42], who reported that peptides with Tyr at the C-terminus and Ile at the N-terminus exhibit highly potent inhibitory activity. In a randomized, double-blind, placebo-controlled human study, a significant depressor effect was observed in mild essential hypertensive volunteers and Val-Tyr was shown to be one of the predominant ACE-I inhibitory peptides involved in this effect [44]. In addition to its high Tyr content, F4-2 may also be absorbed intact into the human circulatory system and induce a reduction in blood pressure. For instance, intravenous and oral administration of Val-Tyr in SHR have shown that this di-peptide caused a long-lasting depressor effect. Val-Tyr is known to be absorbed intact into the human circulatory system, and studies using cross-mated transgenic mice carrying the human renin gene and the human angiotensinogen gene have shown that, as a natural ACE-I inhibitory dipeptide, Val-Tyr regulates the enhanced human renin-angiotensin system and induces a prolonged reduction in blood pressure [45].

Of the fractions purified from F5 by preparative RP-HPLC chromatography (Table 3), F5-2, F5-3 and F5-4 exhibited ACE-I inhibition that was not different among them (P<0.05), but was greater than that of F5-1. The amino acids Arg, Tyr, Met, Ile, Leu, Phe and Lys very probably played a key role in providing greater activity to the first three fractions. As mentioned above, residues such as Leu and Ile are preferred at the amino terminus of tripeptides with ACE-I inhibitory activity due to their low lipophilicity and steric properties or side chain bulk/molecular size values. In addition, Lys and Arg are expected in positions adjacent to the amino terminus due to their low electronic properties and high lipophilicity and steric property values, while residues such as Phe are preferred at the carboxyl terminus due to their low lipophilicity and high steric and electronic property values [41].

Although the precise substrate specificity is not fully understood, ACE-I appears to prefer substrates containing hydrophobic amino acid residues at the three C-terminal positions, suggesting that the higher hydrophobic amino acid content in F5-2 (9.1 g/100g), F5-3 (89.8 g/100g) and F5-4 (34.2 g/100g) versus F5-1 (1.1 g/100g) probably made a substantial contribution to their inhibitory potency. This agrees with Wu et al. [40], who state that aromatic, positively-charged and hydrophobic amino acids are preferred in active tripeptides. Due to substrate specificity differences between the two ACE-I catalytic sites, ACE-I inhibitors may inhibit only one site. Moskowitz [46] proposed a model explaining the clinical superiority of hydrophobic ACE-I inhibitory drugs relative to hydrophilic ones: all ACE-I inhibitors bind to the C-terminal catalytic site, but only hydrophobic ones bind to the occluded N-terminal catalytic site and are therefore better at blocking Ang II production. This would also explain why hydrophobic ACE-I inhibitors have specific local benefits such as organ damage prevention, in addition to reducing blood pressure [46]. The high hydrophobic amino acid (particularly aromatic side-chains) content in the F5 may therefore make a substantial contribution to its fractions' ACE inhibitory activity by blocking Ang II production.

Overall, the highest *in vitro* ACE-I inhibitory activity (IC_{50}) among the cowpea hydrolysate and its derivative fractions was present in the ultrafiltered <1 kDa fraction (0.04 μg/mL), followed by the RP-HPLC F4-2 fraction (0.4704 μg/mL), the gel filtration chromatography fraction F4 (14.195 μg/mL) and finally the hydrolysate (2634.4 μg/mL). Although the IC_{50} values for the hydrolysate, F4 and F4-2 fractions were substantially higher than that of the F<1 kDa fraction, they are still in the same order of magnitude as values reported for many other natural ACE-I inhibitory peptides. Nevertheless, the IC_{50} values for all the studied *V. unguiculata* derivatives are far higher than that of the synthetic ACE-I inhibitor Captopril® (0.0013 μg/mL) [32]. The biological potential in the peptides purified from *V. unguiculata*, and the high ACE-I inhibitory activity in the F<1 kDa fraction, reinforce the need for ACE-I inhibitory peptides to be rich in hydrophobic amino acids (aromatic or branched chains) and peptides rich in Pro. Pro is well-documented as the most favorable amino acid for ACE-I binding (most commercial inhibitors include this residue), but it was not present in the peptides with potential biological activity that had been purified by G-50 gel filtration chromatography and RP-HPLC.

The studied cowpea *V. unguiculata* protein hydrolysate has potential applications in the development of physiologically functional foods aimed at preventing and/or treating hypertension. An added benefit is the balanced amino acid profile of the protein hydrolysate and its peptide fractions, which makes them an appropriate protein source in human nutrition. It should be considered, however, that the *in vitro* ACE-I inhibitory potencies of peptides do not always correlate with their *in vivo* antihypertensive activities as quantified in SHR. This is because they must be absorbed and transported intact from the intestine to the blood stream (in the case of oral administration) and resist plasma peptidase degradation (in the case of oral and intravenous administration) to reach their target sites and exert an antihypertensive effect *in vivo*. Therefore, *in vivo* research is needed to determine to what extent any of the studied ACE-I inhibitory peptides can exercise their antihypertensive activity *in vivo*.

Amino acid	Content (g/100g)*							
	F4-1	F4-2	F4-3	F4-4	F5-1	F5-2	F5-3	F5-4
Asx	9.9 ± 0.063	1.2 ± 0.009	0.8 ± 0.001	12.5 ± 0.242	2.1 ± 0.492	1.2 ± 0.005	2 ± 0.004	10.6 ± 0.731
Glx	5.1 ± 0.214	1.4 ± 0.048	1.1 ± 0.006	12.6 ± 0.329	0 ± 0	1.5 ± 0.046	2 ± 0.071	14.2 ± 0.762
Ser	22.7 ± 0.291	1 ± 0.001	1.2 ± 0.021	8.4 ± 0.177	30.5 ± 0.834	0.9 ± 0.014	1 ± 0.014	8.4 ± 0.009
His	8.2 ± 0.677	0.4 ± 0.015	0.3 ± 0.009	4.1 ± 0.133	16.3 ± 0.194	0.3 ± 0.049	0.4 ± 0.017	2.7 ± 0.957
Gly	19.6 ± 0.983	1.4 ± 0.008	1.7 ± 0.031	9.2 ± 0.226	13.3 ± 0.159	2.9 ± 0.017	2 ± 0.016	10 ± 0.057
Thr	4.3 ± 0.987	0.3 ± 0.005	0.4 ± 0.018	3.5 ± 0.066	5.8 ± 0.069	0.3 ± 0.048	0.4 ± 0.024	2.7 ± 0.957
Arg	3.2 ± 0.065	0.8 ± 0.002	0.8 ± 0.01	7.1 ± 0.151	0 ± 0	0.2 ± 0.236	0.6 ± 0.028	3 ± 0.139
Ala	14.3 ± 0.620	0.4 ± 0.001	0.3 ± 0.005	3.8 ± 0.046	1.1 ± 1.536	0.3 ± 0.048	0.5 ± 0.012	3.4 ± 0.089
Pro	0 ± 0	0 ± 0	0 ± 0	0 ± 0	0 ± 0	0 ± 0	0 ± 0	0 ± 0
Tyr	0 ± 0	71.3 ± 0.012	1.5 ± 0.026	3.9 ± 0.109	0 ± 0	74.7 ± 0.280	0.7 ± 0.109	4.2 ± 0.289
Val	0 ± 0	0 ± 0	0 ± 0	0 ± 0	0 ± 0	0 ± 0	0 ± 0	0 ± 0
Met	0 ± 0	0.5 ± 0.013	0 ± 0	0 ± 0	0 ± 0	0.6 ± 0.071	0.1 ± 0.096	0.5 ± 0.039
Cys	0 ± 0	6.6 ± 0.001	1.3 ± 0.085	3.6 ± 0.288	31 ± 0.369	8.8 ± 0.098	0.9 ± 0.019	6.8 ± 0.070
Ile	0.4 ± 0.555	3.2 ± 0.041	0.2 ± 0.020	2.8 ± 0.059	0 ± 0	1.6 ± 0.035	0.5 ± 0.026	1.8 ± 0.032
Leu	9.3 ± 0.217	10.2 ± 0.006	0.3 ± 0.008	4.7 ± 0.024	0 ± 0	3.1 ± 0.066	0.8 ± 0.093	6.1 ± 0.896
Phe	0 ± 0	0.8 ± 0.003	89.6 ± 0.230	20 ± 0.778	0 ± 0	3.5 ± 0.765	87.9 ± 0.234	22.4 ± 0.304
Lys	3± 0.532	0.3 ± 0.007	0.3 ± 0.008	3.8 ± 0.068	0 ± 0	0.2 ± 0.009	0.3 ± 0.002	2.4 ± 0.073
Trp	0 ± 0	0.2 ± 0.040	0 ± 0	0 ± 0	0 ± 0	0 ± 0	0 ± 0	0 ± 0

Table 3. Amino acid contents of fractions from the F4 and F5 gel filtration chromatography fraction purified by preparative RP-HPLC C-18 chromatography.

4. Conclusions

After modification by Alcalase®, Flavourzyme® and pepsin-pancreatin cowpea *V. unguiculata* proteins proved to be a source of bioactive peptides with ACE-I inhibitory and antioxidant activity. Fractionation of *V. unguiculata* enzymatic hydrolysates by ultrafiltration enhanced their ACE-I inhibitory and antioxidant activity in all the resulting peptides, although the <1 kDa fraction of the Flavourzyme hydrolysate had the highest overall biological activity. Further purification of this fraction by gel filtration chromatography and RP-HPLC produced fractions with different activities, all of which were much higher than the source hydrolysate. Separation of protein hydrolysates by molecular weight and hydrophobicity clearly enhanced peptide ACE-I inhibitory activity, particularly purification by ultrafiltration and chromatography. The highest biological potential among the purified peptides was observed in the ultrafiltered <1 kDa fraction, which supports the importance of high hydrophobic amino acid and proline content in ACE-I inhibitory peptides. These results highlight the promise of controlled protein hydrolysis with a fungal protease complex to isolate bioactive peptides from cowpea *V. unguiculata* proteins, which can then be further purified and/or used as an ingredient in functional foods designed for specific diets.

Author details

Maira R. Segura-Campos, Luis A. Chel-Guerrero and David A. Betancur- Ancona*
Facultad de Ingeniería Química, Campus de Ciencias Exactas e Ingenierías,
Universidad Autónoma de Yucatán, Periférico Nte, Tablaje Catastral 13615,
Col. Chuburná de Hidalgo Inn, Mérida, Yucatán, México

Acknowledgement

This research was supported by the Consejo Nacional de Ciencia y Tecnología (CONACYT) through doctoral scholarship 164186. This research forms part of Project 153012 "Actividad biológica de fracciones peptídicas derivadas de la hidrólisis enzimática de proteínas de frijoles lima (*Phaseolus lunatus*) y caupí (*Vigna unguiculata*)", financed by the CONACYT.

5. References

[1] Miguel M, Alonso M, Salaices M, Aleixandre A, López-Fandiño R. Antihypertensive, ACE inhibitory and vasodilator properties of an egg white hydrolysate: Effect of a simulated intestinal digestion. Food Chemistry 2011; 104: 163-168.

[2] Hooper NM, Turner AJ. An ACE structure. Nature Structural & Molecular Biology 2003; 10: 155-157.

* Corresponding Authors

[3] Tzakos AG, Galanis AS, Spyroulias GA, Cordopatis P, Manessi-Zoupa E, Gerothanassis IP. Structure-function discrimination of the N and C-catalytic domains of human angiotensin-converting enzyme: implications for Cl-activation and peptide hydrolysis mechanisms. Protein Engineering 2003; 16: 993-1003.

[4] Saiga A, Tanabe S, Nishimura T. Antioxidant activity of peptides obtained from porcine myofibrillar proteins by protease treatment. Journal of Agricultural and Food Chemistry 2006; 51: 3661-3667.

[5] Je J Y, Park PJ, Kwon, J Y, Kim AK. A novel angiotensin I converting enzyme inhibitory peptide from Alaska pollack (*Theragra chalcogramma*) frame protein hydrolysate. Journal of Agricultural and Food Chemistry 2004; 52: 7842-7845.

[6] Giasson BI, Ischiropoulos H, Lee VMY, Trojanowski JQ. The relationship between oxidative/nitrosative stress and pathological inclusions in Alzheimer's and Parkinson's diseases.Free Radical Biology & Medicine 2002; 32: 1264-1275.

[7] Pellegrini N, Del Rio D, Colombi B, Bianchi M, Brighenti F. Application of the 2,2′-Azinobis (3-ethylbenzothiazoline-6-sulfonic acid) radical cation assay to a flow injection system for the evaluation of antioxidant activity of some pure compounds and beverages. Journal of Agricultural and Food Chemistry 2003; 51: 260-264.

[8] Moller NP, Scholz-Ahrens KE, Schrezenmeir NR. Bioactive peptides and protein from foods: indication for health effects. European Journal of Nutrition 2008; 47: 171-182.

[9] Korhonen H, Pihlanto A. Bioactive peptides: Production and functionality. International Dairy Journal 2006; 16: 945-960.

[10] Araujo-González F, Chel-Guerrero L, Betancur-Ancona D. Functional properties of hydrolysates from cowpea (*V. unguiculata*) seeds. Advances in Food Science and Food Biotechnology in Developing Countries 2008; 24: 173-183.

[11] Betancur-Ancona D, Gallegos-Tintoré S, Chel-Guerrero L. Wet fractionation of *Phaseolus lunatus* seeds: partial characterization of starch and protein. Journal of the Science of Food and Agriculture 2004; 84: 1193-1201.

[12] Pedroche J, Yust MM, Girón-Calle J, Alaiz M, Millán F, Vioque J. Utilisation of chickpea protein isolates for production of peptides with angiotension I-converting enzyme (ACE) inhibitory activity. Journal of the Science of Food and Agriculture 2002; 82(1): 960-965.

[13] Megías C, Yust MM, Pedroche J, Lquari H, Girón-Calle J, Alaiz M, Millán F, Vioque J. Purification of an ACE inhibitory peptide after hydrolysis of sunflower (*Helianthus annuus* L.) protein isolates. Journal of Agricultural and Food Chemistry 2004; 52: 1928-1932.

[14] Yang Y, Marczak DE, Yokoo M, Usui H, Yoshikawa M. Isolation and antihypertensive effect of Angiotensin I-Converting Enzyme (ACE) inhibitory peptides from spinach rubisco. Journal of Agricultural and Food Chemistry 2003; 51: 4897-4902.

[15] Nielsen P, Petersen D, Dammann C. Improved method for determining food protein degree of hydrolysis. Journal of Food Science 2001; 66(5): 642-646.

[16] Cho MJ, Unklesbay N, Hsieh F, Clarke AD. Hydrophobicity of bitter peptides from soy protein hydrolysate.Journal of Agricultural and Food Chemistry 2004; 52 (19): 5895-5901.

[17] Hayakari M, Kondo Y, Izumi H.A rapid and simple spectrophotometric assay of angiotensin-converting enzyme. Analytical Biochemistry 1978; 84 (1): 361-369.

[18] Pukalskas A, Van Beek T, Venskutonis R, Linssen J, Van Veldhuizen A, Groot A. Identification of radical scavengers in sweet grass (*Hierochloe odorata*), Journal of Agricultural and Food Chemistry 2002; 50: 2914-2919.

[19] Alaiz M, Navarro JL, Giron J, Vioque E. Amino acid analysis by high-performance liquid chromatography after derivatization with diethylethoxymethylenemalonate. Journal of Chromatography 1992; 591: 181-186.

[20] Montgomery D. Diseño y análisis de experimentos (2th Ed.). Mexico: Limusa-Wiley 2004; 686 p.

[21] Vioque J, Sánchez-Vioque R, Clemente A, Pedroche J, Bautista J, Millan F. Production and characterization of an extensive rapeseed protein hydrolysate. Journal of the American Oil Chemists Society 1999; 76(7): 819-823.

[22] Hong LG, Wei L, Liu H, Hui SY. Mung-bean protein hydrolysates obtained with Alcalase® exhibit angiotensin I-converting enzyme inhibitory activity. Food Science and Technology International 2005; 11(4): 281-287.

[23] Doucet D, Otter DE, Gauthier SF, Foegeding EA. Enzyme-induced gelation of extensively hydrolyzed whey proteins by Alcalase: Peptide identification and determination of enzyme specificity Journal of Agricultural and Food Chemistry 2003; 51: 6300-6308.

[24] Pedroche J, Yust M, Girón-Calle J, Vioque J, Alaiz M, Millán F. Plant protein hydrolysates and tailor-made foods. Electronic Journal of Environmental, Agricultural and. Food Chemistry 2003; 2(1): 233-235.

[25] Clemente A, Vioque J, Sánchez-Vioque R, Pedroche J Millán F. Production of extensive chickpea (*Cicer arietinum* L.) Protein hydrolysates with reduced antigenic activity. Journal of Agricultural and Food Chemistry 1999; 47: 3776-3781.

[26] Qi M, Hettiarachychy N, Kalapathy U. Solubility and emulsifying properties of soy protein isolates modified by pancreatin.Journal of Food Science 1997; 62(6): 1110-1115.

[27] Xue-Ying M, Jin-Ren N, Wei-Ling S, Peng-Peng H, Li F. Value-added utilization of yak milk casein for the production of angiotensin-I-converting enzyme inhibitory peptides. Food Chemistry 2007; 103: 1282-1287.

[28] Dávalos A, Miguel M, Bartolomé B, López-Fandiño R. Antioxidant activity of peptides derived from egg white proteins by enzymatic hydrolysis. Journal of Food Protection 2004; 67(9): 1939-1944.

[29] Hui A, Furong W, Chaoliang L. Antioxidant activities of protein-enriched fraction from the larvae of housefly, *Musca domestica*. Natural Product Research 2008; 22(6): 507-515.

[30] Larson M, Donovan SM, Potter SM. Effects of dietary protein source on cholesterol metabolism in neonatal pigs. Nutrition Research 1996; 16(9): 1563-1574.

[31] Ma MS, Bae IY, Lee HG, Yang CB. Purification and identification of angiotensin I-converting enzyme inhibitory peptide from buckwheat (*Fagopyrum esculentum* Moench). Food Chemistry 2006; 96: 36-42.

[32] Hwang JS, Ko WC. Angiotensin I-converting enzyme inhibitory activity of protein hydrolysates from tuna broth. Journal of Food and Drug Analysis 2004; 12(3): 232-237.

[33] Cao W, Zhang C, Hong P, Ji H, Hao J. Purification and identification of an ACE inhibitory peptide from the peptic hydrolysates of *Acetes chinensis* and its antihypertensive effects in spontaneously hypertensive rats. International Journal of Food Science and Technology 2010; 45(1): 959-965.

[34] Wang L, Xu H, He X, Wang X. Studies on the preparation of ACE inhibitory peptides from konjac fly powder by alkaline protease enzymolysis. Journal of Chinese Institute of Food Science and Technology 2010; 1: 42-47.

[35] Ji-Eun K, Ki H, Sam-Pin L. ACE inhibitory and hydrolytic enzyme activities in textured vegetable protein in relation to the solid state fermentation period using *Bacillus subtilis* HA. Food Science and Biotechnology 2010; 19(2): 487-495.

[36] Quirós A, Hernández-Ledesma B, Ramos M, Amigo L, Recio I. Angiotensin- converting enzyme inhibitory activity of peptides derived from caprine kefir. Journal of Dairy Science 2005; 88: 3480-3487.

[37] Bougatef A, Balti R, Nedjar-Arroume N, Ravallec R, Adje EY, Souissi N, Lassoued I, Guillochon D, Nasri M. Evaluation of angiotensin I-converting enzyme (ACE) inhibitory activities of smooth hound (*Mustelus mustelus*) muscle protein hydrolysates generated by gastrointestinal proteases: identification of the most potent active peptide. European Food Research and Technology 2010; 231(1): 127-135.

[38] Lee JK, Lee MS, Park HG, Kim S, Byun HG. Angiotensin I converting enzyme inhibitory peptide extracted from freshwater zooplankton. Journal of Medicinal Food 2010; 13(2): 357-363.

[39] Pozo-Bayón MA, Alcaíde JM, Polo MC, Pueyo E. Angiotensin I converting enzyme inhibitory compounds in white and red wines. Food Chemistry 2007; 100: 43-47.

[40] Kim JM, Whang JH, Suh HJ. Enhancement of angiotensin I converting enzyme inhibitory activity and improvement of the emulsifying and foaming properties of corn gluten hydrolysate using ultrafiltration membranes. European Food Research and Technology 2004; 218: 133-138.

[41] Wu J, Aluko RE, Nakai S. Structural requirements of angiotensin I-converting Enzyme inhibitory peptides: Quantitative structure-activity relationship study of di- and tripeptides. Journal of Agricultural and Food Chemistry 2006; 54: 732-738.

[42] Cheung HS, Wang FL, Ondetti MA, Sabo EF Cushman DW. Binding of peptide substrates and inhibitors of Angiotensin-converting enzyme. Journal of Biological Chemistry 1980; 255(2): 401-407.

[43] Jenn-Shou T, Tai-Jung C, Bonnie-Sun P, Shein-Da G, Mei-Yuh C. Antihypertensive effect of bioactive peptides produced by preotease-facilitated lactic acid fermentation of milk. Food Chemistry 2008; 106(2): 552-558.

[44] Matsufuji H, Matsui T, Seki E, Osajima K, Nakashima M, Osajima Y. Angiotensin I-converting enzyme inhibitory peptides in an alkaline protease hydrolysate derived from sardine muscle. Bioscience, Biotechnology and Biochemistry 1994; 58: 2244–2245.

[45] Matsui T, Hayashi A, Tamaya K, Matsumoto K, Kawasaki T, Murakami K, Kimoto K. Depressor effect induced by dipeptide, Val-Tyr, in part, to the suppression of human circulating renin-angiotensin system. Clinical and Experimental Pharmacology 2003; 30: 262-265.

[46] Moskowitz DW. Is 'somatic' angiotensin I-converting enzyme a mechanosensor? Diabetes Technology and Therapeutics 2003; 4: 841-858.

Chapter 8

Dipeptidyl Peptidase-IV Inhibitory Activity of Peptides in Porcine Skin Gelatin Hydrolysates

Kuo-Chiang Hsu, Yu-Shan Tung, Shih-Li Huang and Chia-Ling Jao

Additional information is available at the end of the chapter

http://dx.doi.org/10.5772/51264

1. Introduction

During a meal, two incretin hormones, glucagon-like peptide 1 (GLP-1) and glucose-dependent insulinotropic polypeptide (GIP), are released from the small intestine into the vasculature and augment glucose-induced insulin secretion from the islet β-cells [1]. It has been estimated that 50-60% of the total insulin secreted during a meal results from the incretin response, mainly the combined effects of GIP and GLP-1 [2]. Previous studies have shown both GIP and GLP-1 stimulate β-cell proliferation, differentiation, and prevent apoptosis [3-6]. GLP-1 also has some actions such as inhibition of glucagon secretion and food intake, glucose homeostasis and slowing the gastric emptying [7-9]. However, GLP-1 has a short half-life of 1-2 min following secretion in response to the nutrients ingestion because of its inactivation by dipeptidyl peptidase-IV (DPP-IV) [10], resulting in loss of insulinotropic activities.

DPP-IV (CD26; E.C. 3.4.14.5) is a 110-kDa plasma membrane glycoprotein ectopeptidase that belongs to the prolyl oligopeptidase family [11]. It acts as a cleaving enzyme with the specificity for removing X-Pro or X-Ala dipeptides from the N terminus of polypeptides and proteins. It has a strong preference for Pro > Ala > Ser as the penultimate amino acid residue [10-12]. This enzyme is also capable of cleavage of N-terminal dipeptides with hydroxyproline (Hyp), dehydroproline, Gly, Val, Thr or Leu [10-14]. GLP-1 has Ala as the N-terminal penultimate amino acid residue, and therefore it is the substrate of DPP-IV. This finding that over 95% of the degradation of GLP-1 is attributed to the action of DPP-IV led to an elevated interest in inhibition of this enzyme for the treatment of type 2 diabetes [15]. Some previous studies have shown that specific DPP-IV inhibition increased the half-life of total circulating GLP-1, decreased plasma glucose, and improved impaired glucose tolerance in animal and human experiments [16-18].

There are several chemical compounds used *in vitro* and in animal models to inhibit DPP-IV activity, such as Val-pyrrolidide [17], NVP-DPP728 [18], Lys[Z(NO_2)]-thiazolidide and

Lys[Z(NO_2)]-pyrrolidide [19]. However, such chemical compounds, which often have to be administered by injection, may result in side effects as chemical drugs often do. Thus, to develop safe and natural DPP-IV inhibitors as the therapeutic agents of type 2 diabetes is necessary.

Proteins are well known as precursors of bioactive peptides. In recent years, peptides have been identified to possess physiological functions, such as immunomodulatory [20], antimicrobial [21], antihypertensive [22], anticancer [23], antioxidative [24] and cholesterol-lowering activities [25]. These bioactive peptides are mostly derived from milk, wheat, soybean, egg and fish proteins by enzymatic hydrolysis or fermentation [26]. Food protein hydrolysates are well-used and natural food ingredients, and therefore they are believed to be safe for consumers when they are served as functional foods. Some studies have reported that bioactive peptides possessed DPP-IV inhibitory activity. Diprotins A and B, isolated from culture filtrates of *Bacillus cereus* BMF673-RF1, are bioactive peptides found to exhibit the DPP-IV inhibitory activity with IC_{50} values of 1.1 and 5.5 μg/mL; they were elucidated to be Ile-Pro-Ile and Val-Pro-Leu, respectively [27]. Two bioactive peptides, Ile-Pro-Ala and Val-Ala-Gly-Thr-Trp-Tyr, derived from β-lactoglobulin hydrolysed by proteinase K and trypsin, showed IC_{50} values of 49 and 174 μM, respectively, against DPP-IV *in vitro* [28,29]. Two patents, WO 2006/068480 and WO 2009/128713 have shown that peptides derived from casein and lysozyme hydrolysates, respectively display DPP-IV inhibitory activity, and the peptides show in particular the presence of at least one Pro within the sequence and mostly as the penultimate N-terminal residue [30,31].

It is well-known that the dominant amino acid in gelatin is Gly, while the imino acids (Pro and Hyp) come second in abundance [32]. The amino acid composition of gelatin is characterized by a repeating sequence of Gly-X-Y triplets, where X is mostly Pro and Y is mostly Hyp. Inside gelatin molecule, Gly constitutes approximately 27% of the total amino acid pool [33]. The total amount of the imino acids is higher in mammalian (20-24%) than in fish (16-20%). In our previous study, we successfully isolated two peptides, Gly-Pro-Ala-Glu and Gly-Pro-Gly-Ala from Atlantic salmon skin gelatin, that showed dose-dependent inhibitory effects on DPP-IV with IC_{50} values of 49.6 and 41.9 μM, respectively [34]. According to the report of previous studies, DPP-IV inhibitory peptides consisted of at least one Pro and mostly as the penultimate N-terminal residue [32]. Therefore, the aim of this study was to examine the DPP-IV inhibitory activity of peptides derived from porcine skin gelatin, which constitutes higher content of imino acids than skin gelatin of Atlantic salmon, a kind of cold-water fish. This is expected to give insight into the possible utilization of porcine skin as a potential source of DPP-IV inhibitors that may be used in the treatment of type 2 diabetes to lower the risk of side effects.

2. Materials and methods

2.1. Materials and reagents

Porcine skin gelatin (G-2500) was purchased from Sigma-Aldrich (St. Louis, MO, USA). Alcalase® 2.4 L FG (from *Bacillus licheniformis*, 2.4 AU/g) (ALA) was the product from Novozymes North America Inc. (Salem, NC, Canada). DPP-IV (D7052, from porcine kidney), Gly-Pro-*p*-nitroanilide hydrochloride, trichloroacetic acid (TCA), L-Leu and Diprotin A were purchased from Sigma-Aldrich. Trinitrobenzenesulfonic acid (TNBS) was from Fluka Biochemika (Oakville, ON, Canada). Other chemicals and reagents used were analytical grade and commercially available.

2.2. Enzymatic hydrolysis

One gram of the gelatin added with 50 mL ddH_2O was incubated at 50°C for 10 min prior to the enzymatic hydrolysis. ALA in liquid form were weighed 10, 30, 50 mg and mixed with 1 mL ddH_2O. The hydrolysis reaction was started by the addition of enzymes at various enzyme/substrate ratios (E/S: 1%, 3%, and 5%). The reaction with ALA was conducted at pH 8.0, respectively, and 50°C for up to 6 h. After hydrolysis, the hydrolysates were heated in boiling water for 15 min to inactivate enzymes and then cooled in cold water at room temperature for 20 min. Hydrolysates were adjusted their pH to 7.0 with 1 M NaOH and centrifuged (Du Pont Sorvall Centrifuge RC 5B, Mandel Scientific Co. Ltd, Guelph, ON, Canada) at 12,000*g* and room temperature for 15 min. The supernatant was lyophilized and stored at -25°C .

2.3. Measurement of degree of hydrolysis

Immediately prior to termination of hydrolysis, 4 mL of the hydrolysate were mixed with an equal volume of 24% TCA solution and centrifuged at 12200*g* for 5 min. The supernatant (0.2 mL) was added to 2.0 mL of 0.05 M sodium tetraborate buffer (pH 9.2) and 1 mL of 4.0 mM TNBS and incubated at room temperature for 30 min in the dark. Then the mixture was added with 1.0 mL of 2.0 M NaH_2PO_4 containing 18 mM Na_2SO_3, and the absorbance was measured at 420 nm using a spectrophotometer (Cary 50 Bio UV-vis spectrophotometer, Varian, Inc., Santa Clara, CA, USA) [35,36]. Degree of hydrolysis (DH) was calculated as % DH=$(h/h_{tot}) \times 100$, where DH=percent ratio of the number of peptide bonds broken (h) to the total number bonds per unit weight (h_{tot}) and h_{tot}=11.1 mequiv/g of gelatin [35]. L-Leu was used for drawing a standard curve.

2.4. Determination of DPP-IV inhibitory activity

DPP-IV activity determination in this study was performed in 96-well microplates measuring the increase in absorbance at 405 nm using Gly-Pro-p-nitroanilide as DPP-IV substrate [37]. The lyophilized hydrolysates were dissolved in 100 mM Tris buffer (pH 8.0) to the concentration of 10 mg/mL and then serially diluted. The hydrolysates (25 μL) were added with 25 μL of 1.59 mM Gly-Pro-p-nitroanilide (in 100 mM Tris buffer, pH 8.0). The mixture was incubated at 37°C °C for 10 min, followed by the addition of 50 μL of DPP-IV (diluted with the same Tris buffer to 0.01 Unit/mL). The reaction mixture was incubated at 37°C for 60 min, and the reaction was stopped by adding 100 μL of 1 M sodium acetate buffer (pH 4.0). The absorbance of the resulting solution was measured at 405 nm with a microplate reader (iEMS reader MF; Labsystems, Helsinki, Finland). Under the conditions of the assay, IC_{50} values were determined by assaying appropriately diluted samples and plotting the DPP-IV inhibition rate as a function of the hydrolysate concentration.

2.5. Ultrafiltration

Hydrolysates were fractionated by ultrafiltration (UF; Model ABL085, Lian Sheng Tech. Co., Taichung, Taiwan) with spiral wound membranes having molecular mass cutoffs of 2.5 and 1 kDa. The fractions were collected as follows: >2.5 kDa, peptides retained without passing through 2.5 kDa membrane; 1-2.5 kDa, peptides permeating through the 2.5 kDa membrane but not the 1 kDa membrane; <1 kDa, peptides permeating through the 1 kDa membrane. All collected fractions were lyophilized and stored in a desiccator until use.

2.6. High Performance Liquid Chromatography (HPLC)

The fractionated hydrolysates by ultrafiltration exhibiting DPP-IV inhibitory activity were further purified using high performance liquid chromatography (Model L-2130 HPLC, Hitachi Ltd., Katsuda, Japan). The lyophilized hydrolysate fraction (100 μg) by gel filtration was dissolved in 1 mL of 0.1% trifluoroacetic acid (TFA) and 90 μL of the mixture, was then injected into a column (ZORBAX Eclipse Plus C18, 4.6 × 250 mm, Agilent Tech. Inc., CA, USA) using a linear gradient of acetonitrile (5 to 15% in 20 min) in 0.1% TFA under a flow rate of 0.7 mL/min. The peptides were detected at 215 nm. Each collected fraction was lyophilized and stored in a desiccator until use.

2.7. Identification of amino acid sequence

An accurate molecular mass and amino acid sequence of the purified peptides was determined using a Q-TOF mass spectrometer (Micromass, Altrincham, UK) coupled with an electrospray (ESI) source. The purified peptides were separately infused into the ESI source after being dissolved in methanol/water (1:1, v/v), and the molecular mass was determined by the doubly charged $(M+2H)^{+2}$ state in the mass spectrum. Automated Edman sequencing was performed by standard procedures using a 477-A protein sequencer chromatogram (Applied Biosystems, Foster, CA, USA).

2.8. Peptide synthesis

Peptides were prepared by the conventional Fmoc solid-phase synthesis method with an automatic peptide synthesizer (Model CS 136, CS Bio Co. San Carlos, CA, USA), and their purity was verified by analytical RP-HPLC-MS/MS.

2.9. Statistical analysis

Each data represents the mean of three samples was subjected to analysis of variance (ANOVA) followed by Tukey's studentized range test, and the significance level of $P<0.05$ was employed.

3. Results and discussion

3.1. Degree of hydrolysis

The DH of porcine skin gelatin hydrolyzed with ALA increased dramatically during the initial 1 h, and then increased gradually thereafter (Fig. 1). Also the highest DH was obtained with the highest E/S ratio. The highest DH (%) for ALA was 16.7% and obtained at the E/S ratio of 5% and 6-h hydrolysis.

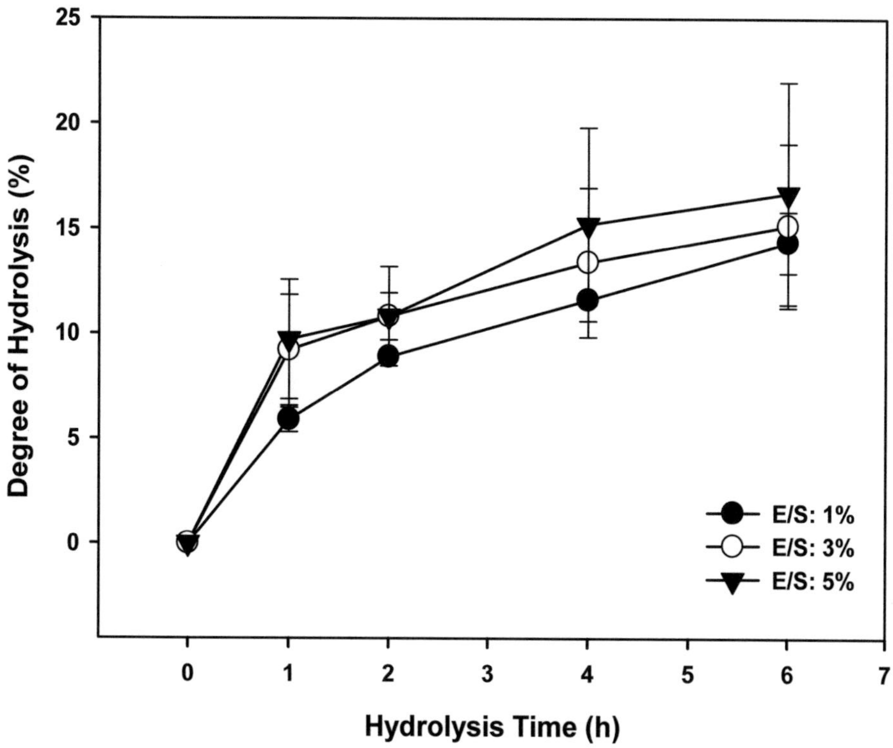

Figure 1. Degree of hydrolysis of porcine skin gelatin hydrolyzed with ALA at various E/S ratio.

3.2. DPP-IV inhibitory activity of hydrolysates

The DPP-IV inhibitory activity of porcine skin gelatin hydrolysates at the concentration of 10 mg/mL was shown in Fig. 2. The gelatin sample without hydrolysis (0 h) showed 9.2% inhibitory rate on DPP-IV. The DPP-IV inhibitory activity of the gelatin hydrolysates increased with E/S ratio and hydrolysis time. The DPP-IV inhibition rates of the 1-h hydrolysates with the E/S ratio of 1, 3 and 5% were 27.2, 44.3 and 48.8%, respectively; while those of 6-h hydrolysates increased to 52.0, 59.7 and 60.0%. The hydrolysates with the E/S ratio of 3% and 5%, and the hydrolysis time of 4 h and 6 h showed the highest DPP-IV inhibition rates between 57.4 to 60.0% among all the samples ($P<0.05$); in the meanwhile, the inhibition rates between the four samples were not significantly different ($P>0.05$). The results showed that the hydrolysates with the smaller size of peptides due to the higher DH possessed greater DPP-IV inhibitory activity. Patent WO 2006/068480 has demonstrated that the hydrolysates possessed great DPP-IV inhibitory activities referred to a mixture of peptides derived from hydrolysis of proteins with the percentage of hydrolysed peptide bonds of most preferably 20 to 40% [30]. As the economic concern for saving time and enzyme used, the hydrolysate with E/S ratio of 3% and 4-h hydrolysis was adopted for furher purification and analysis.

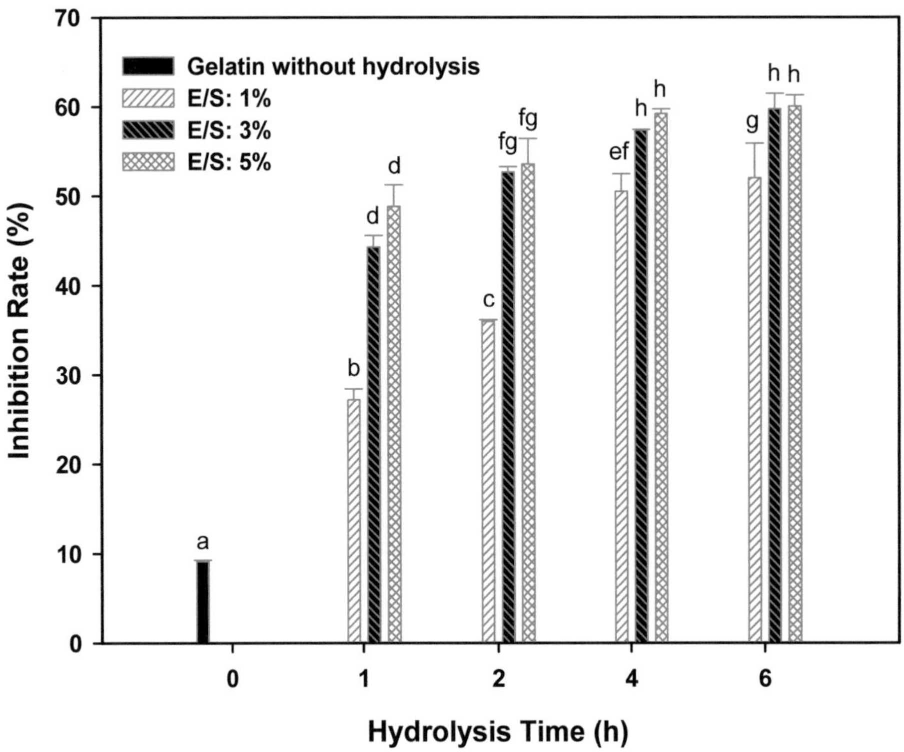

Figure 2. DPP-IV inhibitory rate of porcine skin gelatin hydrolysates. Different letters indicate the significant differences ($P<0.05$).

3.3. DPP-IV inhibitory activtiy of hydrolysates fractionated by UF

The DPP-IV inhibitory activity of hydrolysates with the E/S ratio of 3% and 4-h hydrolysis at the concentration of 1 mg/mL fractionated by UF was shown in Fig. 3A. The result showed the UF fractions of 1-2.5 kDa and < 1 kDa had insignificantly different ($P>0.05$) and higher

DPP-IV inhibition rates of 30.6% and 30.7%, respectively ($P<0.05$), than that within the > 2.5 kDa fraction displaying an inhibition rate of 28.2%. < 1-kDa fraction was selected for further analysis on the basis of the small size peptides may pass through the digestive tract without degradation. The IC_{50} value of the < 1 kDa fraction was determined and found as 1.50 mg/mL (Fig. 3B). The result in this study is in agreement with the former studies using various protein sources that reported the preferable DPP-IV inhibitory peptides derived from food protein consisted of 2-8 amino acid residues [30,31], and their molecular weights were supposed between 200 to 1000 Da.

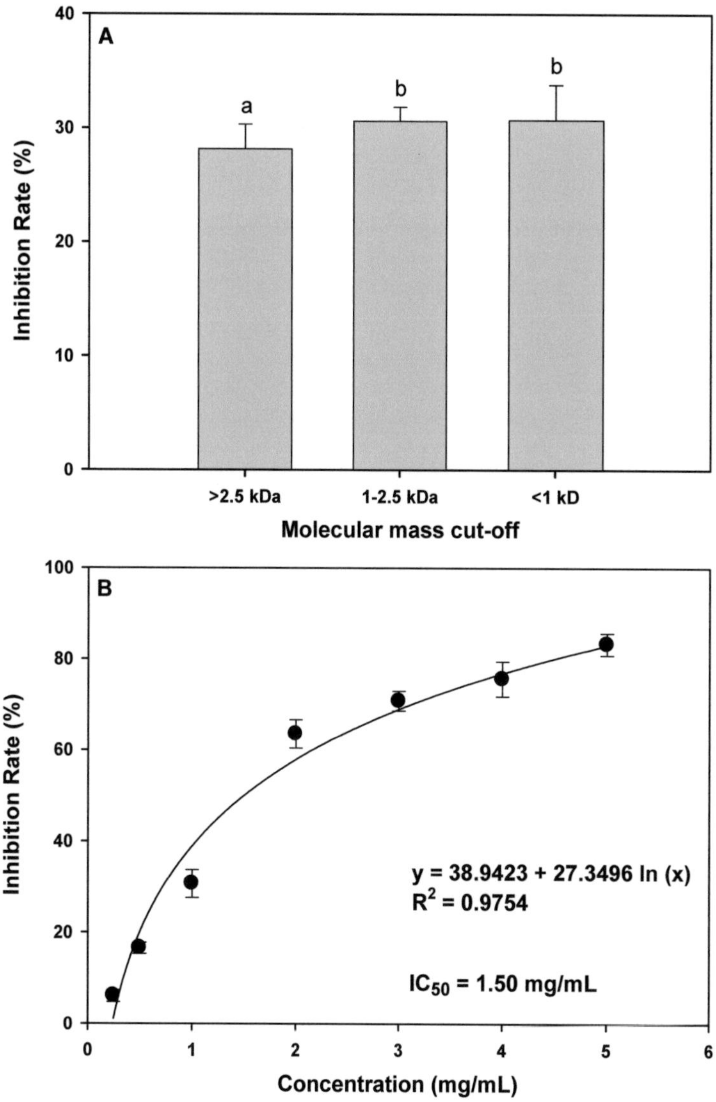

Figure 3. (A) DPP-IV inhibition rate of porcine skin gelatin hydrolysate fractionated by UF at the concentration of 1 mg/mL. (B) DPP-IV inhibition rate of the < 1 kDa UF fraction at various concentrations. Different letters indicate the significant differences ($P<0.05$).

3.4. Purification of DPP-IV-inhibitory peptides by HPLC

The elution profile and DPP-IV inhibitory activity of the peptide fractions from the < 1 kDa UF fraction separated by HPLC were shown in Fig. 4A and B. To obtain a sufficient amount of purified peptide, chromatographic separations were performed repeatedly. Five fractions (F-1 to F-5) were obtained upon HPLC separation of the < 1 kDa UF fraction (Fig. 4A), and they were lyophilized and then used to determine their DPP-IV inhibitory activities at the concentration of 100 μg/mL. The result showed that the fraction F-3 had the highest DPP-IV inhibition rate of 64.6% (P<0.05), as compared to the others which showed the inhibition rates between 26.7 to 45.6%, and its IC_{50} value was also determined as 62.9 μg/mL (Fig. 4C). Therefore, the fraction F-3 was used to identify the amino acid sequences of the peptides.

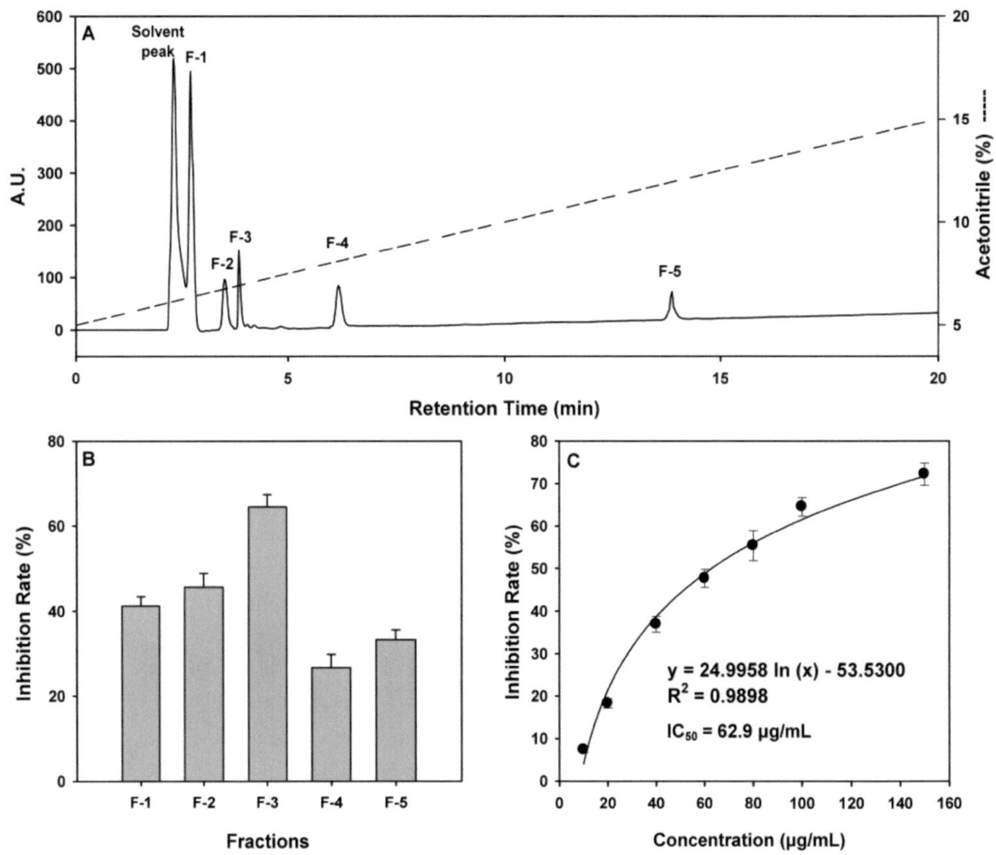

Figure 4. (A) Elution profile and (B) DPP-IV inhibition rate of the peptide fractions from the < 1 kDa UF fraction separated by HPLC. (C) DPP-IV inhibition rate of fraction F-3 at various concentrations. The DPP-IV inhibition rate was determined with each HPLC fraction at the concentration of 100 μg/mL.

3.5. Amino acid sequence of DPP-IV inhibitory peptides

Two peptides were identified in fraction F-3, and their amino acid sequences were Gly-Pro-Hyp (285.3 Da) and Gly-Pro-Ala-Gly (300.4 Da). Patent WO 2006/068480 has reported that 21 peptides which were capable of inhibiting DPP-IV activity showed a hydrophobic character, had a length varying from 3-7 amino acid residues and in particular the presence of Pro residue within the sequence [30]. The Pro residue was located as the first, second, third or fourth N-terminal residue, but mostly as the second N-terminal residue. Besides, the Pro residue was flanked by Leu, Val, Phe, Ala and Gly. In the present study, both peptides comprised Pro as the second N-terminal residue, and the Pro residue was flanked by Ala and Gly. Moreover, the peptides were composed of mostly hydrophobic amino acid residues, such as Ala, Gly and Pro, and one peptide comprised a charged amino acid, Glu, as the C-terminal residue. The present results therefore are consistent with the hypothesis demonstrated in the previous study [30].

3.6. DPP-IV-inhibitory activity of the synthetic peptides

The DPP-IV inhibitory activity of the two synthetic peptides and Diprotin A at various concentrations was determined (Fig. 5). The IC_{50} was calculated for each of the peptides. Diprotin A is well-known as the peptide with the greatest DPP-IV inhibitory activity, and its IC_{50} value was found as 24.7 μM in the present study (Fig. 4C). The IC_{50} values of the two synthetic peptides, Gly-Pro-Hyp and Gly-Pro-Ala-Gly, were 49.6 and 41.9 μM, respectively (Fig. 4A, B). In the previous study, the IC_{50} values against DPP-IV of Diprotin A and Diprotin B isolated from culture filtrates of *B. cereus* BMF673-RF1 were 3.2 and 16.8 μM, respectively [27]. Moreover, Ile-Pro-Ala and Val-Ala-Gly-Thr-Trp-Tyr, both prepared from β-lactoglobulin showed IC_{50} values of 49 and 174 μM against DPP-IV, respectively [28,29]. Patent WO 2006/068480 reported that Diprotin A showed the IC_{50} value of about 5 μM against DPP-IV, and five peptides, His-Pro-Ile-Lys, Leu-Pro-Leu-Pro, Leu-Pro-Val-Pro, Met-Pro-Leu-Trp and Gly-Pro-Phe-Pro, comprised 4 amino acids with Pro as the penultimate N-terminal residue displayed their IC_{50} values between 76 to 120 μM [30]. The results showed that the two peptides obtained in this study showed lower DPP-IV inhibitory activity than only Diprotin A and B, which were composed with 3 amino acid residues. However, they had similar inhibition effect to Ile-Pro-Ala but greater than other peptides comprised 4 or more amino acid residues. It is interesting that the ultimate N-terminal residues of the peptides mentioned above are all hydrophobic amino acids, and therefore we assumed that DPP-IV inhibitory activity of bioactive peptides may be determined by the amino acid length and the two N-terminal amino acid sequence of X-Pro, where X is the hydrophobic amino acid and preferably smaller in size. In conclusion, we found two peptides, Gly-Pro-Hyp and Gly-Pro-Ala-Gly, isolated from porcine skin gelatin hydrolysates having the inhibitory activity against DPP-IV.

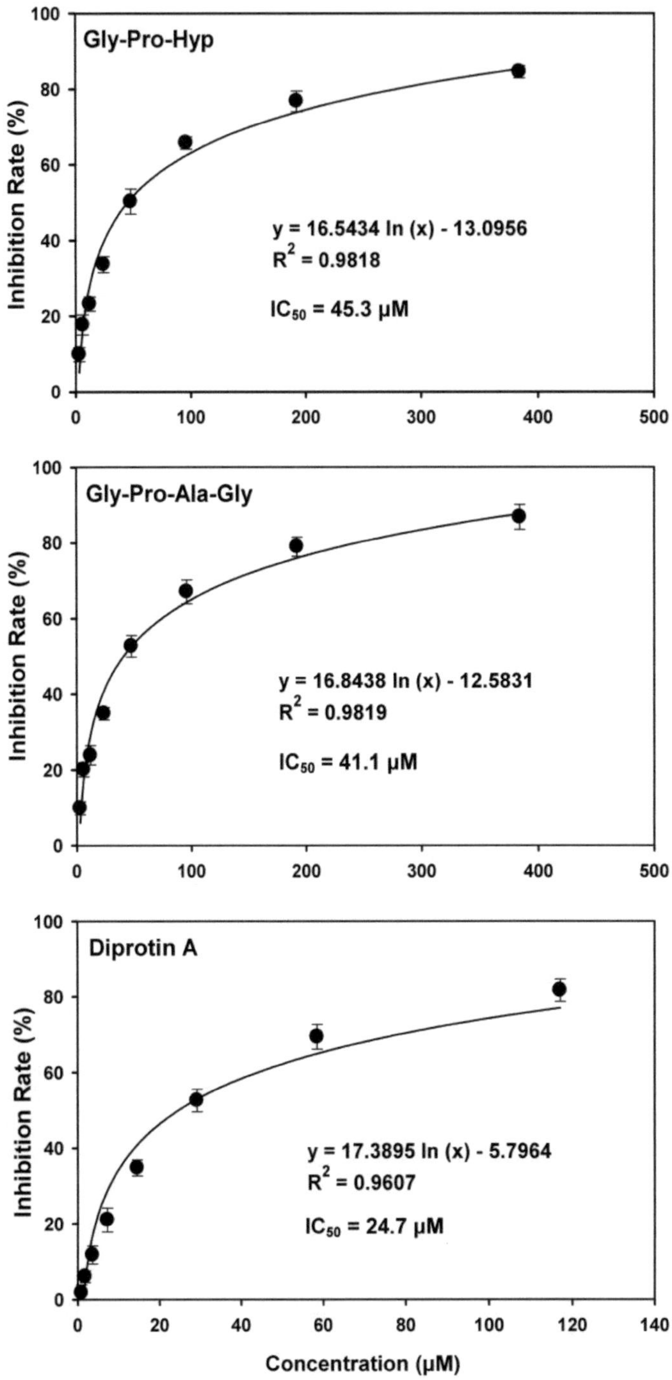

Figure 5. DPP-IV inhibition rates and IC_{50} values of the synthetic peptides and diprotin A.

4. Conclusion

This study has clearly demonstrated that porcine skin gelatin could be a good protein source to produce DPP-IV inhibitory peptides by hydrolysis with ALA. The two peptides identified in this study may have the potential for the therapy or prevention of type 2 diabetes. Further studies using *in vitro* simulated gastrointestinal digestion and Caco-2 cell permeate analysis and *in vivo* animal model systems are therefore necessary to ascertain the required standards of evidence for DPP-IV inhibitory potential of bioactive peptides derived from porcine skin gelatin.

Author details

Kuo-Chiang Hsu and Yu-Shan Tung
Department of Nutrition, China Medical University, Taiwan, Republic of China

Shih-Li Huang
Department of Baking Technology and Management,
National Kaohsiung University of Hospitality and Tourism, Taiwan, Republic of China

Chia-Ling Jao
Department of Food and Beverage Management,
Tung Fang Design University, Taiwan, Republic of China

5. References

[1] Creutzfeldt W. The Entero-Insular Axis in Type 2 Diabetes-Incretins as Therapeutic Agents. Experimental and Clinical Endocrinology & Diabetes 2001;109 288-300.

[2] Creutzfeldt W, Nauck MA. Gut Hormones and Diabetes Mellitus. Diabetes/Metabolism Research and Reviews 1992;8 565-573.

[3] Drucker DJ. Enhancing Incretin Action for the Treatment of Type 2 Diabetes. Diabetes Care 2003;26 2929-2940.

[4] Ehses JA, Casilla V, Doty T, Pospisilik JA, Demuth HU, Pederson RA. Glucose-Dependent Insulinotropic Polypeptide Pormotes β-(INS-1) Cell Survival via Cyclic Adenosine Monophosphate-Mediated Caspase-3 Inhibition and Regulation of p38 Mitogen-Activated Protein Kinase. Endocrinology 2003;144 4433-4455.

[5] Trümper A, Trümper K, Trusheim H, Arnold R, Göke B, Hörsch D. Glucose-Dependent Insulinotropic Polypeptide is a Growth Factor for β-(INS-1) Cells by Pleiotropic Signaling. Molucular Endocrinology 2001;15 1559-1570.

[6] Drucker DJ. Glucagon-Like Peptides: Regulators of Cell Proliferation, Differentiation and Apoptosis. Molecular Endocrinology 2003;17 161-171.

[7] Ahren B. GLP-1 and Extra-Islet Effects. Hormone and Metabolic Research 2004;36 842-845.

[8] De León DD, Crutchlow MF, Ham JYN, Stoffers AS. Role of Glucagon-Like Peptide-1 in the Pathogenesis and Treatment of Diabetes Mellitus. The International Journal of Biochemistry & Cell Biology 2006;38 845-859.

[9] Hansotia T, Drucker DJ. GIP and GLP-1 as Incretin Hormones: Lessons from Single and Double Incretin Receptor Knockout Mice. Regulatory Peptides 2005;128 125-134.

[10] Mentlein R, Gallwitz B, Schmidt WE. Dipeptidyl-Peptidase IV Hydrolyses Gastric Inhibitory Polypeptide, Glucagon-Like Peptide-1 (7-36) Amide, and Peptide Histidine Methionine and is Responsible for Their Degradation in Human Serum. European Journal of Biochemistry 1993;214 829-835.

[11] De Meester I, Lambeir AM, Proost P, Scharpé S. Dipeptidyl Peptidase IV Substrates. Advances in Experimental Medicine and Biology 2003;524 3-13.

[12] Lambeir AM, Durinx C, Scharpé S, De Meester I. Dipeptidyl-Peptidase IV from Bench to Bedside: An Update on Structural Properties, Functions and Clinical Aspects of the Enzyme DPP IV. Critical Reviews in Clinical Laboratory Science 2003;40 209-294.

[13] Mentlein R. Dipeptidyl-Peptidase IV (CD-26)-Role in the Inactivation of Regulatory Peptides. Regulatory Peptides 1999;85 9-24.

[14] Augustyns K, Bal G, Thonus G, Belyaev A, Zhang XM, Bollaert W. The Unique Properties of Dipeptidyl Peptidase IV Inhibitors. Current Medicinal Chemistry 1999;6 311-327.

[15] McIntosh CHS, Demuth HU, Pospisilik JA, Pederson R. Dipeptidyl Peptidase IV Inhibitors: How do They Work as New Antidiabetic Agents. Regulatory Peptides 2005;128 159-165.

[16] Deacon CF, Hughes TE, Holst JJ. Dipeptidyl Peptidase IV Inhibition Potentiates the Insulinotropic Effect of Glucagon-Like Peptide 1 in the Anesthetized Pig. Diabetes 1998;47 764-769.

[17] Deacon CF, Nauck MA, Meier J, Hücking K, Holst JJ. Degradation of Endogenous and Exogenous Gastric Inhibitory Polypeptide in Healthy and in Type 2 Diabetes Subjects as Revealed Using a New Assay for the Intact Peptide. The Journal of Clinical Endocrinology & Metabolism 2000;85 3575-3581.

[18] Mitani H, Takimoto M, Hughes TE, Kimura M. Dipeptidyl Peptidase IV Inhibition Improves Impaired Glucose Tolerance in High-Fat Diet-Fed Rats: Study Using a Fischer 344 Rat Substrain Deficient in Its Enzyme Activity. The Japanese Journal of Pharmacology 2002;88 442-450.

[19] Reinhold D, Vetterb RW, Mnich K, Bühling F, Lendeckel U, Born I, Faust J, Neubert K, Gollnick H, Ansorge S. Dipeptidyl Peptidase IV (DP IV, CD26) is Involved in Regulation of DNA Synthesis in Human Keratinocytes. FEBS Letters 1998;428 100-104.

[20] Gauthier SF, Pouliot Y, Saint-Sauveur D. Immunomodulatory Peptides Obtained by the Enzyme Hydrolysis of Whey Proteins. International Dairy Journal 2006;16 1315-1323.

[21] Galvez A, Abriouel H, Lopez RL, Ben Omar N. Bacteriocin-Based Strategies for Food Biopreservation. International Journal of Food Microbiology 2007;120 51-70.

[22] Hsu KC, Cheng ML, Hwang JS. Hydrolysates from Tuna Cooking Juice as An Anti-Hypertensive Agent. Journal of Food and Drug Analysis 2007;15 169-173.

[23] Kim SE, Kim HH, Kim JY, Kang YI, Woo HJ, Lee HJ. Anticancer Activity of Hydrophobic Peptides from Soy Proteins. BioFactors 2000;12 151-155.

[24] Hsu KC, Lu GH, Jao CL. (2009). Antioxidative Properties of Peptides Prepared from Tuna Cooking Juice Hydrolysates with Orientase (*Bacillus subtilis*). Food Research International 2009;42 647-652.

[25] Fukui K, Tachibana N, Wanezaki S, Tsuzaki S, Takamatsu K, Yamamoto T, Hashimoto Y, Shimoda T. (2002). Isoflavone-Free Soy Protein Prepared by Column Chromatography Reduces Plasma Cholesterol in Rats. Journal of Agricultural and Food Chemistry 2002;50 5717-5721.

[26] Wang W, de Mejia EG. A New Frontier in Soy Bioactive Peptides That May Prevent Age-Related Chronic Diseases. Comprehensive Reviews in Food Science and Food Safety 2005;4 63-78.

[27] Umezawa H, Aoyagi T, Ogawa K, Naganawa H, Hamada M, Takeuchi T. Diprotins A and B, Inhibitors of Dipeptidyl Aminopeptidase IV, Produced by Bacteria. The Journal of Antibiotics 1984;37 422-425.

[28] Tulipano G, Sibilia V, Caroli AM, Cocchi D. Whey Proteins as Source of Dipeptidyl Dipeptidase IV (Dipeptidyl Peptidase-4) Inhibitors. Peptides 2011;32 835-838.

[29] Uchida M, Ohshiba Y, Orie, Mogami. Novel Dipeptidyl Peptidase-4-Inhibiting Peptide Derived from β-lactoglobulin. Journal of Pharmacological Sciences 2011;117 63-66.

[30] Pieter BJW. WO 2006/068480 200, Protein Hydrolysate Enriched in Peptides Inhibiting DPP-IV and Their Use. 2006.

[31] Aart VA, Catharina MJ, Zeeland-Wolbers V, Maria LA, Gilst V, Hendrikus W, Nelissen BJH, Maria JWP. WO 2009/128713, Egg Protein Hydrolysates. 2009.

[32] Tabata Y, Ikada Y. Protein Release from Gelatin Matrices. Advanced Drug Delivery Reviews 1998;31 287-301.

[33] Eastoe JE, Leach AA. Hemical Constitution of Gelatin. In The Science and Technology of Gelatin. Ward AG, Courts A, Eds. Academic Press: New York, NY; 1977. p73-107.

[34] Li-Chan ECY, Huang SL, Jao CL, Ho KP, Hsu KC. Peptides Derived from Atlantic Salmon Skin Gelatin as Dipeptidyl-Peptidase IV Inhibitors. Journal of Agricultural and Food Chemistry 2012;60 973-978.

[35] Alder-Nissen J. Enzymic Hydrolysis of Food Proteins. Elsevier Applied Science Publisher: 1986.

[36] Lo WMY, Li-Chan ECY. Angiotensin I Converting Enzyme Inhibitory Peptides from in vitro Pepsin-Pancreatin Digestion of Soy Protein. Journal of Agricultural and Food Chemistry 2005;53 3369-3376.

[37] Kojima K, Ham T, Kato T. Rapid Chromatographic Purification of Dipeptidyl Peptidase IV in Human Submaxiallry Gland. Journal of Chromatography A 1980;189 233-240.

Chapter 9

Snake Venom Peptides: Promising Molecules with Anti-Tumor Effects

Sameh Sarray, Jose Luis, Mohamed El Ayeb and Naziha Marrakchi

Additional information is available at the end of the chapter

http://dx.doi.org/10.5772/51263

1. Introduction

Tumorigenesis and metastasis are two processes with inter-related mechanisms. These include tumor growth and angiogenesis, detachment of tumor cells from the primary tumor, followed by migration through the local connective tissue and penetration into the circulation (intravasation). Once in the blood stream, tumor cells interact with circulating blood cells, arrest in the microvasculature of target organs, then extravasate and secondary proliferate. During each of these steps, integrin-mediated adhesion, migration, proliferation and survival of tumor cells and angiogenic endothelial cells play crucial roles [1,2].

Integrins are a family of heterodimeric transmembrane receptors that mediate cell-cell and cell-extracellular matrix (ECM) interactions. These cell adhesion molecules are composed by non covalent association of α and β subunits. Although 18 α and 8 β subunits have been described, only 24 different combinations have been identified to date [3]. Specific integrin heterodimers preferentially bind distinct ECM proteins. The repertory of integrins present on a given cell dictates the extent to which that cell will adhere to and migrate on different matrices. Several integrins, among others αv and $\alpha5\beta1$, recognize the RGD sequence on their respective ligands. Other adhesive sequences in ECM proteins have also been observed, including the EILDV and REDV sequences that are recognized by integrin $\alpha4\beta1$ in an alternatively spliced form of fibronectin [3]. On ligation to the ECM, integrins cluster in the plane of the membrane and recruit various signalling and adaptor proteins to form structures known as focal adhesions [4].

Integrin expression can also vary considerably between normal and tumor tissue. Most notably, integrins $\alpha v\beta3$, $\alpha5\beta1$ and $\alpha v\beta6$ are usually expressed at low or undetectable levels in most adult epithelia but can be highly up-regulated in some tumors. Expression levels of some integrins, such as $\alpha2\beta1$, decrease in tumor cells; potentially increasing tumor cell dissemination [5]. The integrin $\alpha v\beta3$ is particularly important for tumor growth and invasiveness [6]. The receptor plays a major role in neo-vessels formation, its expression being strongly up-regulated in endothelial cells and specifically required during angiogenesis stimulated by basic fibroblast growth factor (bFGF) and tumor necrosis factor-α [7,8]. $\alpha v\beta 3$ is functionally involved in the malignant spread of various tumor cell types such as breast carcinoma, prostate carcinoma and melanoma, and supports tumor cell adhesion and migration through endothelium [9] and matrix proteins [10,1]. Blocking $\alpha v\beta 3$ is therefore expected to have a broad impact in cancer therapy and diagnosis. In the last decade, several clinical trials evaluating the efficacy of $\alpha v\beta 3$ blockers have led to encouraging results. Thus, MEDI-522 (Vitaxin), a humanized antibody derived from the mouse LM609 monoclonal antibody, was recently reported to give positive results in a phase II trial enrolling patients with stage IV metastatic melanoma [11]. Cilengitide is an inhibitor of both $\alpha v\beta 3$ and $\alpha v\beta 5$ integrins; it is currently being tested in phase II trials in patients with lung and prostate cancers [12] and in phase II and Phase III trials studying their role against glioblastoma are currently underway.

In addition to their role in tumor cells, integrins are also important for the host cellular response against cancer. Endothelial cells, fibroblasts, pericytes, bone marrow-derived cells, inflammatory cells and platelets all use integrins for various functions, including angiogenesis, desmoplasia and immune response.

Nature has been a source of medicinal products for thousands of years among which snake venoms form a rich source of bioactive molecules such as peptides, proteins and enzymes with important pharmacological activities. International research and development in this area, based on multidisciplinary approaches including molecular screening, proteomics, genomics and pharmacological *in vitro, ex vivo* and *in vivo* assays, allow the identification and characterization of highly specific molecules from snake venom that can potently inhibit integrin functions. These anti-adhesive snake venom proteins belong to different families (phospholipases, disintegrins, C-type lectins and metalloproteinases). By targeting integrins, they exhibit various pharmacological activities such as anti-tumor, anti-angiogenic and/or pro-apoptotic effects.

2. Snake venom protein families

2.1. The Snake Venom Metalloproteinases (SVMP)

Metalloproteinases are among the most abundant toxins in many *Viperidae* venoms. SVMPs are monozinc endopeptidases varying in size from 20 to 100 kDa. They are phylogenetically most closely related to the mammalian disintegrin and metalloproteinase (ADAM) family of proteins. SVMPs are grouped into several subclasses according to their domain organization [13, 14, 15]. P-I SVMPs are the simplest class of enzymes that contain only a metalloproteinase (M) domain. P-II SVMPs contain a M domain followed by a disintegrin (D) domain. P-III SVMPs contain M, disintegrin-like (D) and cysteine-rich (C) domains. Formally called P-IV, the heterotrimeric class of SVMPs that contain an additional snake C-type lectin-like (snaclec) domain [16] is now included in the P-III group as a subclass (P-IIId).

Most of the functional activities of SVMPs are associated with hemorrhage or the disruption of the hemostatic system, which are primarily mediated by the proteolytic activity of the M domain. SVMPs cause hemorrhage by disturbing the interactions between endothelial cells and the basement membrane through the degradation of endothelial cell membrane proteins (e.g., integrin, cadherin) and basement membrane components (e.g., fibronectin, laminin, nidogen, type IV collagen) [17]. Blood coagulation proteins (e.g., fibrinogen, factor X, prothrombin) are also targets of their proteolytic activities.

Echis carinatus venom contains the specific prothrombin activators, ecarin [18,19] and carinactivase [20]. Adamalysin II, a non-hemorrhagic P-I SVMP isolated from *Crotalus adamantus* venom, cleaves and inactivates serum proteinase inhibitors including antithrombin III [21]. Kaouthiagin, isolated from the venom of *Naja kaouthia* specifically binds and cleaves von Willebrand factor (vWF), resulting in loss of both the ristocetin-induced platelet aggregation and collagen-binding activity of vWF [22]. Additionally, a large number of the P-III SVMPs can inhibit platelet aggregation, thus enhancing the hemorrhagic state [23]. The hemorrhagic P-III SVMP jararhagin from the venom of *Bothrops jararaca* has been shown to degrade platelet collagen receptor $\alpha_2\beta_1$ integrin in addition to fibrinogen and vWF, resulting in the inhibition of platelet aggregation [24]. Other platelet receptors are also degraded by SVMPs. GPIbα is cleaved by kistomin; mocarhagin and crotalin [25-27], and GPVI is degraded by alborhagin, crotarhagin and kistomin [28,29].

In the other side, it was reported that several SVMPs inhibited integrin-mediated adhesion of cancer cells on ECM proteins (table 1). BaG, a dimeric PIII class of SVMP from *Bothrops alternatus* with inactivated enzymatic domain but intact D/C domain, has been reported to inhibit fibronectin-mediated K562 cell adhesion *via* α5β1 integrin [30].

Proteins	Snake	Integrins	Effects	References
VAP1, VAP2	*Crotalus atrox*	α3,α6,β1	Induce apoptosis of HUVEC	[31,36]
HV1	*Trimeruserus flavoviridis*	-	Inhibits adhesion of HUVEC and induces apoptosis	[32]
Halysase	*Gloydius halys*	α1β1;α5β1	Inhibits proliferation and Induces apoptosis of HUVEC	[33]
VLAIPs	*Vipera lebetina*	-	Inhibits proliferation and Induces apoptosis of HUVEC	[34]
Graminelysin	*Trimeresurus gramineus*	α1β1;α5β1	Inhibits proliferation and Induces apoptosis of HUVEC	[35]
BaG	*Bothrops alternatus*	α5β1	Inhibits adhesion of K562 cells	[30]
TSV-DM	*Trimeresurus stejnegeri*	-	Inhibits cell proliferation and induces transient cell morphologic changes of endothelial cells.	[113]

Table 1. SVMP affecting tumor cells

Several apoptosis-inducing proteins have been purified from hemorrhagic snake venom, such as VAP1 and VAP2 (*Crotalus atrox*), HV1 (*Trimeresurus flavoviridis*), halysase (*Gloydius halys*), and VLAIPs (*Vipera lebetina*) [31-34], graminelysin [35]. They are members of the SVMP and ADAM family and induce apoptosis of human umbilical vein endothelial cells (HUVECs) [31,36]. The detachment of endothelial cells and resulting apoptosis could be an additional mechanism for the disruption of normal hemostasis by SVMPs. TSV-DM a basic metalloproteinase from *Trimeresurus stejnegeri* venom inhibits cell proliferation and induces cell morphologic changes transiently of ECV304 cells. However, DNA fragmentation and DNA content analysis demonstrated that this metalloproteinase could not induce ECV304 cells apoptosis.

2.2. The disintegrins

Disintegrins are a family of non-enzymatic and low molecular weight proteins derived from viper venom [37-39]. They are able to inhibit platelet aggregation and interact with adhesion molecules in particular integrins in a dose-dependent manner. They have a K / RTS sequence which is known as the RGD adhesive loop [37-39]. Their primary structure shows a strong conservation in the arrangement of cysteines [38]. Most disintegrins represent the C-terminal domain of metalloproteinases PIIa, d and e classes and are released into the venom by proteolytic cleavage [40,37,38]. A minority of these proteins exist as D / C domains from the class of SVMPs PIIIb.

Disintegrins can be conveniently divided into five different groups according to their length and the number of disulfide bridges [41]. The first group includes short disintegrins, single polypeptide composed of 49 - 51 amino acids with four disulfide bridges. The second group comprises medium disintegrins containing about 70 amino acids and six disulfide bridges. The third group includes long disintegrins of 83 residues linked by seven disulfide bridges. The disintegrin domains of PIII snake-venom metalloproteinases, containing approx. hundred amino acids with 16 Cysteine residues involved in the formation of eight disulfide bonds, constitute the fourth subgroup of the disintegrin family. Unlike short-, medium- and long-sized disintegrins, which are single-chain molecules, the fifth subgroup is composed of homo and heterodimers. The dimeric disintegrins subunits contain about 67 residues with four disulfide intra-chain bridges and two interchain bridges [42,43].

Although disintegrins are highly homologous, significant differences exist in their affinity and selectivity for integrins, which explains the multitude of effects of these molecules (Table 2).

Disintegrins were first identified as inhibitors of platelet aggregation and were subsequently shown to antagonize fibrinogen binding to platelet integrin $\alpha IIb\beta 3$ [44,45]. After that, studies on disintegrins have revealed new uses in the diagnosis of cardiovascular diseases and the design of therapeutic agents in arterial thrombosis, osteoporosis, and angiogenesis-related tumor growth and metastasis (table 2). Triflavin from *Trimeresurus flavoviridis* venom was one of the first RGD-disintegrins shown to inhibit angiogenesis both *in vitro* and *in vivo*

[46]. Triflavin strongly inhibited cell migration toward vitronectin and fibronectin nearly thirty orders of magnitude greater than anti-$\alpha v\beta 3$ monoclonal antibodies [46]. Triflavin was also more effective in inhibiting TNF-α-induced angiogenesis in the chicken chorioallantoic membrane (CAM) assay. Similar results were obtained with another RGD-disintegrin, rhodostomin, from *Agkistrodon rhodostoma* venom, which inhibits endothelial cell migration, invasion and tube formation induced by bFGF in Matrigel™ both *in vitro* and *in vivo* [47]. Rhodostomin effects were inhibited by anti-$\alpha v\beta 3$ but not by anti-$\alpha v\beta 5$ antibodies, thus supporting the hypothesis that the effects of RGD-disintegrins are mediated by blockade of the vitronectin receptor.

Proteins	Snake	Integrins	Effects	References
Triflavin	*Trimeresurus flavoviridis*	$\alpha 5\beta 1$,$\alpha v\beta 3$, $\alpha 3\beta 1$	Inhibits adhesion of tumor cells to matrix proteins, cell migration and angiogenesis *in vitro* and *in vivo*	[46]
Rhodostomin	*Agikistrodon rhodostoma*	$\alpha v\beta 3$,$\alpha v\beta 5$	Inhibits cell migration, invasion of endothelial cells; inhibits angiogenesis *in vivo* and *in vitro*	[47]
Contortrostatin	*Agkistrodon contortrix contortrix*	$\alpha v\beta 3$,$\alpha 5\beta 1$, $\alpha v\beta 5$, $\alpha II\beta\beta 3$	Blocks adhesion, migration invasion of different type of tumor cells	[48]
Lebestatin	*Macrovipera lebetina*	$\alpha 1\beta 1$	Inhibits migration and angiogenesis	[56]
Accurhagin-C	*Agkistrodon acutus*	$\alpha v\beta 3$	Prevents migration and invasion of endothelial cells; anti-angiogenic activity *in vitro* and *in vivo*; elicites anoïkis	[58]
Eristostatin	*Eritocophis macmahoni*	$\alpha 4\beta 1$,other integrin not yet determined	Inhibits cell motility; no effect on cell proliferation or angiogenesis	[59,60]
DisBa-01	*Bothrops alternatus*	$\alpha v\beta 3$	Anti-angiogenic and anti-metastatic effect on melanoma cells	[62]
Leberagin-C	*Macrovipera lebetina*	$\alpha v\beta 3$	Inhibits cell adhesion of melanoma tumor cells	[114]
Accutin	*Agkistrodon acutus*	$\alpha v\beta 3$	Inhibits angiogenesis *in vitro* and *in vivo*; induces apoptosis	[115]

Table 2. Effects of disintegrins on cancerous cells

Contortrostatin, a disintegrin isolated from the venom of the southern copperhead snake, exhibits anti-cancer activity in a variety of tumor cells [48-50]. It does not display cytotoxic activity *in vitro* nor animals upon injection. Contortrostatin inhibits adhesion, migration, invasion, metastatic and angiogenesis of tumor and endothelial cells mediated by $\alpha v\beta 3$,$\alpha 5\beta 1$ and $\alpha v\beta 5$ [48,50-54]. Recently, contortrostatin showed an additive inhibitory effect in combination with docetaxel on the growth of xenograft tumors derived from prostate cancer cells [55].

Lebestatin is an example of a non toxic KTS-disintegrin isolated from *Macrovipera lebetina* that inhibits migration and VEGF-induced *in vivo* angiogenesis [56]. The presence of a WGD motif in CC8, a heterodimeric disintegrin from *Echis carinatus*, increases its inhibitory effect on $\alpha v\beta 3$ and $\alpha 5\beta 1$ integrins [57].

There are few reports regarding the effects of ECD-disintegrins on endothelial cell migration. Acurhagin-C, dose-dependently blocked HUVEC migration toward a vitronectin-coated membrane. Furthermore, acurhagin-C elicited endothelial anoïkis *via* disruption of the $\alpha v\beta 3$/FAK/PI3K survival cascade and subsequent initiation of the procaspase-3 apoptotic signaling pathway [58].

Eristostatin, an RGD-disintegrin from *Eristocophis macmahoni* was tested on individual metastasis steps such as cell arrest, extravasation and migration [59]. Eristostatin treatment did not prevent tumor cell extravasation or migration [60]. However, it was shown later that eristostatin inhibited melanoma cell motility, an effect mediated by fibronectin-binding integrins [61]. Interestingly, this disintegrin, contrary to other RGD-disintegrins, did not inhibit angiogenesis, as stated before [61]. DisBa-01, a $\alpha v\beta 3$ integrin-blocking RGD-disintegrin, inhibits not only migration of endothelial cells *in vivo* [62] but also *in vitro* migratory ability of fibroblasts and two tumor cell lines.

Since integrin receptors are also quite indiscriminate as they support cell adhesion to several substrates, it seems highly reasonable that the general RGD-disintegrin scaffold of the integrin-binding motif could be employed as a prototype for drug design for new anti-metastatic therapies *via* blocking both tumor cell adhesion and tumor angiogenesis.

2.3. The snake venom phospholipases

Snake venom is one of the most abundant sources of secretory phospholipases A2 (PLA2), which are one of the potent molecules in snake venoms [63-65].

PLA2 (EC 3.1.1.4)—are enzymes that catalyze the hydrolysis of sn-2-acyl bond of sn-3-phospholipids, generating free fatty acids and lysophospholipids as products [66]. They are currently classified in 15 groups and many subgroups that include five distinct types of enzymes, namely secreted PLA2 (sPLA2), cytosolic PLA2 (cPLA2), Ca^{2+} independent PLA2s (iPLA2), platelet-activating factor acetyl-hydrolases (PAF-AH), lysosomal PLA2, and a recently identified adipose-specific PLA2 [65,67]. PLA2 are low molecular weight proteins with molecular masses ranging from 13-19 kDa that generally require Ca^{2+} for their activities

[69,70]. Snake venom sPLA2 are secreted enzymes belonging to only two groups that are based on their primary structure and disulfide bridge pattern [68,71,72]. Those of group I are similar to pancreatic sPLA2 present in mammals, were found in venom of *Elapidae* snakes, while group II PLA2s belong to the *Viperidae* and are similar to mammals nonpancreatic, inflammatory sPLA2s [73,74]. The group II can be subdivided mainly in two subgroups, depending on the residue at position 49 in the primary structure: Aspartic acid-49 PLA2s are enzymatically active, while Lysine 49 present low or no enzymatic activity [75]. There are other subgroups, such as Asparagine-49, Serine-49, Glutamine-49 and Arginine -49 [76-83]. Studies have found that catalytic activity is reduced or even abolished when an Aspartic acid of native PLA2 is replaced by another amino acid [80,84].

Despite a high identity of their amino acid sequences, sPLA2 exhibit a wide variety of pharmacological properties such as anticoagulant, haemolytic, neurotoxic, myotoxic, oedema-inducing, hemorrhagic, cytolytic, cardiotoxic, muscarinic inhibitor and antiplatelet activities [63,85-92].

Recently, PLA2s have been shown to possess anti-tumor and anti-angiogenic properties (Table 3). CC-PLA2-1 and CC-PLA2-2 from *Cerastes cerastes* viper are non-toxic and acidic proteins. They have high inhibitory effects on platelet aggregation and coagulation. In addition, CC-PLA2-1 and CC-PLA2-2 inhibit the adhesion of the human fibrosarcoma (HT1080) and melanoma (IGR39) cells to fibrinogen and fibronectin. In the same direction, CC-PLA2-1 and CC-PLA2-2 potently reduces HT1080 cell migration to fibrinogen and fibronectin with nearly similar IC_{50} values [93]. This anti-adhesive effect was due to the inhibition of $\alpha5\beta1$ and αv-containing integrins [94]. A recent report demonstrated that Bth-A-I, a non-toxic PLA2 isolated from *Bothrops jararacussu* venom display an anti-tumoral effect upon breast adenocarcinoma as well as upon human leukaemia T and Erlich ascetic tumor [95].

Proteins	Snake	Integrins	Effects	References
CCPLA2-1; CCPLA2-2	*Cerastes cerastes*	$\alpha5\beta1,\alpha v$	Inhibits migration and adhesion of fibrosarcoma and melanoma cells	[93,94]
Bth-A-I-PLA2	*Bothrops jararacussu*	-	Anti-tumor activity on adenocarcinoma and leukaemia cells	[95]
MVL-PLA2	*Macrovipera lebetina*	$\alpha5\beta1,\alpha v$	Inhibits adhesion and migration of human microvascular cells and inhibits angiogenesis *in vivo* and *in vitro*.	[96]
BP II	Prothobotrops flavoviridis	-	Induces cell death in human leukaemia cells	[97]

Table 3. PLA2s targeting tumor cells

MVL-PLA2 is a snake venom phospholipase purified from Macrovipera lebetina venom that inhibited adhesion and migration of human microvascular endothelial cells (HMEC-1) without being cytotoxic. Using Matrigel™ and chick chorioallantoic membrane assays, MVL-PLA2, as well as its catalytically inactivated form, significantly inhibited angiogenesis both *in vitro* and *in vivo*. Also, the actin cytoskeleton and the distribution of $\alpha v\beta 3$ integrin, a critical regulator of angiogenesis and a major component of focal adhesions, were disturbed after MVL-PLA2 treatment. The enhancement of microtubule dynamics of HMEC-1 cells, in consequence of treatments by MVL-PLA2, may explain the alterations in the formation of focal adhesions, leading to inhibition of cell adhesion and migration [96].

A cell death activity was discovered in Lysine 49-PLA2 called BPII. It induces caspase-independent cell death in human leukaemia cells regardless of its depressed enzymatic activity [97].

2.4. The C-type lectins

The C-type lectins are abundant components of snake venom with various function. Typically, these proteins bind calcium and sugar residues. However, the C-type lectin like proteins from snake venom (termed actually snaclec) does not contain the classic calcium/sugar binding loop and have evolved to bind a wide range of physiologically important proteins and receptors [98].

Snaclecs have a basic heterodimeric structure with two subunits, nearly always linked covalently, *via* a disulphide bond. The heterodimers are often further multimerized either non-covalently or covalently *via* additional disulphide bonds, to form larger structures [99]. The two subunits form a concave surface between them [100] thus constituting the main site of ligand binding [101,102]. The subunits have a high structural degree of homology between them and with other snaclecs [103]. Despite their highly conserved primary structure, the snaclecs are characterized by various biological activities. They were and are still considered as modulators of platelet aggregation by targeting vWF, GPIb-IX-V, GPVI and possibly other platelet receptors.

Recently, novel activities of snaclecs were highlighted. They were described for their potential anti-tumor effect by blocking adhesion, migration, proliferation and invasion of different cancer cell lines (Table 4). Among these proteins, EMS16, a heterodimer isolated from the venom of *Echis multisquamatus*, inhibits the adhesion of HUVECs cells on ECM proteins and their migration by inhibiting the binding of integrin $\alpha 2\beta 1$ to collagen [104].

Lebecetin and lebectin, purified from *Macrovipera lebetina* venom, are the only snaclecs, until today, with an evident anti-tumor effect in addition to their anti-aggregation activity on platelets. Indeed, these two non cytotoxic proteins inhibit the adhesion of various cancer cell lines: melanoma (IGR39), adenocarcinoma (HT29-D4), fibrosarcoma (HT1080) and leukemia cells (K562) on different ECM proteins. They also inhibit the proliferation, migration and invasion of HT1080 cells [105,106]. Lebectin also displays anti-angiogenic activity at very low concentrations both *in vitro* and *in vivo* [107]. Thus, lebectin presents the best anti-angiogenic efficacy yet described for snake venom-derived peptides [108,109]. These observed effects are mediated by $\alpha5\beta1$ and αv integrins [107].

Extensive researches have been shown that cell adhesion activities in cancer disease are deregulated. According to this idea, it was also reported that lebectin inhibits these alterations by promoting N-cadherin/catenin complex reorganisation at cell-cell contacts, inducing a strengthening of intercellular adhesion [110].

Another snaclec, BJcuL isolated from *Bothrops jararacussa* venom, was also described for its anti-tumor, but the receptor or integrin implicated has not been determined yet. This homodimeric protein inhibits proliferation of several cell lines of renal, pancreatic, prostate and melanoma origin, but no effect was observed on colon or breast cancer cells [111]. BJcuL also affects the viability of some tumor cell lines of different origins, but has no effect on the growth of K562 and T24 cells, suggesting that these cells do not express the receptor recognized by the lectin. BJcuL induces apoptosis in human gastric carcinoma cells accompanied by inhibition of cell adhesion and actin cytoskeleton disassembly [112].

Proteins	Snake	Integrins	Effects	References
Lebecetin, lebectin	*Macrovipera lebetina*	$\alpha5\beta1,\alpha v$	Inhibits adhesion, migration and invasion of human tumor cells; inhibits angiogenesis	[106]
BJcuL	*Bothrops jararacussu*	–	Inhibits tumor cell and endothelial cell growth; induces apoptosis of human gastric carcinoma cells; inhibits cell adhesion and actin cytoskeleton disassembly	[111,112]
EM16	*Echis multisquamatus*	$\alpha2\beta1$	Inhibits adhesion and migration of HUVEC cells	[104]

Table 4. Snaclecs and their effects on tumor cells

3. Potential application of snake venom compounds

Venoms are a rich source of molecules endowed with diverse pharmacological effects. Most part of these molecules act *via* the adhesion molecules. The intervention of the scientists and the clinicians in the pharmaceutical development field would employ these molecules as therapeutic agents for several pathologies such as cancer, thrombosis, diabetes....

Until now, no medicine was produced from a native molecule purified from venom. However, several peptidomimetics were designed by basing on the structure of these molecules. The benefits of these peptidomimetics compared to antibodies that can be used for the treatment of certain diseases are: a shorter half-life, reversible inhibition, easier to control a problem and very low immunogenicity. For example, the antihypertensive drug captopril, modelled from the venom of the Brazilian arrowhead viper (*Bothrops jaracusa*); the anticoagulant Integrilin (eptifibatide), a heptapeptide derived from a protein found in the venom of the American southeastern pygmy rattlesnake (*Sistrurus miliarius barbouri*); Ancrod, a compound isolated from the venom of the Malaysian pit viper (*Agkistrodon rhodostoma*) for use in the treatment of heparin-induced thrombocytopenia and stroke and alfimeprase, a novel fibrinolytic metalloproteinase for thrombolysis derived from southern copperhead snake (*Agkistrodon contortrix contortrix*) venom (Table 5). Two venom proteins from the Australian brown snake, *Pseudonaja textilis*, are currently in development as human therapeutics (QRxPharma). The first is a single agent procoagulant that is a homolog of mammalian Factor Xa prothrombin activator, whereas the other is a plasmin inhibitor, named Textilinin-1, with antihemorrhagic properties.

Name	Snake	Target and function/treatment	Clinical stage
Capoten ® (Captropil)	*Bothrops jaracusa*	Angiotensin converted enzyme (ACE) inhibitor/ high blood pressure	Granted FDA approval
Integrilin ® (Eptifibatide)	*Sisturus miliarus barbouri*	Platelet aggregation inhibitor/acute coronary syndrome	Granted FDA approval
Aggrastat ® (tirofiban)	*Echis carinatus*	GPIIb-IIIa inhibitor/ myocardial infarct, refractory ischemia	Approved for use with heparin and aspirin for the treatment of acute coronary syndrome
Exanta	*Cobra*	Thrombin inhibitor/ arterial fibrillation and blood	Seeking FDA approval
Alfimeprase	(Agkistrodon contortrix contortrix)	Thrombolytic/ Acute ischemic stroke, acute peripheral arterial occlusion	Phase III
Ancrod ® (viprinex)	*Agkistrodon rhodostoma*	Fibrinogen inhibitor/ stroke	Phase III
hemocoagulase	*Bothrops atrox*	Thrombin-like effect and thromboplastin activity/ prevention and treatment of haemorrhage	Phase III
Protac/ Protein C activator	*Agkistrodon contortrix contortrix*	Protein C activator/clinical diagnosis of haemostatic disorder	Granted FDA approval
Reptilase	*Bothrops jaraca*	Diagnosis of blood coagulation disorder	Granted FDA approval
Ecarin	*Echis carinatus*	Prothrombin activator/ diagnostic	Granted FDA approval

Table 5. Drugs and clinical diagnostic kits from snake venom

Actually, most of the current anticancer therapies (radiotherapy, chemotherapy) are not specific and are targeting at both tumor cells and healthy cells. However, in recent years, new treatments tend to focus on the tumor microenvironment and particularly on the inhibition of tumor angiogenesis. These treatments are based on several active and non toxic proteins from snake venom, as for example contortrostatin from *Agkistrodon contortrix contortrix* and eristostatin from *Eristocophis macmahoo.* Although all these molecules are still currently in clinical trials, they could in the future open new ways of healing and could be used as drugs.

4. Conclusions

From the initial discovery of captopril, the first oral ACE inhibitor, to the recent application of disintegrins for the potential treatment of cancer, the various components of snake venoms have never failed to reveal amazing new properties. While the original native snake venom compounds are usually unsuitable as therapeutics, interventions by medicinal chemists as well as scientists and clinicians in pharmaceutical R&D have made it possible to use the snake venom proteins as potential drugs for multiple disorders or scaffolds for drug design.

Author details

Sameh Sarray*
Pasteur Institute of Tunis, Tunis Belvedere, Tunisie
Faculty of Science, University of Tunis El Manar, Tunisie

Jose Luis
Research Center of Biologic Oncology and Oncopharmacology (CRO2), INSERM UMR 911, Marseille, France
University of Aix-Marseille, Marseille, France

Mohamed El Ayeb and Naziha Marrakchi
Pasteur Institute of Tunis, Tunis Belvedere, Tunisie

5. References

[1] Felding-Habermann, B. Integrin adhesion receptors in tumor metastasis. Clinical and Experimental Metastasis 2003;20(3) 203–213.

[2] Fidler IJ. Biological behavior of malignant melanoma cells correlated to their survival in vivo. Cancer Research 1975; 35(1) 218–224.

[3] Barczyk M, Carracedo S, Gullberg D. Integrins. Cell Tissue Research 2010;339(1) 269-280.

[4] Berrier AL and Yamada KM. Cell Matrix. Journal of cellular physiology 2007;213(3) 565-573

---

* Corresponding Author

[5] Kren A, Baeriswyl V, Lehembre F, Wunderlin C, Strittmatter K, Antoniadis H, Fässler R, Cavallaro U, Christofori G. Increased tumor cell dissemination and cellular senescence in the absence of β1-integrin function. The EMBO Journal 2007;26(12) 2832–2842.

[6] Albelda SM, Mette SA, Elder DE, Stewart R, Damjanovich L, Herlyn M, Buck CA. Integrin distribution in malignant melanoma: association of the beta 3 subunit with tumor progression. Cancer Research 1990;50(20) 6757–6764.

[7] Brooks PC, Clark RA, Cheresh DA. Requirement of vascular integrin alpha v beta 3 for angiogenesis. Science 1994;264(5158) 569–571

[8] Enenstein J, Waleh NS, Kramer RH. Basic FGF and TGFbeta differentially modulate integrin expression of human microvascular endothelial cells. Experimental Cell Research 1992;203(2) 499–503.

[9] Voura EB, Ramjeesingh RA, Montgomery AM, Siu CH. Involvement of integrin alpha(v)beta(3) and cell adhesion molecule L1 in transendothelial migration of melanoma cells. Molecular Biology of the Cell 2001; 12(9) 2699–2710.

[10] Cooper CR, Chay CH, Pienta KJ. The role of alpha(v)beta (3) in prostate cancer progression. Neoplasia 2002;4(3) 191–194.

[11] Hersey P, Sosman J, O'Day S. A phase II, randomized, open-label study evaluating the antitumor activity of MEDI-522, a humanized monoclonal antibody directed against the human alpha v beta 3 (αvβ3) integrin, ± dacarbazine (DTIC) in patients with metastatic melanoma (MM). ASCO Annual Meeting Proceedings, Journal of Clinical Oncology 2005;

[12] Beekman KW, Colevas AD, Cooney K, Dipaola R, Dunn RL, Gross M, Keller ET, Pienta KJ, Ryan CJ, Smith D, Hussain M. Phase II evaluations of cilengitide in asymptomatic patients with androgen independent prostate cancer: scientific rationale and study design. Clinical Genitourinary Cancer 2006;4(4) 299–302.

[13] Fox JW and Serrano SM. Structural considerations of the snake venom metalloproteinases, key members of the M12 reprolysin family of metalloproteinases, Toxicon 2005;45(8) 969–985.

[14] Fox JW and Serrano SM. Insights into and speculations about snake venom metalloproteinase (SVMP) synthesis, folding and disulfide bond formation and their contribution to venom complexity, FEBS Journal 2008;275 (12) 3016–3030.

[15] Takeya H, Miyata T, Nishino N, Omori-Satoh T, Iwanaga S. Snake venom hemorrhagic and non hemorrhagic metalloendopeptidases, Methods in Enzymolgy 1993;223 365–378.

[16] Clemetson KJ, Morita T, Manjunatha Kini R. Scientific and standardization committee communications: classification and nomenclature of snake venom C-type lectins and related proteins, Journal of Thrombosis Haemostasis 2009;7(2) 360.

[17] Baramova EN, Shannon JD, Bjarnason JB, Fox JW. Degradation of extracellular matrix proteins by hemorrhagic metalloproteinases. Archives of Biochemistry and Biophysics 1989;275(1) 63–71.

[18] Morita T and Iwanaga S. Purification and properties of prothrombin activator from the venom of Echis carinatus. Journal of Biochemistry 1978;83(2) 559-570.

[19] Nishida S, Fujita T, Kohno N, Atoda H, Morita T, Takeya H, Kido I, Paine MJ, Kawabata S, Iwanaga S. cDNA cloning and deduced amino acid sequence of prothrombin

activator (ecarin) from Kenyan Echis carinatus venom. Biochemistry 1995;34 (5) 1771–1778.

[20] Yamada D, Sekiya F, Morita T. Isolation and characterization of carinactivase, a novel prothrombin activator in Echis carinatus venom with a unique catalytic mechanism. Journal Biological Chemistry 1996;271(9) 5200–5207.

[21] Kress LF and Paroski EA. Enzymatic inactivation of human serum proteinase inhibitors by snake venom proteinases. Biochemical and Biophysics Research Communications 1978;83(2) 649–656.

[22] Hamako J, Matsui T, Nishida S, Nomura S, Fujimura Y, Ito M, Ozeki Y, Titani K. Purification and characterization of kaouthiagin, a von Willebrand factor-binding and –cleaving metalloproteinase from Naja kaouthia cobra venom, Thrombosis and Haemostasis 1998;80(3) 499–505.

[23] Laing GD, Moura-da-Silva AM. Jararhagin and its multiple effects on hemostasis. Toxicon 2005; 45 (8) 987–996.

[24] Kamiguti AS. Platelets as targets of snake venom metalloproteinases. Toxicon 2005;45(8) 1041–1049.

[25] Huang TF, Chang MC, Teng CM. Antiplatelet protease, kistomin, selectively cleaves human platelet glycoprotein Ib. Biochemica et Biophysica Acta 1993;1158(3) 293–299.

[26] Ward CM, Andrews RK, Smith AI, Berndt MC. Mocarhagin, a novel cobra venom metalloproteinase, cleaves the platelet von Willebrand factor receptor glycoprotein Ib alpha. Identification of the sulfated tyrosine/anionic sequence Tyr-276-Glu-282 of glycoprotein Ib alpha as a binding site for von Willebrand factor and alpha-thrombin, Biochemistry 1996;35(15) 4929–4938.

[27] Wu WB, Peng HC, Huang TF. Crotalin, a vWF and GP Ib cleaving metalloproteinase from venom of Crotalus atrox. Thrombosis and Haemostasis 2001;86(6) 1501–1511.

[28] Hsu CC, Wu WB, Huang TF. A snake venom metalloproteinase, kistomin, cleaves platelet glycoprotein VI and impairs platelet functions, Journal of Thrombosis and Haemostasis 2008;6(9) 1578–1585.

[29] Wijeyewickrema LC, Gardiner EE, Moroi M, Berndt MC, Andrews RK. Snake venom metalloproteinases, crotarhagin and alborhagin, induce ectodomain shedding of the platelet collagen receptor, glycoprotein VI. Thrombosis and Haemostasis 2007;98(6) 1285–1290.

[30] Cominetti MR, Ribeiro JU, Fox JW, Selistre-de-Araujo HS. BaG, a new dimeric metalloproteinase/disintegrin from the Bothrops alternatus snake venom that interacts with alpha5beta1 integrin. Archives of Biochemistry and. Biophysics 2003;416(2) 171–179.

[31] Masuda S, Araki S, Kaji K, Hayashi H. Purification of a vascular apoptosis-inducing factor from hemorrhagic snake venom. Biochemical and Biophysics Research Communications 1997;235(1) 59–63.

[32] Masuda S, Hayashi H, Atoda H, Morita T, Araki S. Purification, cDNA cloning and characterization of the vascular apoptosis-inducing protein, HV1, from Trimeresurus flavoviridis. European Journal of Biochemistry 2001;268(11) 3339–3345.

[33] You WK, Seo HJ, Chung KH, Kim DS. A novel metalloprotease from Gloydius halys venom induces endothelial cell apoptosis through its protease and disintegrin-like domains. Journal of Biochemistry 2003;134(5) 739–749.

[34] Trummal K, Tonismagi K, Siigur E, Aaspollu A, Lopp A, Sillat T, Saat R, Kasak L, Tammiste I, Kogerman, P, Kalkkinen, N, Siigu, J. A novel metalloprotease from Vipera lebetina venom induces human endothelial cell apoptosis. Toxicon 2005;46(1) 46–61.

[35] WU WB, Shin CC, Ming-Yi L, Tur-Fu H. Purification, molecular cloning and mechanism of action of graminelysin I, a snake-venom-derived metalloproteinase that induces apoptosis of human endothelial cells. Biochemical Journal 2001;357(Pt 3) 719-728.

[36] Masuda S, Hayashi H, Araki S. Two vascular apoptosis-inducing proteins from snake venom are members of the metalloprotease/disintegrin family. European Journal of Biochemistry 1998;253(1) 36–41.

[37] McLane MA, Sanchez EE, Wong A, Paquette-Straub C, Perez JC. Disintegrins. Current Drugs Targets. Cardiovascular & Haematological Disorders 2004; 4 327-355.

[38] Calvete JJ. Structure-function correlations of snake venom disintegrins. Current Pharmaceutical Design 2005;11(7) 829-835.

[39] Calvete JJ, Marcinkiewicz C, Monleon D, Esteve V, Celda B, Juarez P, Sanz L. Snake venom disintegrins: evolution of structure and function. Toxicon 2005;45(8) 1063-1074.

[40] Kini R M and Evans HJ. Structural domains in venom proteins: evidence that metalloproteinases and nonenzymatic platelet aggregation inhibitors (disintegrins) from snake venoms are derived by proteolysis from a common precursor. Toxicon 1992;30(3) 265-293

[41] Calvete JJ, Moreno-Murciano MP, Theakston RD, Kisiel DG, Marcinkiewicz, C. Snake venom disintegrins: novel dimeric disintegrins and structural diversification by disulphide bond engineering. Biochemical Journal 2003;372(Pt 3) 725-734.

[42] Calvete JJ, Jurgens M, Marcinkiewicz C, Romero A, Schrader M, Niewiarowski S. Disulphide-bond pattern and molecular modelling of the dimeric disintegrin EMF-10, a potent and selective integrin alpha5beta1 antagonist from Eristocophis macmahoni venom. Biochemical Journal 2000a;345 (Pt 3) 573-581.

[43] Bilgrami S, Tomar S, Yadav S, Kaur P, Kumar J, Jabeen T, Sharma S, Singh TP. Crystal structure of schistatin, a disintegrin homodimer from saw-scaled viper (Echis carinatus) at 2.5 A° resolution. Journal of Molecular Biology 2004;341(3) 829-837.

[44] Gould RJ, Polokoff MA, Friedman PA, Huang TF, Holt JC, Cook JJ, Niewiarowski S. Disintegrins: a family of integrin inhibitory proteins from viper venoms. Proceeding of the Society for Experimental Biology and Medicine 1990;195(2) 168–171.

[45] Ouyang C, Yeh HI, Huang TF. A potent platelet aggregation inhibitor purified from Agkistrodon halys (mamushi) snake venom. Toxicon 1983;21(6) 797 – 804.

[46] Sheu JR, Yen MH, Kan YC, Hung WC, Chang PT, Luk HN. Inhibition of angiogenesis in vitro and in vivo: comparison of the relative activities of triflavin, an Arg-Gly-Asp-containing peptide and anti-alpha(v)beta3 integrin monoclonal antibody. Biochemica et Biophysica Acta 1997;1336(3) 445–454.

[47] Yeh CH; Peng HC; Yang RS, Huang TF. Rhodostomin, a snake venom disintegrin, inhibits angiogenesis elicited by basic fibroblast growth factor and suppresses tumor

growth by a selective alpha(v)beta(3) blockade of endothelial cells. Molecular Pharmacology 2001;59(5) 1333–1342
[48] Swenson S, Costa F, Ernst W, Fujii G, Markland FS. Contortrostatin, a snake venom disintegrin with anti-angiogenic and anti-tumor activity. Pathophysiology of Haemostasis and Thrombosis 2005;34(4-5) 169–176.
[49] Swenson S, Costa F, Minea R, Sherwin RP, Ernst W, Fujii G, Yang D, Markland FS Jr. Intravenous liposomal delivery of the snake venom disintegrin contortrostatin limits breast cancer progression. Molecular Cancer Therapeutics 2004;3(4) 499–511.
[50] Trikha M, De Clerck YA, Markland FS. Contortrostatin, a snake venom disintegrin, inhibits beta 1 integrin-mediated human metastatic melanoma cell adhesion and blocks experimental metastasis. Cancer Research 1994;54(18) 4993–4998.
[51] Zhou Q, Sherwin RP, Parrish C, Richters V, Groshen SG, Tsao-Wei D, Markland FS. Contortrostatin, a dimeric disintegrin from Agkistrodon contortrix contortrix, inhibits breast cancer progression. Breast Cancer Research and Treatment 2000a;61(3) 249 – 260.
[52] Zhou Q, Nakada MT, Brooks PC, Swenson SD, Ritter MR, Argounova S, Arnold C, Markland FS. Contortrostatin, a homodimeric disintegrin, binds to integrin alphavbeta5. Biochemical Biophysical Research Communications 2000b;267(1) 350 – 355.
[53] Trikha M, Rote WE, Manley PJ, Lucchesi BR, Markland FS. Purification and characterization of platelet aggregation inhibitors from snake venoms. Thrombosis Research 1994;73(1) 39–52.
[54] Ritter MR, Zhou Q, Markland FS Jr. Contortrostatin, a snake venom disintegrin, induces alphavbeta3-mediated tyrosine phosphorylation of CAS and FAK in tumor cells. Journal of Cellular Biochemistry 2000;79(1) 28–37.
[55] Lin E, Wang Q, Swenson S, Jadvar H, Susan G, Ye W, Markland F, Pinski J. The disintegrin contortrostatinin combination with docetaxel is a Potent Inhibitor of Prostate cancer in vitro and in vivo. The Prostate 2010;70(12) 1359-1370
[56] Olfa KZ; Jose L; Salma D, Bazaa A, Srairi N, Nicolas A; Maxime L, Zouari R, Mabrouk K, Marvaldi J, Sabatier JM, El Ayeb M; Marrakchi N. Lebestatin, a disintegrin from Macrovipera venom, inhibits integrin-mediated cell adhesion, migration and angiogenesis. Laboratory Investigation 2005;85(12) 1507–1516.
[57] Calvete JJ; Fox JW; Agelan A; Niewiarowski S; Marcinkiewicz C. The presence of the WGD motif in CC8 heterodimeric disintegrin increases its inhibitory effect on alphaII(b)beta3, alpha(v)beta3, and alpha5beta1 integrins. Biochemistry 2002;41(6) 2014–2021.
[58] Morris VL; Schmidt EE, Koop S, MacDonald IC, Grattan M; Khokha R, McLane MA; Niewiarowski S, Chambers AF, Groom AC. Effects of the disintegrin eristostatin on individual steps of hematogenous metastasis. Experimental Cell Research 1995;219(2) 571–578.
[59] Wang WJ. Acurhagin-C, an ECD disintegrin, inhibits integrin alphavbeta3-mediated human endothelial cell functions by inducing apoptosis via caspase-3 activation. British Journal of Pharmacology 2010;160(6) 1338–1351.

[60] Sohn YD, Cho KS, Sun SA, Sung HJ, Kwak KW, Hong SY, Kim DS, Chung KH. Suppressive effect and mechanism of saxatilin, a disintegrin from Korean snake (Gloydiussaxatilis), in vascular smooth muscle cells. Toxicon 2008;52(3) 474–480.

[61] Tian J, Paquette-Straub C, Sage EH, Funk SE, Patel V, Galileo D, McLane MA. Inhibition of melanoma cell motility by the snake venom disintegrin eristostatin. Toxicon 2007; 49(7) 899–908.

[62] Ramos OH, Kauskot A, Cominetti MR, Bechyne I, Salla Pontes CL, Chareyre F, Manent J, Vassy R, Giovannini M, Legrand C, Selistre-de-Araujo HS, Crepin M, Bonnefoy AA. Novel alpha(v)beta (3)-blocking disintegrin containing the RGD motive, DisBa-01, inhibits bFGF-induced angiogenesis and melanoma metastasis. Clinical & Experimental Metastasis 2008;25(1) 53–64.

[63] Kini RM. Excitement ahead: structure, function and mechanism of snake venom phospholipase A2 enzymes. Toxicon 2003;42(8) 827–840.

[64] Burke JE and Dennis EA. Phospholipase A2 Biochemistry. Cardiovascular drugs and Therapy 2009a;23(1) 1-22.

[65] Ramar PS, Gopalakrishnakone P, Bow H, Puspharaj PN, Chow VT. Identification and characterization of a phospholipase A2 from the venom of the Saw-scaled viper: Novel bactericidal and membrane damaging activities. Biochimie 2010;92(12) 1854-1866.

[66] Ritonja A and Gubensek F. Ammodytoxin A, a highly lethal phospholipase A2 from Vipera ammodytes ammodytes venom. Biochemica et Biophysica Acta 1985; 828(3) 93 306-312.

[67] Maung-Maung T, Gopalakrishnakone P, Yuen R, Tan CH. A major lethal factor of the venom of Burmese Russell's viper (Daboia russelli siamensis): isolation, N-terminal sequencing and biological activities of daboiatoxin. Toxicon 1995;33(1) 63-76.

[68] Chakrabarty D, Datta K, Gomes A, Bhattacharyya D. Haemorrhagic protein of Russell's viper venom with fibrinolytic and esterolytic activities. Toxicon 2000; 38(11) 1475-1490.

[69] Kini RM. Phospholipase A2-a complex multifunctional protein puzzle. In: Kini, R M (ed) Enzymes: Structure, Function and Mechanism. John Wiley and Sons, Chichester, England;1997. p 1-28.

[70] Valentin E and Lambeau G. Increasing molecular diversity of secreted phospholipases A(2) and their receptors and binding proteins. Biochemica et Biophysica Acta 2000;1488(1-2) 59-70.

[71] Six DA and Dennis EA. The expanding superfamily of phospholipase A(2) enzymes: classification and characterization. Biochemica et Biophysica Acta 2000;1488(1-2) 1-19.

[72] Rouault M, Bollinger JG, Lazdunski M, Gelb MH, Lambeau G. Novel mammalian group XII secreted phospholipase A2 lacking enzymatic activity. Biochemistry 2003;42(39) 11494-11503.

[73] Lambeau G and Lazdunski M. Receptors for a growing family of secreted phospholipases A2. Trends in Pharmacological Sciences 1999;20(4) 162-170.

[74] Dennis EA. Phospholipase A2 in ecicosanoid generation. American Journal of Respiratory and Critical Care Medicine 2000 61(2 Pt 2) S32-S35.

[75] Lomonte B, Angulo Y, Calderon L. An overview of lysine-49 phospholipase A2 myotoxins from crotalid snake venoms and their structural determinants of myotoxic action. Toxicon 2003;42(8) 885-901

[76] Tsai IH, Wang YM, Chen YH, Tsai TS. Venom phospholipases A2 of bamboo viper (Trimeresurus stejnegeri): molecular characterization, geographic variations and evidence of multiple ancestries. Biochemical Journal 2004; 77(Pt 1) 215–223.

[77] Wei JF, Wei XL, Chen QY, Huang T, Qiao LY, Wang WY, Xiong YL, He SH. N49 phospholipase A2, a unique subgroup of snake venom group II phospholipase A2. Biochemica et Biophysica Acta 2006;1760(3) 462–471.

[78] Krizaj I, Bieber AL, Ritonja A, Gubensek F. The primary structure of ammodytin L, a myotoxic phospholipase A2 homologue from Vipera ammodytes venom. European Journal of Biochemistry 1991;202(3) 1165–1168.

[79] Polgar J, Magnenat EM, Peitsch MC, Wells TN, Clemetson KJ. Asp-49 is not an absolute prerequisite for the enzymic activity of low-M(r) phospholipases A2: purification, characterization and computer modelling of an enzymically active Ser- 49 phospholipase A2, ecarpholin S, from the venom of Echis carinatus sochureki (saw-scaled viper). Biochemical Journal 1996;319(Pt 3) 961–968.

[80] Bao Y, Bu P, Jin L, Wang H, Yang Q, An L. Purification, characterization and gene cloning of a novel phospholipase A2 from the venom of Agkistrodon blomhoffii ussurensis. The International Journal of Biochemistry & Cell Biology Cell Biol 2005;37(3) 558–565.

[81] Chijiwa T, Tokunaga E, Ikeda R, Terada K, Ogawa T, Oda-Ueda N, Hattori, S, Nozaki, M, Ohno M. Discovery of novel [Arg49] phospholipase A2 isozymes from Protobothrops elegans venom and regional evolution of Crotalinae snake venom phospholipase A2 isozymes in the southwestern islands of Japan and Taiwan. Toxicon 2006;48(6) 672–682.

[82] Mebs D, Kuch U, Coronas FIV, Batista CVF, Gumprecht A, Possani LD. Biochemical and biological activities of the venom of the Chinese pitviper Zhaoermia mangshanensis, with the complete amino acid sequence and phylogenetic analysis of a novel Arg49 phospholipase A2 myotoxin. Toxicon 2006;47(7) 797–811.

[83] Wei JF, Li T, Wei XL, Sun QY, Yang FM, Chen QY, Wang WY, Xiong YL, He SH. Purification, characterization and cytokine release function of a novel Arg-49 phospholipase A2 from the venom of Protobothrops mucrosquamatus. Biochimie 2006;88(10) 1331–1342.

[84] Li Y, Yu BZ, Zhu H, Jain MK, Tsai MD. Phospholipase A2 engineering. Structural and functional roles of the highly conserved active site residue aspartate-49. Biochemistry 1994;33(49)14714-14722.

[85] Kini RM and Evans HJ. Correlation between the enzymatic activity, anticoagulant and antiplatelet effects of phospholipase A2 isoenzymes from Naja nigricollis venom. Thrombosis and Haemostasis 1988;60(2) 170-173.

[86] Kasturi S and Gowda TV. Purification and characterization of a major phospholipase A2 from Russell's viper (Vipera russelli) venom. Toxicon 1989;27(2) 229-237.

[87] Stefansson S, Kini, RM, Evans HJ. The inhibition of clotting complexes of the extrinsic coagulation cascade by the phospholipase A2 isoenzymes from Naja nigricollis venom. Thrombosis Research 1989; 55(4) 481-491.

[88] Maung-Maung T, Gopalakrishnakone P, Yuen R, Tan CH. A major lethal factor of the venom of Burmese Russell's viper (Daboia russelli siamensis): isolation, N-terminal sequencing and biological activities of daboiatoxin. Toxicon 1995; 33(1) 63-76.

[89] Huang MZ, Gopalakrishnakone P, Kini RM. Role of enzymatic activity in the antiplatelet effects of a phospholipase A2 from Ophiophagus hannah snake venom. Life Sciences 1997; 61(22) 2211-2217.

[90] Kole L, Chakrabarty D, Datta K, Bhattacharyya D. Purification and characterization of an organ specific haemorrhagic toxin from Vipera russelli russelli (Russell's viper) venom. Indian Journal of Biochemistry & Biophysics 2000; 37(2) 114-120.

[91] Chakrabarty AK, Hall RH, Ghose AC. Purification and characterization of a potent hemolytic toxin with phospholipase A2 activity from the venom of Indian Russell's viper. Molecular and Cellular Biochemistry 2002; 237(1-2) 95-102.

[92] Dong M, Guda K, Nambiar PR, Rezaie A, Belinsky GS, Lambeau G, Giardina C, Rosenberg DW. Inverse association between phospholipase A2 and COX-2 expression during mouse colon tumorigenesis. Carcinogenesis 2003; 24(2) 307-315.

[93] Zouari-Kessentini R, Luis J, Karray A, Kallech-Ziri O, Srairi-Abid N, Bazaa A, Loret E, Bezzine S, El Ayeb M, Marrakchi N. Two purified and characterized phospholipases A2 from Cerastes cerastes venom that inhibits cancerous cell adhesion and migration. Toxicon 2009; 53(4) 444–453

[94] Zouari-Kessentini R, Jebali J, Taboubi S, Srairi-Abid N, Morjen M, Kallech-Ziri O, Bezzine S, Marvaldi J, El Ayeb M, Marrakchi N. CC-PLA2-1 and CC-PLA2-2, two cerastes cerastes venom derived phospholipases A2, inhibit angiogenesis both in vitro and in vivo. Laboratory Investigation 2010; 90(4) 510-519.

[95] Roberto PG, Kashima S, Marcussi S, Pereira JO, Astolfi-Filho S, Nomizo A, Giglio JR, Fontes MR, Soares AM, Fança SC. Cloning and identification of a complete cDNA coding for a bactericidal and anti-tumoral acidic phospholipase A2 from Bothrops jararacussu venom. Protein Journal 2004; 23(4) 273-285.

[96] Bazaa A, Pasquier E, Defilles C, Limam I, Kessentini-Zouari R, Kallech-Ziri O, El Battari A, Braguer D, El Ayeb M, Marrakchi N, Luis J. MVL-PLA2, a snake venom phospholipase A2, inhibits angiogenesis through an increase in microtubule dynamics and disorganization of focal adhesions. PLoS One 2010; 5(4):e10124.

[97] Murakami T, Kamikado N, Fujimoti R, Hamaguchi K, Nakamura H, Chijiwa T, Ohno M, Oda-Ueda. A [Lys49] phospholipase A2 from Protobothrops flavoviridis venom induces caspase-independent apoptotic cell death accompanied by rapid plasma-membrane rupture in human leukemia cells. Biosciences, Biotechnology and Biochemistry 2011; 75(5) 864-870.

[98] Lu Q, Navdaev A, Clemetson JM. Clemetson KJ. Snake venom C-type lectins interacting with platelet receptors: structure-function relationships and effects on haemostasis. Toxicon 2005; 45(8) 1089–1098.

[99] Eble JA, Niland S, Bracht T, Mormann M, Peter-Katalinic J, Pohlentz G, Stetefeld J.. The alpha2beta1 integrin-specific antagonist rhodocetin is a cruciform, heterotetrameric molecule. FASEB Journal 2009; 23(9) 2917-2927.

[100] Mizuno H, Fujimoto Z, Koizumi M, Kano H, Atoda H, Morita T. Crystal structure of coagulation factor IX-binding protein from habu snake venom at 2.6 A: implication of central loop swapping based on deletion in the linker region. Journal of Molecular Biology 1999; 289(1) 103-112.

[101] Horii K, Okuda D, Morita T, Mizuno H. Crystal structure of EMS16 in complex with the integrin alpha2-I domain. Journal of Molecular Biology 2004; 341(2) 519-527.

[102] Maita N, Nishio K, Nishimoto E, Matsui T, Shikamoto Y, Morita T, Sadler JE, Mizuno H. Crystal structure of von willebrand factor A1 domain complexed with snake venom, bitiscetin: insight into glycoprotein Ibα binding mechanism induced by snake venom proteins. The Journal of Biological Chemistry 2003; 278(39) 37777–37781.

[103] Runhua WR, Manjunatha K, Max CMC. Rhodocetin, a novel platelet aggregation inhibitor from the venom of Calloselasma rhodostoma (Malayan pit viper): synergistic and non covalent interaction between subunits. Biochemistry 1999; 38(23) 7584-7593.

[104] Marcinkiewicz C, Lobb RR, Marcinkiewicz MM, Daniel JL, Smith JB, Dangelmaier C, Weinreb PH., Beacham DA, Niewiarowski S. Isolation and characterization of EMS16, a C-lectin type protein from Echis multisquamatus venom, a potent and selective inhibitor of the $\alpha 2\beta 1$ integrin. Biochemistry 2000; 39(32) 9859–9867.

[105] Sarray S, Berthet V, Calvete JJ, Secchi J, Marvaldi J, El Ayeb M, Marrakchi N, Luis J. Lebectin, a novel C-type lectin from Macrovipera lebetina venom, inhibits integrin mediated adhesion, migration and invasion of human tumour cells. Laboratory Investigation 2004; 84(5) 573-581.

[106] Sarray S, Delamarre E, Marvaldi J, El Ayeb M, Marrakchi N, Luis J. Lebectin and lebecetin, two C-type lectins from snake venom, inhibit alpha5 beta1 and alphaV-containing integrins. Matrix Biology 2007; 26(4) 306-313.

[107] Pilorget A, Conesa M, Sarray S, Michaud-Levesque J, Daoud S, Kim KS, Demeule M, Marvaldi J, El Ayeb M, Marrakchi N, Beliveau R, Luis J. Lebectin, a Macrovipera lebetina venom-derived C-type lectin, inhibits angiogenesis both in vitro and in vivo. Journal of Cellular Physiology 2007 211(2) 307-315.

[108] Golubkov V, Hawes D, Markland FS. Anti-angiogenic activity of contortrostatin, a disintegrin from Agkistrodon contortrix contortrix snake venom. Angiogenesis 2003; 6(3) 213–224.

[109] Marcinkiewicz C, Weinreb PH, Calvete JJ, Kisiel DG, Mousa SA, Tuszynski GP, Lobb RR. Obtustatin: A potent selective inhibitor of alpha1beta1 integrin in vitro and angiogenesis in vivo. Cancer Research 2003; 63(9) 2020–2023.

[110] Sarray S, Siret C, Lehmann M, Marrakchi N, Luis J, El Ayeb M, Andre F. Lebectin increases N-cadherin-mediated adhesion through PI3K/AKT pathway. Cancer Letters 2009; 285(2) 174–181

[111] Pereira-Bittencourt M, Carvalho DD. Gagliard AR, Collins DC. The effect of a lectin from the venom of the snake, Bothrops jararacussu, on tumor cell proliferation. Anticancer Research 1999; 19(5B) 4023-4025.

[112] Nolte S, de Castro Damasio D, Baréa AC, Gomes J, Magalhães A, Mello Zischler LF, Stuelp-Campelo PM, Elífio-Esposito SL, Roque-Barreira MC, Reis CA, Moreno-Amaral AN. BJcuL, a lectin purified from Bothrops jararacussu venom, induces apoptosis in human gastric carcinoma cells accompanied by inhibition of cell adhesion and actin cytoskeleton disassembly. Toxicon 2012; 59(1) 81-85.

[113] Wan SG, Jin Y, Lee WH, Zhang Y. A snake venom metalloproteinase that inhibited cell proliferation and induced morphological changes of ECV304 cells. Toxicon 2006; 47(4) 480-489.

[114] Limam I, Bazaa A, Srairi-Abid N, Taboubi S, Jebali J, Zouari-Kessentini R, Kallech-Ziri O, Mejdoub H, Hammami A, El Ayeb M, Luis J, Marrakchi N. Leberagin-C, A disintegrin-like/cysteine-rich protein from Macrovipera lebetina transmediterranea venom, inhibits alphavbeta3 integrin-mediated cell adhesion. Matrix Biology 2010; 29(2)117-126.

[115] Yeh CH, Peng HC, Yih JB, Huang TF A new short chain RGD-containing disintegrin, accutin, inhibits the common pathway of human platelet aggregation. Biochemica Biophysica Acta 1998; 1425(3) 493-504.

Section 3

Production of Bioactive Food Peptides

Chapter 10

Advancements in the Fractionation of Milk Biopeptides by Means of Membrane Processes

Claudia Muro, Francisco Riera and Ayoa Fernández

Additional information is available at the end of the chapter

http://dx.doi.org/10.5772/53674

1. Introduction

Nowadays the most common way to obtain bioactive peptides is by enzymatic hydrolysis of protein solutions. The most studied substrates used to produce bioactive peptides are milk proteins in the form of co-products from dairy industries: caseins, cheese whey, buttermilk, whey protein concentrates and isolates or even pure single proteins that can be obtained at a reasonable price on an industrial scale (e.g. β-lactoglobulin, β-Lg).

Different specific and non-specific enzymes are used to obtain hydrolysates (trypsin, pepsin, pancreatin and alcalase). The catalytic activity of some of them is quite specific and the composition of the hydrolysate is predictable when substrates are quite pure [1]. In other cases, the activity of the enzyme is non-specific and produces a complex mixture of peptides and amino acids in which individual effect of each molecule in the subsequent fractionation process is difficult to demonstrate and quantify. The design of an efficient fractionation methodology is then of paramount importance for peptides separation and even more, when the process must applied on an industrial scale. Separation technologies, which discriminate small differences in charge, size and hydrophobicity, can be employed to fractionate protein hydrolysates and obtain peptide fractions with higher functionality or higher nutritional value in a more purified form. Membrane separation techniques seem to be well suited for this purpose. These processes are based upon selective permeability of one or more of the liquid constituents through the membrane according to the driving forces.

2. Overview of techniques used for peptide fractionation

Due to the demonstration of their impact on human health, the market for functional food and nutraceuticals containing bioactive peptides is increasing very rapidly and, consequently, the food and bio-pharmaceutic industries are looking for processes allowing the production of this kind of products from natural sources. Considering that most functional peptides are present in complex mixtures containing a large number of hydrolysed protein fractions, their separation and purification are required.

The methodologies commonly used for peptide fractionation and enrichment include: selective precipitation, membrane filtration, ion exchange, gel filtration technologies and liquid chromatography [1]. However, significant differences concerning the number and type of extracted peptides occur among extraction procedures. Additionally, undesired peptides, such as allergenic or bitter-tasting peptides, could be enriched in the process when using some of those techniques [2].

Fractionation methods involving precipitation steps are carried out by means of the addition of organic solvents like ethanol, methanol or acetone; adding acids like trichloroacetic acid (TCA), sulphosalicylic acid or phosphotungstic acid (PPTA); by means of the addition of salts (ammonium sulphate) or just by adjusting the pH to the isoelectric point. Precipitation often results in a selective fractionation of peptides depending on their solubility in the precipitating agent [3]; however the addition of chemical compounds causes in some cases peptide degradation and changes in the biological and physical properties.

Chromatographic methods for peptide separation are currently used at lab-scale: high performance liquid chromatography (HPLC), fast protein liquid chromatography (FPLC), isoelectric focusing (IEF) and ion exchange chromatography (IEC) are some of them. In most cases, one or two cycles of successive HPLC separation had been adequate to isolate peptides one by one. In the same way, IEC has been used for the enrichment of casein phosphopeptides from casein hydrolysates or for the isolation of cationic antibacterial peptides from lactoferrin. However, although chromatographic processes can provide good separation selectivity, the low productivity and high production costs involved in these processes make impossible its use at industrial scale.

Size exclusion chromatography (SEC) and more frequently Ultrafiltration-Nanofiltration (UF-NF) are the main techniques used to isolate peptides according to their molecular size [4-10]. In addition it is possible to obtain more purified hydrolysate samples by removing salts and other interfering components by means of UF membranes [11]. In fact, investigations into these methodologies under optimized conditions to reduce time and cost are ongoing [12].

Pressure-driven membrane-based processes, such as UF and NF, are used to fractionate peptide mixtures and amino acids [13]. These types of membrane have been widely used to fractionate milk protein hydrolysates with the aim of enhancing their biological or functional properties [14-15]. It has been shown that variations in operating conditions may favor the permeation of bioactive peptides [16-17].

Membrane technology has become an important separation technology in recent decades probably because their main advantages (it works without the addition of chemicals, with a relatively low use of energy, it has low processing costs, the scale-up is an easy subject and the process lines are well arranged) make it the ideal technology for use on an industrial scale. In addition, membrane processes are especially suitable for the food industry, because of the mild working conditions, relatively easy scale up and low processing costs in comparison to chromatographic techniques.

The separation of peptides by UF mainly depends on the molecular weight (MW) cut-off (MWCO) of the membrane. However, when the MW of the peptides involved in the process is quite similar, their isolation is a hard subject; in these cases, NF is the best membrane separation technique [18]. The fact that NF membranes are usually charged offers the possibility of separating solutes through a combination of size and charge mechanisms.

3. Membrane technology applied to peptide fractionation

Membrane processes are now viewed as efficient tools for the development of new value-added products by separating minor compounds such as bioactive peptides [19]. These separation processes are based upon selective permeability of one or more of the liquid constituents through the membrane according to the pressure difference. Amongst the pressure-driven membrane techniques, which main features are summarized in Figure 1, UF and NF have been tested for the fractionation of protein hydrolysates due to the fact that the molecular weight of most bioactive peptides is within the normal pore size range of these membranes.

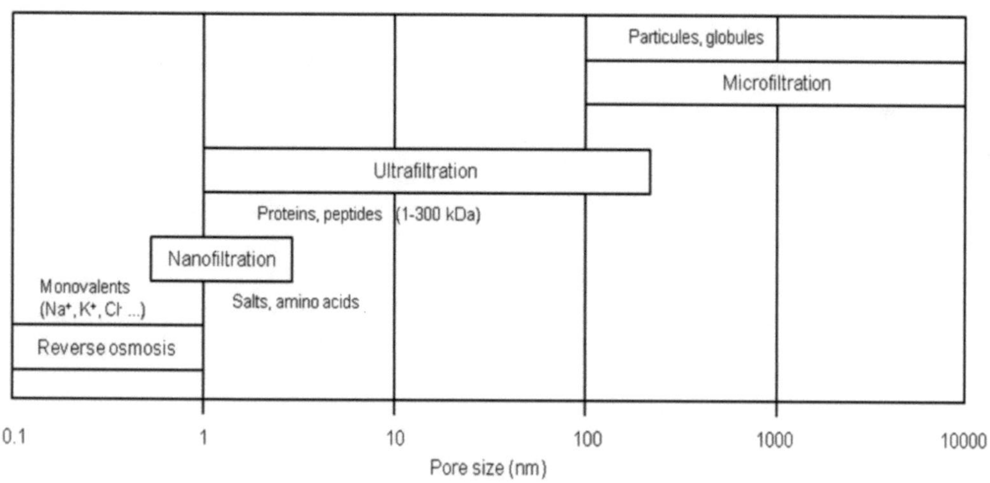

Figure 1. Pressure-driven membrane processes

UF is commonly applied to prepare enriched bioactive solutions from protein hydrolysates and improve the bioactivity of peptides. This process is also used to separate peptides with a size lower than 7 kDa [20]. The fractions are collected by subsequently filtration in two or three streams to obtain peptides with different size [21]. For example, amino acids and small peptides can be separated at pH 4.6 into four ranges of molecular mass (I<30 kDa, II>30 kDa (protein fraction), III>10 kDa (protein fraction), IV>0.3 kDa) [22]. Recent results on fractionation peptides by UF-membranes show that crude yoghurt fractions obtained after ion exchange can be separated into four fractions by successive UF using membranes with molecular cut off sizes of 30, 10 and 3 kDa [23]; whereas UF membranes < 1 kDa are efficient for peptides fractionation from milk hydrolysates if the last permeate contains free amino acids [21].

The combination of membrane processes (UF and NF) is also often used to separation of peptides. The first step of these processes consists in the UF of the hydrolysate in order to obtain complete rejection of intact proteins and intermediate peptides. The resulting permeate fractions is then subjected to a fractionation by NF and a peptide fraction having a molar mass < 1 kDa is isolated of the mixture by means of these membranes.

In this case, permeates obtained after UF could be adjusted at two pH values (9.5 and 3.0) that corresponded to the different charged states of the membrane and of the peptides to improve of separation of polypeptides of molar mass < 1 kDa [23-24].

Recently a method that couple UF and HPLC has also been applied on milk hydrolysate samples for enhance the peptides separation. A current study showed that an UF-membrane was enough to concentrate peptides and subsequently, both permeate and retentate were fractioned by SE-HPLC to obtain small peptides with biological activity [25].

There are also other important UF-processes to separate specific compounds of whey as caseinomacropeptide (CMP). A first method was designed to obtain CMP fractions trough UF membranes with MWCO 20-50 kDa by two diafiltration steps [26]. The method is based on the ability of CMP to form non-covalent linked polymers with a molecular weight up to 50 kDa at neutral pH, which dissociate at acid conditions. The dissociated form of CMP permeates through the UF-membrane at pH 3.5, whereas the majority of whey proteins such as β-Lg, α-lactalbumin (α-La), immunoglobulins (IGs) and bovine serum albumin (BSA) are held back. At pH 7.0, permeate containing CMP can be concentrated by means of the same membrane; however a low permeation rate is obtained with this technique. A second method for separation of CMP can be seen in [27]. Thermal stability of CMP is used in comparison to that of the rest of whey proteins. Complete denaturation and aggregation of proteins is obtained by treating whey at 90°C for 1h; with this method, the denatured proteins can be removed by centrifugation at 5200 g and 4°C for 15 min and the supernatant containing CMP can be concentrated by UF with MWCO 10 kDa after pH adjustment to 7.0; however whey proteins lose part of their functionality due to the denaturation.

Another method for separation of CMP consists in the pretreatment of whey protein concentrate with the enzyme transglutaminase (Tgase) followed by microfiltration [28]. The amino acid sequence of CMP includes two glutamine and three lysine residues, whereby this peptide can be cross-linked by tranglutaminase. The covalent linked CMP aggregates can be removed be means of microfiltration or diafiltration to obtain CMP-free whey protein.

3.1. Enzymatic membrane reactor equipped with membranes: first step to peptide fractionation

Enzymatic membrane reactor (EMR) consists on a coupling of a membrane separation process with an enzymatic reaction. EMR allows the continuous production and separation of specific peptide sequences by means a selective membrane, which is used to separate the biocatalyst from the reaction products and the peptides fractionation [29]. At present, EMR is used when working on an industrial scale. This technology for peptides separation is gaining interest, because it is a specific mode for running batch or continuous processes in which enzymes are separated from end products with the help of a selective membrane. By that way, it is possible to obtain complete retention of the enzyme without deactivation problems typical of enzyme immobilization. Furthermore, EMR have been shown to improve the efficiency of enzyme-catalyzed bioconversion and to increase product yields [13, 30-31].

EMR technology has been investigated for the production and separation of peptides since the 90's. Antithrombotic peptides derived from hydrolysed CMP can be recovered by UF membranes [32-33] and Lactorphin have been successfully produced through continuous hydrolysis of whey in an UF-reactor [34-35]. Multicompartment EMR has also been designed for the continuous hydrolysis of milk proteins. Nowadays, this technique is operated under an electric field for continuous harvesting of some biologically active peptides, such as phosphopeptides and precursors of casomorphins from the tryptic digest of β-casein [36]. Special attention had also had the study of the hydrolysis of whey protein isolates (WPI) using a tangential flow filter membrane (TFF) of 10 kDa in EMR [37]. The factors influencing on the operation of the EMR (substrate concentration, ionic strength, and transmembrane pressure) have been studied and discussed in other research works [30, 38]. In recent years, the use of EMR has emerged as an exciting area of research due to their low production cost, product safety and easy scaled up [39]. Table 1 summarizes some examples of processes for the separation or concentration of bioactive peptides by means of UF membranes. UF offers possibilities for a large-scale production of bioactive peptides but seems limited because of fouling and poor selectivity. Another drawback of UF membranes is their pore size, because the large pores are not selective enough to fractionate small peptides MW of bioactive peptides is usually smaller than 1 kDa). To sum up, with the use of an EMR equipped with UF membranes, the first peptide fractionation is achieved but if a more purified permeate is required; NF membranes should be used as an additional step instead of UF membranes.

3.2. NF membranes and peptide fractionation

NF is a pressure-driven membrane technique in which the pore size of the membrane is in the nanometers range. As can be observed in Figure 1, this technique is an intermediate step between reverse osmosis (RO) and UF and it is useful to separate/fractionate solutes with MW lower than 5 kDa. Transmembrane pressure in NF is lower than in RO and the permeate flux is usually higher, which represents an important energetic advantage in industrial applications. NF membranes of cut-off < 1 kDa are particularly useful for the filtration of the smaller peptides from hydrolysates solutions.

The selectivity of NF membranes is based on both size and charge characteristics of the solutes and on the interaction between charged solutes and membrane surface. Hydrodynamic parameters (mainly transmembrane pressure and linear velocities) and membrane material exert influence on membrane selectivity too. NF membranes have a slightly charged surface; because the dimensions of the pores are less than one order of magnitude larger than the size of ions [54].

Protein Hydrolysate Source	Biological Activity	References
Bovine caseinomacropeptide	Antithrombotic	[32-33]
	Calcium bioavailability improvement	[40]
	Bone and teeth mineralization	
Bovine whey β-lactoglobulin	ACE inhibitor	[41]
	Opioid	[42]
	Anti microbial	[43]
	Muscular contraction	[44]
Bovine whey α-lactalbumin	ACE inhibitor	[4]
Fish protein	ACE inhibitor	[45]
Alfalta white protein	ACE inhibitor	[46]
Alfalta leaf protein	Antioxidant	[47]
Wheat gluten	ACE inhibitor	[48]
Soybean protein	ACE inhibitor	[49]
Soybean β-conglycinin	ACE inhibitor	[50]
Sea cucumber gelatin	ACE inhibitor	[51]
Potato	Antimicrobial	[52]
Potato	Antimicrobial	[53]

Table 1. Bioactive peptides obtained by means of UF membranes

3.2.1. NF transport mechanism

The mechanism behind the selectivity of membrane processes is generally the size of the component. This mainly applies in the case of UF membranes and in the case of NF membranes with uncharged solutes. Charge effects are minimized in this case and the transmission of the solutes depends largely on the size exclusion effects of the membrane. This sieving effect is usually modeled and corrected [55] using continues hydrodynamic models such as originally proposed by Ferry. In this model, the membrane is assumed to be a network of perfectly cylindrical and parallel pores in which solvent velocity follows Poiseuille's law with a parabolic profile and solutes are assimilated to hard spheres. The transmission coefficient (*Tr*) of a given solute can be calculated according to equation (1) However, the selectivity of NF membranes is based on both size and charge characteristics of the solutes and on the interaction between charged solutes and membrane surface [56].

$$Tr = (1-(\lambda(\lambda-2))^2 \exp(-0.7146 \lambda^2) \quad (1)$$

Where λ is the relation between the radius of the solute and the radius of the pore.

The selectivity of the separation when using NF membranes is based on the following factors: a) Solute (peptide) size, shape and charge. b) Membrane pore size and surface charge (sign and surface charge density). c) Hydrodynamic conditions of the fractionation process (transmembrane pressure, lineal velocities and solute concentration). d) Membrane characteristics (manufacture process, surface roughness, porosity, film layer material and hydrophilic/hydrophobic surface). All these aspects must be considered in order to estimate the viability of a peptide fractionation process.

Especially in NF membranes involving peptide fractionation from mixtures, charge exclusion mechanisms are predominant in the separation. The charge effects affect membrane-peptide and peptide-peptide interactions in the mixture or at the membrane surface. The transport mechanism through the pores is governed by convective and diffusive fluxes as well as by electromigrative flux. These phenomena make the prediction of the separation selectivity a difficult objective.

The current state of science the knowledge of the NF process is not sufficient to make a model fulfilling the requirements. The difficulties in modeling permeate flow rates and solute rejection come from the scale at which the different phenomena takes place at the membrane surface and through the membrane pores, where most of the hydrodynamic and macroscopic interactions begin to break down. However, simplified approaches could be used to explain qualitatively the experimental results obtained, as can be seen below.

The solute transfer through the membrane follow two main steps: distribution of ionic species at the selective interface according to their charge (both solutes and membrane) and transfer by a complex combination among diffusion, convection and electrophoretic mobility through the membrane, at least at low feed concentrations [13]. According to Donnan theory, the passage of charged solutes through a charged NF membrane is likely to be different whether they are considered to be co-ions, i.e. with the same charge of the membrane, or counter-ions, i.e. with a charge of opposite sign. In fact, due to electrostatic repulsive/attractive forces between the membrane and the solutes the concentration of co-ions will be lower in the membrane than in the solutions. On the contrary, the counter-ions have a higher concentration in the membrane than in the solution. This concentration difference of the ions generates a potential difference at the interface between the membrane and the solution, which is called Donnan potential. Under equilibrium conditions, electro-neutrality and equality of electrochemical potentials are maintained through the system. The Donnan equilibrium depends on the ion concentration, the fixed charge concentration in the membrane and the valences of the co-ions and counter-ions. Figure 2 shows an adapted schematic representation [57] of the influence of the electrostatic interactions in the transmission of charged peptides through a charged NF membrane.

Because of the electro-neutrality principle, and on the assumption that the charge density of the membrane is quite higher than the net charge of the co-ions, is possible to calculate the distribution of the co-ion resulting from a binary electrolyte $AB \rightarrow A^{z_A} + B^{z_B}$ between the membrane surface and the solution as a function of the charge density of the membrane.

$$K = \frac{C_B^m}{C_B} = \frac{(z_B \,.C_B)^{z_B/Z_A}}{(Z_B.C_B^m + Z_X\,.C_X^m)} \tag{2}$$

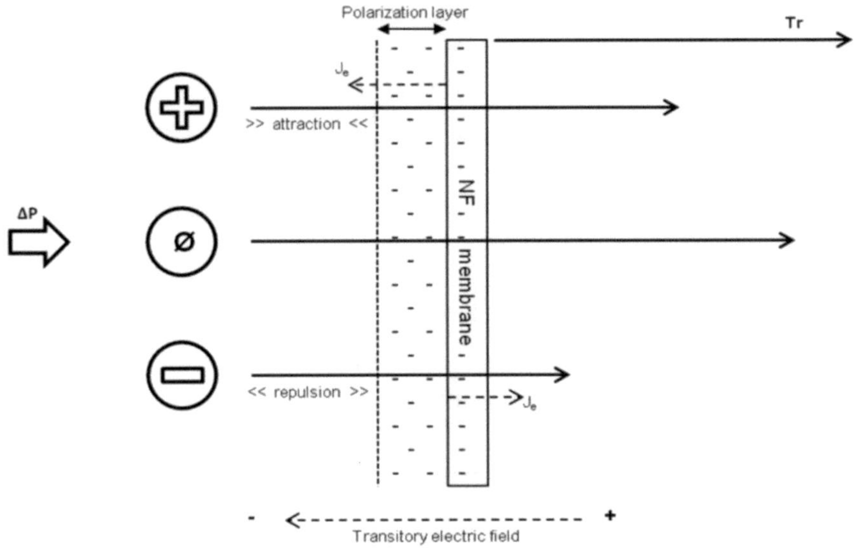

Figure 2. Schematic representation of solute flows across a negatively charged NF membranes. J_e: electromigrative flow as a consequence of the transitory electric field. Tr: transmission of the solute. Attractive (>> <<) and repulsive (<< >>) electrostatic interactions between charged solutes and the membrane are also represented.

C_B^m and C_B represent the concentration of co-ions B in the membrane and in the solution respectively. The coefficient of distribution, K, can be used to predict the rejection value of a binary electrolyte if the ionic transport is mainly due to convection and size exclusion effects are negligible. Under these conditions, K will mainly depend on: the co-ion valence (z_B), the counter-ion valence (z_A), the membrane charge (C_x^m), its valence (z_x) and the concentration of the co-ion in the solution (C_B).

According to equation 2, Donnan equilibrium predicts that an increase in the concentration of co-ions in the global solution and/or a decrease in the membrane charge density lead to a decrease in the exclusion of co-ions from the membrane surface (K is increased) and to a decrease in the retention of the binary salt (co-ion and cointer-ion) in order to maintain electroneutrality in both sides of the membrane [58-59]. The concentration of co-ions in the membrane will change according to the valence of the co-ion and counter-ions present in the solution. Thus, if the valence of the co-ion (z_B) has a lower value and the valence of the counter-ion (z_A) is increased, the concentration of co-ions in the membrane will be favored. For example, the retention of some common salts by descending order (Na_2SO_4 > NaCl > $CaCl_2$) through a negatively charged NF membrane can be predicted according to these principles [60-63].

Donnan theory is generally used to describe the permeability and selectivity of NF membranes using solutions containing only one amino acid. For example, for an amino acid co-ion and its associated counter-ion, in accordance with the Donnan equilibrium, the amino acid is electrically rejected by the charged active layer of the membrane. Simultaneously, the counter-ion is retained to ensure the balance of charges as the consequence of the electromigrative flow that opposes the convective one. However unfortunately, the extrapolation of Donnan theory to predict the behavior of individual solutes in mixed solutions containing several negative, neutral and positive solutes is very limited, mainly because of coupling and competitive effects. For this reason, NF process of complex mixtures of amino acids and peptides is a difficult object for mathematical modeling [64].

3.2.2. *NF Applied to amino acid and peptide fractionation: Review*

To clarify the mechanisms involved in the separation of biomolecules by NF membranes several fundamental researches have been published. Table 2 shows relevant NF studies involving amino acids and peptides. The data obtained are relative at different factors affecting the separation of single amino acid (AA) solutions, peptides mixtures and protein hydrolysates. For example, the influence of pH in the retention of amino acids through NF membranes was studied to analyse the separation of small peptides (only two amino acids). In this case, different isoelectric points (pI) by adjusting the pH of the mixture were considered in peptides rejection [65]. Another report showed the separation of a mixture of nine amino acids on the basis of electrostatic interactions of solutes-membrane [66]. According to results, pH has the greater influence on membrane selectivity. In addition the content of inorganic ions compared to the content of ionized amino acids affects also the separation. Therefore these variables are crucial for optimization of membrane selectivity.

Reference	**Solution**	**Experiments**	**Membrane**
[65]	Single AA solutions Mixtures of dipeptides	pH variation experiments Separation experiments of mixed dipeptides	Flat-sheet membranes **Materials:** Phosphatidic Acid (PA), Thin Film Composite (TFC), Sulfonated Polyethersulfone SPES) and Sulfonated Polystyrene (SPE) **MWCO**: 0.2-3 kDa **Charge at pH 7**: negative (SPES, SPE and TFC) or amphoteric (PA)
[66]	Mixtures of AA	Separation of a mixture of 9 AAs on the basis of differential electrostatic interactions with the membrane Membrane selectivity as a function of pH, AA concentration and Ionic Strength	**Material:** Inorganic membrane, chemical modification of the ZrO_2 layer of a UF membrane with cross linked Polyetherimide (PEI) **Charge:** positive

[13]	Single AA solutions AA mixtures Peptides (from protein hydrolysate)	NF of charged AA (single solutions and mixtures) and peptides(similar MW but different pI)	**Material:** ZrO2 filtering layer on a mineral support **Charge:** weakly negative charge at pH 8.0
[67]	Single AA solutions AA mixtures	Influence of concentration and ionic composition (salt concentration and kind of salt added) on single AA retention. Separation of AA mixtures	**Material:** cellulose acetate, SPES, SPS and Polysulfunate (PS) **MWCO:** 35-45% (NaCl retention), 1 kDa, 3, 6 kDa respectively
[68]	Protein hydrolysate	Separation of a mixture of 10 small peptides Influence of physicochemical conditions (ionic strength and pH) on the fractionation (permeate flux and *Tr*)	**M5+PEI:** ZrO_2 modified with PEI **Kerasep Solgel:** microporous active layer of ZrO_2
[16]	Protein hydrolysate	Effect of adjusting pH and ionic strength in the fractionation of the hydrolysate.	Flat sheet TFC membranes. **Material and MWCO:** PA (2.5 kDa), cellulose acetate (0.5, 0.8, 1-5 and 8-10 kDa). **Charge:** anionic characteristics
[69]	Single AA solutions AA mixtures	Influence of experimental conditions on the steady-state regime pH effect on retention coefficients of single AA solutions and AA mixtures Influence of ionic strength and transmembrane pressure on retention coefficients of an AA mixture	Cross-flow NF membrane **Material:** ceramic alumina γ with an average pore radius of 2.5 nm. **Charge:** zero point charge in the range of pH 8-9. Positively charged in the pH range tested.
[70]	AA mixtures	Separation performance of two different NF membranes. Influence of pH and operation pressure on the selectivity of the separation. Simulation NF process system for separation and concentration of L-Phe and	CTF membranes with asymmetric structure **Material:** aromatic PA and SPS

		L-Asp	
[71]	Protein hydrolysate	Concentration polarization phenomena: effect of hydrodynamic conditions on the *Tr* of selected peptides from the hydrolysate	Flat sheet membrane **Material:** cellulose acetate **MWCO:** 2.500 kDa **Charge:** anionic charge characteristics at basic pH
[72]	Single AA solutions Fermentation broth	Effect of pH, concentration and physicoquemical environment (ionic strength and kind of salt added) on single AA rejection Effect of operating pressure and concentration of fermentation broth on NF (selectivity and AA rejection)	**Material:** SPES **Charge:** high negative charge at neutral pH
[14]	Protein hydrolysate	Effect of feed concentration, pH, transmembrane pressure and feed velocity in the ability of a "loose" composite NF membrane to fractionate acid, neutral and basic peptides. Evaluation of the effect of peptides fouling on sieving and electrostatic characteristics of the membrane: PEG and Effect of aggregating peptides on the fractionation of a protein hydrolysate	Flat sheet membrane **Material:** PA (proprietary) **MWCO:** 2.5 kDa **Charge:** negatively charged at alkaline pH
[57]	Protein hydrolysate	NaCl retention measurements.	Flat sheet membrane **Material:** PA (proprietary) **MWCO:** 2.5 kDa **Charge:** negatively charged at alkaline pH
[23]	Peptide mixture	Selectivity estimation in the separation peptides from lactose and effect of pH in fouling	**Material:** SPES **MWCO:** 1 kDa **Charge:** negatively charged at neutral pH
[73]	AA mixtures	Separation of neutral AA using multilayer polyelectrolyte NF	**Material:** Bilayers of Phosphatidylserine synthase (PSS) on porous

		membranes	alumina support
[74]	Protein hydrolysate	Fractionation of small peptides using a 1 kDa NF membrane. Influence of pH and ionic strength on *Tr*	Cross-flow filtration **Material:** cellulose acetate **MWCO:** 1 kDa
[75]	Single AA solutions	Permeation of single AA solutions in the whole range of their solubility with a stepwise pH scan ranging from 0 to -1 total net charge	Membrane discs **MWCO and material:** 0.15-0.30 kDa (proprietary), 1 kDa (proprietary), 2.5 kDa (proprietary), 0.15-0.30 kDa (permanently hydrophilic PES) and 1kDa (permanently hydrophilic PES)
[76]	Single AA solutions	Study solute rejection versus concentration of 5 different AA. Comparison of experimental data against a combined steric and charge rejection model.	**Material:** SPES **MWCO:** 1kDa

Table 2. NF studies involving amino acids and peptides

Influence of concentration and ionic composition (salt concentration and type of salt added) on single amino acid retention and on the separation of amino acid mixtures was also studied to explain peptides rejection [67]. The different results show that both parameters have a negative impact on the selectivity of the membrane when size effects are not dominant. Under these conditions, the membrane seems to be more permeable to charged components due to saturation of its charged sites which makes that repulsive/attractive force between the membrane and the charged peptides become weaker.

Other studies have showed that the mixture of amino acids and their concentration affect also the behavior of NF membranes. However very few works have focused on concentrated amino acid or peptide mixtures. The most NF studies involve highly diluted amino acid solutions, which are the most likely to be found in industrial processes, and the results obtained to date are not completely understood due to at the difference in the data. For example, the results of the separation of l-glutamine (l-Gln) from Gln fermentation broth by NF, showed the effects of various experimental parameters such as transmembrane pressure, pH and concentration of broth on the rejection of l-Gln and l-glutamate (l-Glu). However, the rejection of fermentation broth from a single l-Gln or l-Glu solution was mainly caused by the complex ionic composition of the real fermentation broth [72]. Increase of I-GIn rejection was reported as a function of concentration in a concentration range from 0.3 to 3% (w/v) while the rejection of I-Glu decreased in the range from 0.1 to 0.85%.

Permeation experiments of aqueous solutions of diprotic amino acids (L-glutamine and glycine) showed different data [75]. Amino acid rejection became more concentration dependant at higher pH values due to the increased net charge of the solutes. In this high concentration regime (up to 2 M of glycine) and under alkaline conditions, an important decrease in amino acid rejection was observed in all tests.

Recent results were also found in the experiments of rejection of five amino acids by NF membranes, where experimental data were compared against a combined steric and charge rejection model [76]. Only positive charged amino acids showed good agreement with the model in all the concentration range studied while the behavior of negatively charged peptides only agree with the model at the highest concentration values and rejection of neutral amino acids was decreased due to its smaller net charge. Despite these data, the separation of bioactive peptides from natural sources and the prediction of their individual behavior require previous NF studies of complex mixed solutions.

At the other hand, the study of separation of tryptic β-casein peptides trough UF membranes showed that the separation of peptides is also affected by ionic strength by means a controlled dual mechanism: size exclusion and electrostatic repulsion [77]. Electrostatic interactions affect the peptides transport, especially if the ionic strength of the solution is low.

Another subsequent work reported the interesting potential of NF membranes for separating peptides in the range of 0.3-1 kDa [68]. Specific conditions of ionic strength and especially pH promoted the separation of peptides because the membrane and peptides showed amphoteric properties. Three categories of peptides (acid, basic, neutral) were separated according to their pI. At optimum pH 8 this led to high transmissions of basic peptides (even over 100%), intermediate transmissions for neutral peptides, and low transmissions for acid peptides. The addition of multicharged cationic and anionic species in the hydrolysate induced a markedly enhanced selectivity when the polyelectrolyte was a membrane co-ion and a complete reversion of selectivity when it was a membrane.

An additional research was later performed in order to understand the separation of peptide mixtures through NF membranes [13]. In this case, the solution tested was a mixture of 4 small peptides (4-7 residues) obtained by trypsin hydrolysis of caseinomacropeptide. From above results, it was proposed the first comprehensive approach concerning at filtration of mixtures of peptides, under two principles: (i) electro-neutrality of the solutions is always recovered, which means that all charged solute transmission are interdependent, and (ii) the number of charges along the peptide sequence, rather than the global net charge, has to be considered in order to explain the transmission of a given peptide.

Afterwards, it was investigated the potential of organic NF membranes with a MWCO between 1 and 5 kDa for the fractionation of whey protein hydrolysates. The effect of adjusting pH and ionic strength on the separation properties of the membranes was also characterized in these tests [16]. Highest selectivity between basic and acidic peptides was found at alkaline conditions without the addition of NaCl. In addition the authors demonstrated that two peptides differing by only one amino acid are transmitted differently. Consequently a single change in the amino acid sequence can affect peptides transmission.

Influence of peptide interactions on peptide separation was also established in some studies of NF membranes. The data show that the same peptide could be transmitted differently when issued from different hydrolysates, reflecting the importance of surrounding peptides, and, hence, the possible occurrence of peptide-peptide interactions [78]. Therefore hydrophobic interactions between peptides when the pH of the solution is close to their pI can lead to their aggregation and subsequent fouling of the NF membrane.

By means of NF experiments on fractionation of β-Lg tryptic hydrolysate, it was shown that peptide-peptide interactions are mainly driven by hydrophobic interactions and that some peptides are aggregated at acidic pH [14]. The morphology of these aggregates avoids the neutralization of the negative charge of the membrane surface with the alkaline peptides in the bulk. Therefore, higher permeability and higher transmission of small positive peptides is obtained under these conditions.

Furthermore peptide aggregates contribute at the polarization concentration on the membrane surface. In this case, the peptides can interact in the polarized layer during the filtration process and their transmission decreases with the time under specific conditions [79].

Other successive tests demonstrated that although physico-chemical parameters such as pH and ionic strength are the dominant ones in the case of NF membranes, operational parameters which determine permeate flux through the membrane, and in particular transmembrane pressure, have also an important influence on the retention of peptides and therefore on the selectivity of the membrane [71]. Furthermore, it should be noted that the resulting sieving properties of some NF membranes could depend on the fouled peptide layer and the composition of this layer interacting with the membrane is pH dependant [28].

The combination of membrane processes (UF and NF) was also recently used in the fractionation of whey hydrolysates to study peptides transmission [29]. The first step of this process consisted in the UF of the hydrolysate in order to obtain complete rejection of intact proteins and intermediate peptides. The resulting permeate fractions were then subjected to a fractionation by NF and a peptide fraction having a molar mass range of 5-2 kDa was isolated in this step. Transmission of peptides, amino acids and lactose were found to be mainly affected by the permeability of the fouling layer showing the effect of peptide aggregates.

Comparison of results of NF peptides using a single amino acid solutions, amino acid mixtures and peptide mixtures, had enabled to conclude that whatever the complexity of the solution: the charge is the most important criterion for the separation of peptides having similar molecular weight. The pH value of the solution is the parameter, which has the greatest effect on the separation. Addition of salts (increase of ionic strength) could decrease the intensity of charge effects. The determination of both the membrane and the mixture characteristics are of paramount importance in order to predict and optimize the performance of NF membranes for the fractionation of complex peptide mixtures.

3.2.3. *Main parameters influencing peptide fractionation using NF membranes*

The interactions of peptide-peptide and peptide-membrane affect the separation process performance and thus it is difficult to predict the selectivity of the membranes when the objective is the fractionation of complex peptide mixtures. According the literature, the most important parameters that cause effect on membrane selectivity are pH, ionic strength, polarization layer and fouling.

1. The pH of solution is an important control variable in NF processes for the fractionation of complex peptide mixtures, because peptides are molecules that have at least one carboxylic group ($R-COOH \leftrightarrow R-COO^-$ and one amine group ($R-NH_3^+ \leftrightarrow R-NH_2$). The total number of acid and basic groups depends on its primary structure (amino acid sequence) and it determines the pH value at which the peptides have the same number of negative than positive charges, i.e., its pI. Peptides can be classified in three different groups according to their pI: acidic peptides ($pI \leq 5$), neutral peptides ($5 < pI \leq 7$) and basic peptides ($pI > 7$). Their net charge depends on the pH of the solution, as well as the charge density of the NF membrane. This last value will vary because of the ionization of its functional groups (acidic and basic).

In NF, the transmission of amino acids and peptides reaches its maximum value when the pH is equal to the pI. Under these conditions, repulsive electrostatic interactions are minimized. That way, the modification of NF membranes transmission is possible by changing the pH of the mixture.

In the case of protein hydrolysates, which composition is more complex, there will be a pH value at which the fractionation of acid, neutral and basic peptides is maximized. For example, it has been shown that the separation factor between basic and acid peptides reaches its maximum value when the pH of the mixture is alkaline [16]. However, literature published on this topic only describes the behavior of "tracer" peptides in the hydrolysate and this limits the scope of the separation factor calculated.

2. The ionic strength of peptides solution affects the selectivity of NF membranes. In an aqueous medium the increase of the ionic strength, for example by the addition of NaCl, results in a decrease of zeta potential of the NF membrane [80-83] as well as a decrease in the electrophoretic mobility of proteins and peptides [84]. According to these observations, electrostatic interactions between the membrane and the peptides become less intense, which usually leads to better transmission values of the peptides.

Several authors have demonstrated the preponderance of a selectivity based on electrostatic interactions at low ionic strength values [16, 68, 85]. The fact that electrostatic interactions membrane-peptide lose significance at high ionic strength values results in a decrease of the double selectivity size/charge in processes involving NF membranes. In addition to the effect over the charge density of the membrane, ionic strength also influences the effective hydrodynamic volume of charged proteins and peptides [86]. A charged protein is surrounded by a diffuse ion cloud, typically called the electrical double layer, and the thickness of this layer is characterized by the Debye length (L_D):

$$L_D = 0.304\, I^{-1/2} \tag{3}$$

Where I is the ionic strength (mol/L) and L_D is in nm. According to equation 3 the higher the ionic strength the narrower the Debye length.

In addition, the effect of the electrical double layer could be described in terms of an increase in the effective protein radius R_{eff}:

$$R_{eff} = r_s + 0.045\, Z^2 \frac{L_D}{r_s} \tag{4}$$

Where r_s is the hard-sphere radius of the uncharged protein or peptide (in nm) and z is the surface charge of the protein (in electronic charge units).

Equations (3) and (4) indicate that relatively low salt concentration is needed in order to enhance the magnitude of the electrostatic interactions. However, the increase in the ionic strength leads to an increase in the transmission of charged peptides though the membrane.

This last observation, which is well known and it has been applied to explain the selectivity of several protein separation processes using UF membranes, is not usually mentioned in works involving the separation of peptides by NF membranes. The effects of the ionic strength over the charge density of the membrane and over the effective hydrodynamic volume of charged peptides are complementary and both of them contribute in the explanation of experimental results.

Variation of these parameters has been applied by some authors [86-88] to obtain good selectivity values in the fractionation of different proteins with similar sizes. The wise combination between membranes, pH and ionic strength is called HPTFF (High-performance-tangential flow filtration) and it is effective when proteins or peptides to be fractionated show different pI and when low or medium protein concentration is processed.

3. Concentration polarization and fouling is also a condition affecting the peptides separation. Physico-chemical parameters such as pH and ionic strength are of paramount importance in NF processes because they modulate the electrostatic interactions on which the selectivity of these membranes is supported. In addition electrostatic interactions may partly explain the distribution of a peptide between the whole solution and the membrane interface [89-90]. However, when using porous membranes, peptides are involved in a convective transport flux and its rejection is therefore the result of (i) electrostatic interactions between the membrane and the peptides plus (ii) a steric mechanism through the porous. In this sense, hydrodynamic parameters have influence in peptide rejection [91]. Thus, for example, when the MWCO of the membrane and the molecular weight of the peptide have similar values or in the presence of electrostatic interactions, an increase in transmembrane pressure will result in an increase of amino acid retention.

Concentration polarization is one of the consequences of selective solute transport through membranes. The constituents of the solution that are retained by the membrane tend to accumulate over its surface and this creates a concentration gradient in the area called polarization boundary layer. This phenomenon is quickly established at the beginning of the process and leads to a modification in the efficiency of membrane processes as well as a change in the composition of the permeate stream. The management of hydrodynamic conditions could minimize its effects. In addition size exclusion properties related to the pore size of the membrane could be completely modified due to pore blocking by the peptides. Fouling is a general term for any accumulation of deposits and materials over the membrane surface or within the pores. Two kinds of fouling can be defined: reversible fouling, the one which can be reduced by adjusting hydrodynamic conditions (velocity or transmembrane pressure), and irreversible fouling, which effect can't be avoid by cleaning procedures.

In practice, the series resistance model is widely used for fouling quantification in membrane processes. This approach derives from Darcy's phenomenological equation. The clean water flux rate (Jw) through a membrane is defined by equations (5).

$$J_w = \frac{P_T}{\mu_w R_M} \tag{5}$$

Where P_T is the transmembrane pressure, μ_W the water viscosity and R_M the intrinsic resistance of the membrane.

The measurement of water flux rate through the same membrane after being used (J_W') can be expressed as:

$$J_W' = \frac{P_T}{\mu_w (R_M + R_f)} \tag{6}$$

Equation 6 allows the calculation of the resistance associated to fouling (R_f).

Studies involving peptides transmission or retention don't usually take into account the polarization and fouling phenomena but it has been demonstrated that these phenomena are crucial in the case of protein hydrolysates, especially at acid pH values [16, 68]. Complex peptide mixtures contain peptides, which with different physicochemical characteristics (pI, hydrophobicity, charge) promote the creation of strong interactions with filtration membranes [92-93].

4. Future potential of peptides fractionation by means of membrane techniques

Currently, conventional membrane separation techniques can be employed to obtain peptide fractions in purified form with higher functionality and higher nutritional value. Special properties of the NF membranes make possible novel peptide separations. However, the specific separation of one or more peptides from a raw hydrolysate is a difficult subject because ionic interactions between peptides and membranes can markedly influence on peptides fractionation. In addition these pressure-driven processes involve the accumulation of particles on membrane leading formation of a fouling and to the modification of the membrane transport selectivity. Therefore, it is clear that NF still has to grow more in terms of understanding, materials, and process control. In addition modeling studies are necessary to predict of the process performance in all circumstances.

Alternatively the application of an external electrical field, which acts as an additional driving force to the pressure gradient, can be seen as a technique that could improve the efficiency of the conventional membrane processes for the separation of charged bioactive molecules. In this sense, two different configurations can be distinguished: electrically-enhanced filtration, which can be used with conventional pressure driven membrane filtration, and forced-flow membrane electrophoresis, which is conducted in an electrophoretic cell. Intensive researches on these membrane processes have been carried out including electromembrane filtration (EMF) [94-95], electrodialysis with UF membranes (EDUF) [96-99] and forced-flow electrophoresis (FFE) [100] for the separation of charged bioactive molecules.

EDUF couples size exclusion capabilities of UF membranes with the charge selectivity of electrodyalysis (ED) allowing separation of molecules according to their electric charges and to their molecular mass (membrane filtration cut-off). The feasibility of peptide fractionation by EDUF was demonstrated notably with β-Lg tryptic hydrolysate solutions and was suggested to improve the separation between basic and neutral peptides [97]. Actually, EDUF process also allowed a selective and a simultaneous separation of anionic and cationic peptides presents in an uncharacterized concentrated polypeptide mixture of snow crab by-products hydrolysate [101].

Recently a comparative study on NF and EDUF was performed in terms of flux and mass balance [102]. The results showed that NF provides a greater mass flux while when using EDUF a wider range of peptides and more polar amino acids are recovered. EDUF can be seen to be a promising separation technology, but further scale-up developments will be necessary to confirm its feasibility at large scale.

EMF combines the separation mechanisms of membrane filtration and electrophoresis. Ion exchange membranes are replaced by UF in a conventional electrodialysis cell. In electrophoretic separators, a porous membrane is used to put into contact two flowing liquids between which an electrically driven mass transfer takes place. During this process the mass transport is affected by electrostatic interactions taking place at the membrane solution interface. The perspectives in the field of peptide fractionation will be the complete understanding of the interactions of peptides and membrane as well as the development of new membrane materials of gels limiting or increasing these interactions to improve the selectivity and the yield of production of specific peptides [100].

Author details

Claudia Muro
Institute Technological of Toluca, México

Francisco Riera and Ayoa Fernández
University of Oviedo, Spain

5. References

[1] Muro-Urista C, Álvarez-Fernández R, Riera-Rodríguez F, Arana-Cuenca A, Téllez-Jurado. A. Review: Production and Functionality of Active Peptides from Milk. Food Science and Technology International 2011; 17(4) 293-317.

[2] Korhonen H, Pihlanto-Leppälä A, Rantamäki P, Tupasela T. Impact of Pocessing on Bioactive Proteins and Peptides. Trends in Food Science and Technology 1998; 9(8) 307-319.

[3] Korhonen H. Technology Options for New Nutritional Concepts. Society Dairy Technology 2002; 55(2) 79-88.

[4] Pihlanto-Leppälä A, Koskinen P, Piilola K, Korhonen H. Angiotensin I Converting Enzyme Inhibitory Properties of Whey Protein Digest: Concentration and Characterization of Active Peptides. Journal Dairy Research 2000; 67(1) 53-64.

[5] Saito T, Nakamura T, Kitazawa H, Kawai Y, Itoh T. Isolation and Structural Analysis of Antihypertensive Peptides that Exist Natural in Gouda Cheese. Journal Dairy Science 2000; 83(7) 1434-1440.

[6] Toll H, Oberacher H, Swart R, Huber CG. Separation, Detection and Identification of Peptides by Ion-Pair Reversed-Phase 5 High-Performance Liquid Chromatography-Electrospray Ionization Mass Spectrometry at High and Low pH. Journal Chromatography 2005; 1079(1) 274-286.

[7] Vermeirssen V, Van Camp J, Verstraete W. Fractionation of Angiotensin I Converting Enzyme Inhibitory Activity from Pea and Whey Protein in Vitro Gastrointestinal Digests. Journal Science Food Agriculture 2005; 85(3) 399-405.

[8] Lopez-Fandiño R, Otte J, Van Camp J. Physiological, Chemical and Technological Aspects of Milk-Protein-Derived Peptides with Antihypertensive and ACE-Inhibitory Activity. International Dairy Journal 2006; 16(11) 1277-1293.

[9] Hernández-Ledesma B, Quiros A, Amigo L, Recio I. Identification of Bioactive Peptides After Digestion of Human Milk and Infant Formula with Pepsin and Pancreatin. International Dairy Journal 2007; 17(1) 42-49.

[10] Otte J, Shalaby S, Zakora M, Nielsen MS. Fractionation and Identification of ACE-Inhibitory Peptides from α-Lactalbumin and β-Casein Produced by Thermolysin-Catalysed Hydrolysis. International Dairy Journal 2007; 17(11)1460-1472.

[11] Wijers MC, Pouliot Y, Gauthier SF, Pouliot, M, Nadeau L. Use of Nanofiltration Membranes for the Desalting of Peptide Fractions from Whey Protein Enzymatic Hydrolysates. Le Lait 1998; 78(6) 621-632.

[12] Agyei D, Danquah MK. Industrial-Scale Manufacturing of Pharmaceutical-Grade Bioactive Peptides. Biotechnology Advances 2011; 29(3) 272-277.

[13] Martin-Orue C, Bouhallab S, Garem A. Nanofiltration of Amino Acid and Peptide Solutions: Mechanisms of Separation. Journal Membrane Science 1998; 142(2) 225-233.

[14] Groleau PE, Lapointe JF, Gauthier SF, Pouliot Y. Effect of Aggregating Peptides on the Fractionation of β-Lg Tryptic Hydrolysate by Nanofiltration Membrane. Journal Membrane Science 2004; 234(1) 121-129.

[15] Konrad G, Lieske B, Faber W. A Large-Scale Isolation of Native β-Lactoglobulin: Characterization of Physicochemical Properties and Comparison with Other Methods. International Dairy Journal 2000; 10(10) 713-721.

[16] Pouliot Y, Wijers MC, Gauthier SF, Nadeau L. Fractionation of Whey Hydrolysates Using Charged UF/NF Membranes. Journal Membrane Science 1999; 158(1) 105-114.

[17] Lapointe JF, Bouchard C, Gauthier S, Pouliot Y. Selective Separation of a Cationic Peptide from a Tryptic Hydrolysate of β-Lactoglobulin by Electrofiltration. Biotechnology and Bioengineering 2005; 94(2) 223-233.

[18] Futselaar H, Schonewille H, Van Der Meer W. Direct Capillary Nanofiltration-a New High-Grade Purification Concept. Desalination 2002; 145(1) 75-80.

[19] Pouliot Y. Membrane Processes in Dairy Technology-from a Simple Idea to Worldwide Panacea. International Dairy Journal 2008; 18(7) 735-740.

[20] Mehra R, Kelly PM. Whey Protein Fractionation Using Cascade Membrane Filtration. In: Brussels, Belgium Revue (Ed) 1985: Processes for Novel Dairy Applications, Advances in Fractionation and Separation: proceedings of the IDF World Dairy Summit 2003, Sep 2003, Bruges, Belgium: International Dairy Federation Bulletin; 2004.

[21] Hogan S, Zhang L, Li J, Wang H, Zhou K. Development of Antioxidant Rich Peptides from Milk Protein by Microbial Proteases and Analysis of Their Effects on Lipid Peroxidation in Cooked Beef. Food Chemical 2009; 117(3) 438-443.

[22] Herraiz T. Sample Preparation and Reversed Phase-High Performance Chromatography Analysis of Food-Derived Peptides Liquid. Analytica Chimica Acta 1997; 352(1) 119-139.

[23] Butylina S, Luque S, Nyström M. Fractionation of Whey-Derived Peptides Using a Combination of Ultrafiltration and Nanofiltration. Journal Membrane Science 2006; 280(1) 418-426.

[24] Farvin KH, Baron CP, Nielsen SN, Otte J, Jacobsen C. Antioxidant Activity of Yoghurt Peptides: Part 2-Characterisation of Peptide Fractions. Food Chemistry 2010; 123(41) 1090-1097.

[25] Kapel R, Klingenberg F, Framboisier X, Dhulster P, Marc I. An Original Use of Size Exclusion-HPLC for Predicting the Performances of Batch Ultrafiltration Implemented to Enrich a Complex Protein Hydrolysate in a Targeted Bioactive Peptide. Journal Membrane Science 2011; 383(1) 26-34.

[26] Kawasaki Y, Kawakami M, Tanimoto M, Dosako S, Tomizawa A, Kotake M. pH-Dependent Molecular Weight Changes of J-Casein Glycomacropeptide and its Preparation by Ultrafiltration. Milchwissenschaft 1996; 48(4) 191-196.

[27] Martín-Diana AB, Fontecha MJF. Isolation and Characterisation of Caseinomacropeptide from Bovine, Ovine and Caprine Cheese Whey. European Food Research and Technology 2002; 214(4) 282-286.

[28] Tolkach A, Kulozik U. Fractionation of Whey Proteins and Caseinomacropeptide by Means of Enzymatic Crosslinking and Membrane Separation Techniques. Journal Food Engineering 2005; 67(1) 13-20.

[29] Pouliot Y, Gauthier SF, Groleau PE. Membrane-Based Fractionation and Purification Strategies for Bioactive Peptides. In: Mine Y, Shahidi F, (eds) Nutraceutical Proteins and Peptides in Health and Disease. Marcel Dekker Inc: New York, USA: Tayloy & Francis Group, LLC; 2005. p 639-658.

[30] Prata-Vidal M, Bouhallab S, Henry G, Aimar P. An Experimental Study of Caseinomacropeptide Hydrolysis by Trypsin in a Continuous Membrane Reactor. Biochemical Engineering Journal 2001; 8(3) 195-202.

[31] Guadix A, Camacho F, Guadix EM. Production of Whey Protein Hydrolysates with Reduced Allergenicity in a Stable Membrane Reactor. Journal Food Engineering 2006; 72(4) 398-405.

[32] Bouhallab S, Mollé D, Léonil J. Tryptic Hydrolysis of Caseinomacropeptide in Membrane Reactor: Preparation of Bioactive Peptides. Biotechnology Letters 1995; 104(1) 805-810.

[33] Bouhallab S, Touzé C. Continuous Hydrolysis of Caseinomacropeptides in a Membrane Reactor: Kinetic Study and Gram-Scale Production of Antithrombotic Peptides. Lait 1995; 75(3) 251-258.

[34] Bordenave S, Sannier F, Ricart G, Piot JM. Continuous Hydrolysis of Goat Whey in an Ultrafiltration Reactor: Generation of Alpha-Lactorphin. Preparative Biochemistry Biotechnology 1999; 29(2) 189-202.

[35] Sannier F, Bordenave S, Piot JM. Purification of Goat β-Lactoglobulin from Whey by an Ultrafiltration Membrane Enzymic Reactor. Journal Dairy Research 2000; 67(1) 43-51.

[36] Righetti PG, Nembri F, Bossi A, Mortarino M. Continuous Enzymatic Hydrolysis of Beta-Casein and Isoelectric Collection of Some of the Biologically Active Peptides in an Electric Field. Biotechnology Progress 1997; 13(3) 258-264.

[37] Cheison SC, Wang Z, Xu SY. Use of Response Surface Methodology to Optimise the Hydrolysis of Whey Protein Isolate in a Tangential Flow Filter Membrane Reactor. Journal Food Engineering 2007; 80(4) 1134-1145.

[38] Trusek-Holownia A, Noworyta A. Peptides Removing in Enzymatic Membrane Bioreactor. Desalination 2008; 221(1) 543-551.

[39] Sharma AK, Sharma MK. Plants as Bioreactors: Recent Developments and Emerging Opportunities. Biotechnology Advances 2009; 27(6) 811-832.

[40] Brulé G, Roger L, Fouquant J, Plot M. Procédé de Traitement d´une Matiére á Base de Caséine Contenant des Phosphocaséinates de Cations Monovaients ou Leurs Dérivés, Products Obtenus et Applications. French Patent, 1982; 2, 474,829.

[41] Mullally MM, Meisel H, Fitzgerald RJ. Angiotensin-I-Converting Enzyme Inhibitory Activities of Gastric and Pancreatic Proteinase Digests of Whey Proteins. International Dairy Journal 1997; 7(5) 299-303.

[42] Antila P, Paakkari I, Järvinen A, Mattila MJ, Laukkanen M, Pihlanto-Leppälä A, Mänrsälä P, Hellman J. Opioid Peptides Derived from In-Vitro Proteolysis of Bovine Whey Proteins. International Dairy Journal 1991; 1(4) 215-229.
[43] Pellegrini ACU, Dettling U, Thomas N, Hunziker P. Isolation and Characterization of Four Bactericidal Domains in the Bovine β-Lactoglobulin. Biochimica et Biophysica Acta 2001; 1526(1) 247-259.
[44] Pihlanto-Leppälä A, Paakkari I, Rinta-Koski M, Antila P. Bioactive Peptide Derived from in Vitro Proteolysis of Bovine β-Lactoglobulin and its Effect on Smooth Muscle. Journal Dairy Research 1997; 64(1)149-155.
[45] Fujita H, Yamagami T, Oshima K. Effects of an ACE-Inhibitory Agent, Katsuobushi Oligopeptide, in the Spontaneously Hypertensive Rat and in Borderline and Mildly Hypertensive Subjects. Nutrition Research 2001; 21(8) 1149-1158.
[46] Kapel R, Rahhou E, Lecouturier D, Guillochon D, Dhulster P. Characterization of an Antihypertensive Peptide from an Alfalfa White Protein Hydrolysate Produced by a Continuous Enzymatic Membrane Reactor. Process Biochemistry 2006; 41(9) 1961-1966.
[47] Xie Z, Huang J, Xu X, Jin A. Antioxidant Activity of Peptides Isolated from Alfalfa Leaf Protein Hydrolysate. Food Chemistry 2008; 111(2) 370-376.
[48] Kong X, Zhou H, Hua Y. Preparation and Antioxidant Activity of Wheat Gluten Hydrolysates (WGHS) Using Ultrafiltration Membranes. Journal of the Science of Food and Agriculture 2008; 88(5) 920-926.
[49] Wu J, Ding X. Characterization of Inhibition and Stability of Soy-Protein-Derived Angiotensin I-Converting Enzyme Inhibitory Peptides. Food Research International, 2002; 35(4) 367-375.
[50] Kuba M, Tana C, Tawata S, Yasuda M. Production of Angiotensin I-Converting Enzyme Inhibitory Peptides from Soybean Protein with Monascus Purpureus Acid Proteinase. Process Biochemistry 2005; 40(6) 2191-2196.
[51] Yuanhui Z, Bafang L, Zunying L, Shiyuan D, Xue Z, Mingyong A. Anhihypertensive Effect and Purification of an ACE Inhibiroty Peptide from Sea Cucumber Gelatine Hydrolysate. Process Biochemistry 2007; 42(12) 1586-1591.
[52] Kim J-Y, Park S-C, Kim M-H, Lim H-T, Park Y, Hahm K-S. Antimicrobial Activity Studies on a Trypsin-Chymotrypsin Protease Inhibitor Obtained from Potato. Biochemical and Biophysical Research Communications 2005; 330(3) 921-927.
[53] Kim M-H, Park S-C, Kim J-Y, Lee S-Y, Lim H-T, Cheong H, Hahm K-S, Park Y. Purification and Characterization of a Heat-Stable Serine Protease Inhibitor from the Tubers of New Potato Variety "Golden Valley". Biochemical Biophysical Research Communications, 2008; 346(3) 681-686.
[54] Fievet P, Labbez C, Szymczyk A, Vidonne A, Foissy A, Pagetti J. Electrolyte Transport Through Amphoteric Nanofiltration Membranes. Chemical Engineering Science 2002; 57(15) 2921-2931.
[55] Zeman L, Wales M. Steric Rejection of Polymeric Solutes by Membrane with Uniform Pore Sizes Distribution. Separation Science and Technology 1981; 16(3) 275-290.

[56] Van Der Bruggen B, Schaep J, Wilms D, Vandecasteele C. Influence of Molecular Size, Polarity and Charge on the Retention of Organic Molecules by Nanofiltration. Journal Membrane Science 1999; 156(2) 29-41.

[57] Lapointe JF, Gauthier SF, Pouliot Y, Bouchard C. Fouling of a Nanofiltration Membrane by a β-Lactoglobulin Tryptic Hydrolysate: Impact on the Membrane Sieving and Electrostatic Properties. Journal Membrane Science 2005; 253(1) 89-102.

[58] Jeantet R, Maubois JL. Sélectivités de Membranes de Nanofiltration: Effect du pH, de la Nature et de la Concentration des Solutes. Le Lait 1995; 7(3) 595-610.

[59] Levenstein R, Hasson D, Semiat R. Utilization of the Donnan Effect for Improving Electrolyte Separation with Nanofiltration Membranes. Journal Membrane Science 1996; 116(1) 77-92.

[60] Lapointe JF, Gauthier SF, Pouliot Y, Bouchard C. Characterization of Interactions between_β-Lactoglobulin Tryptic Peptides and a Nanofiltration Membrane: Impact on the Surface Membrane Properties as Determined by Contact Angle Measurements. Journal Membrane Science 2005; 261(1) 36-48.

[61] Schaep J, Van Der Bruggen B, Vandecasteele C, Wilms D. Influence of Ion Size and Charge in Nanofiltration. Separation and Purification Technology 1998; 14(1) 155-162.

[62] Schaep J, Vandecasteele C, Mohammad AW, Bowen WR. Analysis of the Salt Retention of Nanofiltration Membranes Using the Donnan-Steric Partitioning Pore Model. Separation Science and Technogy 1999; 34(15) 3009-3030.

[63] Childress AE, Elimelech M. Relating Nanofiltration Membrane Performance to Membrane Charge (electrokinetic) Characteristics. Environmental Science and Technology 2000; 265(1) 3710-3716.

[64] Van Der Bruggen B, Mäanttäri M, Nyström M. Drawbacks of Applying Nanofiltration and how to Avoid Them: A Review. Separation and Purification Technology 2008; 63(2) 251-263.

[65] Tsuru T, Shutou T, Nakao SI, Kimura S. Peptide and Amino Acid Separation with Nanofiltration Membranes. Separation Science and Technology 1994; 29(8) 971-984.

[66] Garem A, Daufin G, Maubois JL, Léonil J. Selective Separation of Amino Acids with a Charged Inorganic Nanofiltration Membrane: Effect of Physicochemical Parameters on Selectivity. Biotechnology and Bioengineering 1997; 54(4) 291-302.

[67] Timmer JMK, Speelmans MPJ, Van Der Horst HC. Separation of Amino Acids by Nanofiltration and Ultrafiltration Membranes. Separation and Purification Technology 1998; 14(1)133-144.

[68] Garem A, Daufin G, Maubois JL, Chaufer B, Léonil J. Ionic Interactions in Nanofiltration of β-Casein Peptides. Biotechnology and Bioengineering 1998; 142(2), 109-116.

[69] Grib H, Persin M, Gavach C, Piron DL, Sandeaux J, Mameri N. Amino Acid Retention with Alumina Nanofiltration Membrane. Journal Membrane Science 2000; 172(1) 9-17.

[70] Wang XL, Ying AL, Wang WN. Nanofiltration of L-Phenylalanine and L-Aspartic Acid Aqueous Solutions. Journal Membrane Science 2002; 196(1) 59-67.

[71] Lapointe JF, Gauthier SF, Pouliot Y, Bouchard C. Effect of Hydrodynamic Conditions on Fractionation of β-Lactoglobulin Tryptic Peptides Using Nanofiltration Membranes. Journal Membrane Science 2003; 212(1) 55-67.

[72] Li SL, Li C, Liu YS, Wang XL, Cao ZA. Separation of L-Glutamine from Fermentation Broth by Nanofiltration. Journal Membrane Science 2003; 222(1) 191-201.

[73] Hong SU, Bruening ML. Separation of Amino Acid Mixtures Using Multilayer Polyelectrolyte Nanofiltration Membranes. Journal Membrane Science 2005; 280(1)1-5.

[74] Tessier B, Harscoat-Schiavo C, Marc I. Contribution of Electrostatic Interactions during Fractionation of Small Peptides Complex Mixtures by UF/NF Membranes. Desalination 2003; 200(2) 333-343.

[75] Kovacs Z, Samhaber W. Nanofiltration of Concentrated Amino Acid Solutions. Desalination, 2009; 240(1) 78-88.

[76] Shirley J, Mandale S, Williams PM. Amino Acid Rejection Behavior as a Function of Concentration. Advances in Colloid and Interface Science 2011; 164(1) 118-125.

[77] Nau F, Kerhervé FL, Léonil J, Daufin G, Aimar P. Separation of β-Casein Peptides Through UF Inorganic Membranes. Bioseparation 1993; 46(3) 205-215.

[78] Pouliot Y, Gauthier SF, L'Heureux J. Effect of Peptide Distribution on the Fractionation of Whey Protein Hydrolysates by Nanofiltration Membranes. Le Lait 2000; 80(1) 113-122.

[79] Lapointe JF, Gauthier SF, Pouliot Y, Bouchard C. Effect of Hydrodynamic Conditions on Fractionation of β-Lactoglobulin Tryptic Peptides Using Nanofiltration Membranes. Journal Membrane Science 2002; 212(1) 55-67.

[80] Peeters JMM, Mulder MHV, Strathmann H. Steaming Potential Measurements as a Characterization Method for Nanofiltration Membranes. Colloid and Surfaces A 1999; 150(1) 247-259.

[81] Afonso MD, Hagmeyer G, Gimbel R. Steaming Potential Measurements to Assess the Variation of Nanofiltration Membranes Surface Charge with the Concentration of Salt Solutions. Separation and Purification Technology 2001; 22(23) 529-541.

[82] Schaep J, Vandecasteele C, Mohammad AW, Bowen WR. Modeling the Retention of Ionic Components for Different Nanofiltration Membranes. Separation and Purification Technology 2001; 22(23) 169-179.

[83] Schaep J, Vandecasteele C, Evaluating the Charge of Nanofiltration Membranes. Journal Membrane Science 2001; 188(1) 129-136.

[84] Adamson NJ, Reynolds EC. Rules Relating Mobility, Charge and Molecular Size of Peptides and Proteins. Journal Chromatography B 1997; 699(1) 133-147.

[85] Nau F, Kerhervé FL, Léonil J, Daufin G. Selective Separation of Tryptic β-Casein Peptides through Ultrafiltration Membranes: Influence of Ionic Interactions. Biothechnology and Bioengineering 1995; 46(3) 246-253.

[86] Zydney AL. Protein Separations Using Membrane Filtration: New Opportunities for Whey Fractionation. International Dairy Journal 1998; 8(3) 243-250.

[87] Pujar, NS, Zydney AL. Electrostatic and Electrokinetic Interactions during Protein Transport through Narrow Pore Membranes. Industrial and Engineering Chemistry Research 1994; 33(10) 2473-2482.

[88] Van Reis R, Gadam S, Frautschy LN, Orlando S, Goodrich EM, Saksena S, Kuriyel R, Simpson CM, Pearl S, Zydney AL. High Performance Tangential Flow Filtration. Biotechnology and Bioengineering 1997; 159(1) 71-82.

[89] Groleau PE, Morin P, Gauthier SF, Pouliot Y. Effect of Physicochemical Conditions on Peptide-Peptide Interactions in a Tryptic Hydrolysate of α-Lactoglobulin and Identification of Aggregating Peptides. Journal of Agricultural Food Chemistry 2003; 51(15) 4370-4375.

[90] Jarusutthirak C, Mattaraj S, Jiraratananon R. Factors Affecting Nanofiltration Performances in Natural Organic Matter Rejection and Flux Decline. Separation and Purification Technology 2007; 58(1) 68-75.

[91] Ben-David A, Oren Y, Freger V. Thermodynamic Factors in Partitioning and Rejection of Organic Compounds by Polyamide Composite Membranes. Environmental Science and Technology 2006; 40(22) 7023-7028.

[92] Gourley L, Britten M, Gauthier SF, Pouliot Y. Identification of Casein Peptides Interacting with Polysulfone Ultrafiltration Membrane. Journal Membrane Science 1998; 78(6) 633-646.

[93] Noiseux I, Gauthier S, Fand SF, Turgeon SL. Interactions Between Bovine β-Lactoglobulin and Peptides under Different Physicochemical Conditions. Journal of Agricultural Food Chemistry 2002; 50(6) 1587-1592.

[94] Bargeman GJ, Houwing I, Recio GH, Koops C, Van Der Horst. Electro-Membrane Filtration for the Selective Isolation of Bioactive Peptides from an alpha S_2-Casein Hydrolysate. Biotechnology and Bioengineering 2002; 80(6) 599-609.

[95] Bargeman GJ, Dohmen-Speelmans I, Recio M, Timmer C, Van Der Horst. Selective Isolation of Cationic Amino Acids and Peptides by Electro-Membrane Filtration. Le Lait 2002; 80(1) 175-185.

[96] Lapointe JF, Gauthier SF, Pouliot Y, Bouchard C. Selective Separation of Cationic Peptides from a Tryptic Hydrolysate of β-Lactoglobulin by Electrofiltration. Biotechnology and Bioengineering 2006; 94(2) 223-233.

[97] Poulin JF, Amiot J, Bazinet L. Simultaneous Separation of Acid and Basic Bioactive Peptides by Electrodialysis with Ultrafiltration Membrane. Journal Biotechnology 2006; 123(3) 314-328.

[98] Poulin JF, Amiot J, Bazinet L. Impact of Feed Solution Flow Rate on Peptide Fractionation by Electrodialysis with Ultrafiltration Membrane. Journal of Agricultural Food Chemistry 2007; 56(6) 2007-2011.

[99] Poulin JF, Amiot J, Bazinet L. Improved Peptide Fractionation by Electrodialysis with Ultrafiltration Membrane: Influence of Ultrafiltration Membrane Stacking and Electrical Field Strength. Journal Membrane Science 2007; 299(1) 83-90.

[100] Bazinet L, Firdaous L. Membrane Processes and Devices for Separation of Bioactive Peptides. Recent Patents on Biotechnology 2009; 3(1) 61-72.

[101] Doyen A, Baulieu L, Sauciera L, Pouliot Y, Bazinet L. Demonstration of in Vitro Anticancer Properties of Peptide Fractions from a Snow Crab by-Products Hydrolysate after Separation by Electrodialysis with Ultrafiltration Membranes. Separation and Purification Technology 2011; 78(3) 321-329.

[102] Langevin ME, Roblet C, Moresoli C, Ramassamy C, Bazinet L. Comparative Application of Pressure and Electrically-Driven Membrane Processes for Isolation of Bioactive Peptides from Soy Protein Hydrolysate. Journal Membrane Science 2012; 403-404 (1) 15-24.